Nuclear Development

Forty Years of Uranium Resources, Production and Demand in Perspective

"The Red Book Retrospective"

© OECD 2006
NEA No. 6096

NUCLEAR ENERGY AGENCY
ORGANISATION FOR ECONOMIC CO-OPERATION AND DEVELOPMENT

ORGANISATION FOR ECONOMIC COOPERATION AND DEVELOPMENT

The OECD is a unique forum where the governments of 30 democracies work together to address the economic, social and environmental challenges of globalisation. The OECD is also at the forefront of efforts to understand and to help governments respond to new developments and concerns, such as corporate governance, the information economy and the challenges of an ageing population. The Organisation provides a setting where governments can compare policy experiences, seek answers to common problems, identify good practice and work to co-ordinate domestic and international policies.

The OECD member countries are: Australia, Austria, Belgium, Canada, the Czech Republic, Denmark, Finland, France, Germany, Greece, Hungary, Iceland, Ireland, Italy, Japan, Korea, Luxembourg, Mexico, the Netherlands, New Zealand, Norway, Poland, Portugal, the Slovak Republic, Spain, Sweden, Switzerland, Turkey, the United Kingdom and the United States. The Commission of the European Communities takes part in the work of the OECD.

OECD Publishing disseminates widely the results of the Organisation's statistics gathering and research on economic, social and environmental issues, as well as the conventions, guidelines and standards agreed by its members.

* * *

This work is published on the responsibility of the Secretary-General of the OECD. The opinions expressed and arguments employed herein do not necessarily reflect the official views of the Organisation or of the governments of its member countries.

NUCLEAR ENERGY AGENCY

The OECD Nuclear Energy Agency (NEA) was established on 1st February 1958 under the name of the OEEC European Nuclear Energy Agency. It received its present designation on 20th April 1972, when Japan became its first nonEuropean full member. NEA membership today consists of 28 OECD member countries: Australia, Austria, Belgium, Canada, the Czech Republic, Denmark, Finland, France, Germany, Greece, Hungary, Iceland, Ireland, Italy, Japan, Luxembourg, Mexico, the Netherlands, Norway, Portugal, Republic of Korea, the Slovak Republic, Spain, Sweden, Switzerland, Turkey, the United Kingdom and the United States. The Commission of the European Communities also takes part in the work of the Agency.

The mission of the NEA is:

– to assist its member countries in maintaining and further developing, through international co-operation, the scientific, technological and legal bases required for a safe, environmentally friendly and economical use of nuclear energy for peaceful purposes, as well as

– to provide authoritative assessments and to forge common understandings on key issues, as input to government decisions on nuclear energy policy and to broader OECD policy analyses in areas such as energy and sustainable development.

Specific areas of competence of the NEA include safety and regulation of nuclear activities, radioactive waste management, radiological protection, nuclear science, economic and technical analyses of the nuclear fuel cycle, nuclear law and liability, and public information. The NEA Data Bank provides nuclear data and computer program services for participating countries.

In these and related tasks, the NEA works in close collaboration with the International Atomic Energy Agency in Vienna, with which it has a Cooperation Agreement, as well as with other international organisations in the nuclear field.

Publié en français sous le titre :
Ressources, production et demande de l'uranium : un bilan de quarante ans

Credit cover photograhs: AREVA, France; NEI, United States; COGEMA, France.

FOREWORD

Since 1965, with the co-operation of their member countries and states, the OECD Nuclear Energy Agency (NEA) and the International Atomic Energy Agency (IAEA) have jointly prepared periodical updates (currently every two years) on world uranium resources, production and demand. These updates have been published by the OECD/NEA in what is commonly known as the Red Book.

This publication was undertaken to collect, collate, analyse and publish all of the key data and information collected in the 20 editions of the Red Book published between 1965 and 2004. The *Red Book Retrospective* gives a historical profile of the world uranium industry in the areas of uranium resources, exploration, production, installed nuclear capacity, annual uranium requirements, uranium stocks, price, environmental activities and relevant uranium policies. It provides in-depth information relating to the histories of the major uranium-producing countries including Australia, Canada, France, Germany (including the former German Democratic Republic), the Russian Federation (including the former Union of Soviet Socialist Republics) and the United States. Expert analyses provide fresh insights into important aspects of the industry including: the cost of discovery, resources to production ratios and the time to reach production after discovery, among others. Over 100 countries have provided official government data for this publication which is now available in an easy-to-use form, including electronically to allow others to analyse the information and further the understanding of this important industry.

This report is published on the responsibility of the Secretary-General of the OECD.

Acknowledgements

The Secretariat would like to acknowledge the valuable contributions of Fritz Barthel (Germany), Jean-René Blaise (France) and Jay M. McMurray (United States) to this report.

3

TABLE OF CONTENTS

EXECUTIVE SUMMARY

When the first Red Book was published in 1965, there were 29 reactors in operation worldwide with generating capacity totalling about 4 500 MWe. By 2003, 435 reactors were in operation with generating capacity totalling about 359 400 MWe. From 1965 to 2004, 20 Red Books were published, which over time tracked the growth of nuclear power and provided comprehensive official government data on uranium resources, exploration and production to the public. The 1965 Red Book included information relating to uranium resources in 16 countries with Reasonably Assured Resources (RAR) totalling 993 000 tU. By 2003 RAR totalling 3 169 000 tU were reported by 43 countries. The history of the Red Book has paralleled the growth of nuclear energy but has also been influenced by world events. Foremost among these was the Cold War, during which military requirements for uranium were a major influence on the uranium market. Other significant events included the oil crisis in 1973 that increased public awareness of the potential of nuclear energy, the Three Mile Island and Chernobyl reactor accidents that slowed the growth of nuclear power and the end of the Cold War in 1989 that led to introduction of significant secondary sources of uranium to the world market as well as the inclusion of new information on the uranium industries of Central and Eastern European countries beginning in 1991 and countries of the former USSR beginning in 1993.

This *Red Book Retrospective* was undertaken to collect, collate, analyse and publish all of the information collected over the 40 years of the existence of the Red Book. In addition to capturing information included in the Red Books published between 1965 and 2004, every effort has been made to fill in gaps in information to ensure as complete a perspective as possible on the history of uranium supply and demand. With the inclusion of this supplementary information, this report should be the most complete record of the uranium industry publicly available dating from the birth of civilian nuclear energy through the dawn of the 21st century. Some key findings are highlighted below with further details provided in 12 chapters of the text while the appendices contain the raw data that will allow others to continue this work.

Installed nuclear capacity and reactor-related uranium requirements (demand)

Civilian nuclear generating capacity has grown from 50 MWe, the capacity of the world's first civilian reactor, which began operating in 1957, to about 359 400 MWe, the installed capacity of 435 reactors operating in the beginning of 2003 operating in 31 countries. The leading countries in terms of nuclear generating capacity in 2003 are listed in Table ES.1.

Table ES.1. Worldwide leadership in nuclear generating capacity (as of 1 January 2004)

Country	Generating capacity (MWe net)	Number of operating reactors
United States	98 357	104
France	63 180	58
Japan	43 197	52
Germany	21 283	19
Russian Federation	20 793	30
Others (total)	112 623	172
World Total	359 433	435

Between 1956 and 2003, the 33 countries that have used commercial nuclear reactors had reactor-related requirements that are estimated to have totalled 1 513 327 tU. The top five users of uranium during that period are shown in Table ES.2.

Table ES.2. Uranium requirements of selected countries (1956-2003)

Country	Reactor-related requirements	Percentage of world total
United States	364 180	24.1
France	173 837	11.5
Japan	163 520	10.8
Russian Federation	94 925	6.3
Germany	73 842	4.9
Others (total)	643 023	42.4
World total	1 513 327	100

Market price

The price of uranium reached its all-time peak in the 1970s driven by a combination of military requirements and growth of civilian nuclear power. After this peak prices rapidly dropped and then began a steady decline over the next 20 years driven in large part by slower than expected growth in nuclear power, a result of the Three Mile Island and Chernobyl accidents, and a supply over capacity that resulted in the build-up of large inventories. The price hit a historic low in 2000 and began a rebound that continued through 2005, as the market adjusted to the reality of potential near to mid-term supply shortfalls.

Fluctuations in the uranium market price have impacted several aspects of the uranium industry over the years including exploration expenditures, exploration objectives (e.g. exploration for higher grade deposits in the 1980s and 1990s when the market price was low), uranium resource estimates and production capacity as well as changing emphasis on production methods (e.g. development of *in situ* leaching (ISL) and non-entry mining methods for very high-grade deposits). Though today's market price generally reflects the perceived balance between supply and demand, in the past military demand for uranium has distorted its behaviour as a commodity. Since the end of the Cold War, however, uranium has increasingly behaved as a typical commodity, with prices responding to perceptions as to the balance between supply and demand.

Exploration

A total of 81 countries have reported exploration expenditures related to uranium with cumulative worldwide exploration expenditures between 1945 and 2003 of about USD 13 400 million. The modern era of uranium exploration began in the early 1940s, largely motivated by the need to satisfy military requirements. Exploration for uranium to fuel civilian power reactors began to gain importance in the 1950s. The world leaders in total exploration expenditures during this time are listed in Table ES.3.

Table ES.3. Countries with highest exploration expenditures (1945-2003)[1]

Country	USD million	Percentage of world total
USSR[1]	3 692	27.6
United States	2 507	18.7
Germany[2]	2 003	14.9
Canada	1 289	0.6
France	907	6.8
Others (total)	3 002	22.4
World total	13 400	100

1. Does not include expenditures by Kazakhstan, the Russian Federation, Ukraine and Uzbekistan since 1991.
2. Includes the German Democratic Republic (GDR).

The trend in worldwide exploration expenditures has closely paralleled uranium market prices, with the peak in expenditures lagging the 1978 market price peak by only one year. As one would expect, the parallelism between market price and exploration expenditures was very close in market-based economies, whereas expenditures in centrally planned economies showed little relationship to price.

Exploration data from 13 countries were used to quantify exploration effectiveness. The cost of discovery for uranium resources based on historical data in these countries is estimated to have been approximately USD 1.82/kgU.

Resources

Uranium resources are reported in categories of confidence level and production cost. Reporting of resources has evolved in response to changes in the uranium market and to growing sophistication in resource calculation. The 1965 Red Book reported resources in 16 countries totalling 3.21 million tU; in 2003, 56 countries reported total resources in all confidence and cost categories of 14.38 million tU. Several factors contributed to the increase in resources including:

- discovery of additional resources as the result of ongoing exploration;

- more countries participating in the reporting process including those of the former Soviet Union;

- several countries, including China, Iran and India began reporting resources according to Red Book classification criteria;

- new resource categories having been added over time, along with changes in production cost categories.

The countries with the largest reported uranium resources are listed in Table ES.4.

Table ES.4. Countries with largest Known Conventional Resources recoverable at <USD 130/kgU (2003)[1]

Country	tU	Percentage of world total
Australia	1 058 000	23.1
Kazakhstan	847 620	18.5
Canada	438 544	9.6
South Africa	395 670	8.6
United States[2]	345 000	7.5
Others (total)	1 503 166	32.7
World total	4 588 000	100

1. Includes Reasonably Assured Resources (RAR) and Estimated Additional Resources - Category I (EAR-I) at <USD 130/kgU.
2. The United States does not report resources in the EAR-I category.

Market price indirectly affects resources because it drives exploration expenditures and cut-off grades in market-based economies, as well as other parameters used in resource calculations. However, because of the time lag between exploration and resource reporting, the relationship is seldom readily apparent.

Depletion of resources through production has not yet become a factor as far as the adequacy of supply is concerned. Despite cumulative production of more than 2.2 million tU through 2003, additions to resource totals have kept pace with production so that overall resource levels have remained level or have increased over time. The ratio between Known Conventional Resources and reactor-related uranium requirements in 2003 was 52 compared to an average of 47 since 1985.

Production

Uranium production in 1945 is estimated to have totalled 507 tU. By 1965, when the first Red Book was published, production totalled 31 564 tU. Production peaked in 1980 at 69 692 tU from 22 countries. In 2003, uranium production was reported by 19 countries with output totalling 35 492 tU. Cumulative worldwide uranium production between 1945 and 2003 totalled 2 204 732 tU with production having been reported or estimated from 35 different countries since 1945. The leading countries in cumulative uranium production from 1945-2003 are listed in Table ES.5.

Table ES.5. Leading uranium producer countries based on cumulative production (1945-2003)

Country	tU	Percentage of world total
USSR[1]	377 613	17.1
Canada	374 548	17.0
United States	356 485	16.2
Germany[2]	219 239	9.9
South Africa	157 618	7.1
Others (total)	719 229	32.7
World total	2 204 732	100

1. Only includes production until 1991.
2. Includes production of GDR (1946-1989) and Federal Republic of Germany (FRG) (1961-2003).

Primary supply exceeded reactor-related uranium requirements until 1991, when that relationship was reversed. Since 1991, the gap between primary supply and uranium requirements has been filled by secondary supply (e.g. uranium enriched for military use declared excess to national security needs by the Russian Federation and the United States, reprocessed uranium and plutonium), with requirements in 2003 being met almost equally by primary and secondary supply.

Historically, uranium production has averaged about 76% of capacity, and this share has varied within a range of 57-88%. The fact that production has never matched capacity is largely attributable to the uranium industry having to lower output to match demand for primary supply. Slower growth of nuclear power and competition from secondary supply significantly reduced primary supply demand.

The time to bring a deposit into production after its discovery has steadily increased over the years from just a few years in the early 1950s to as much as 20-30 years at the end of the 20th century.

Natural and enriched uranium stocks and inventories

Cumulative worldwide production is estimated to have exceeded requirements by approximately 690 000 tU between 1945 and 2003. This total can be seen as an estimate of the amount of already-mined uranium, some of which may become available for commercial use in the future. However, it more probably represents an upper bound of total already-mined uranium and not a true inventory of excess material that will be available for civilian use. Nevertheless, natural and enriched uranium stocks and inventories represent a significant resource that has the potential to continue to supplement primary supply in the future.

Inventories of natural and enriched uranium have been reduced over time as seen by the ratio between inventory and requirements, which decreased from 5.7 in 1983 to 1.5 in 2001. The ratio has varied within a narrow range at about 1.5 since 1999, suggesting that utility-held inventories are approaching strategic levels, with much less discretionary material available to the market.

Some key messages

While there are many lessons available from the facts presented in the text and tables of this report, several key messages deserve highlighting as the world considers whether to expand the use of nuclear energy in a manner not seen since the 1970s. First, past exploration for uranium has resulted in the discovery of uranium deposits and ultimately led to recoverable resources. A period of low levels of exploration expenditures ended in late 2000 having lasted for over 20 years. Since 2001, exploration expenditures have steadily increased as they follow the market price upward from historic lows. It can be expected that this new period of exploration will result in the discovery of new deposits of uranium and in an increased resource base. Despite these low levels of exploration and the cumulative uranium production of over 2 200 000 tU since 1945, reported uranium resources have steadily increased since the mid-1980s. The analysis of annual reactor-related requirements to reported resources shows a forward looking reserves ratio that has averaged about 45 over the past 20 years despite steadily increasing requirements (Figure ES.1). Taken together the lessons of the past provide confidence that uranium resources will remain adequate to meet projected demands even were requirements to significantly increase.

Figure ES.1. Ratio between uranium resources and uranium requirements

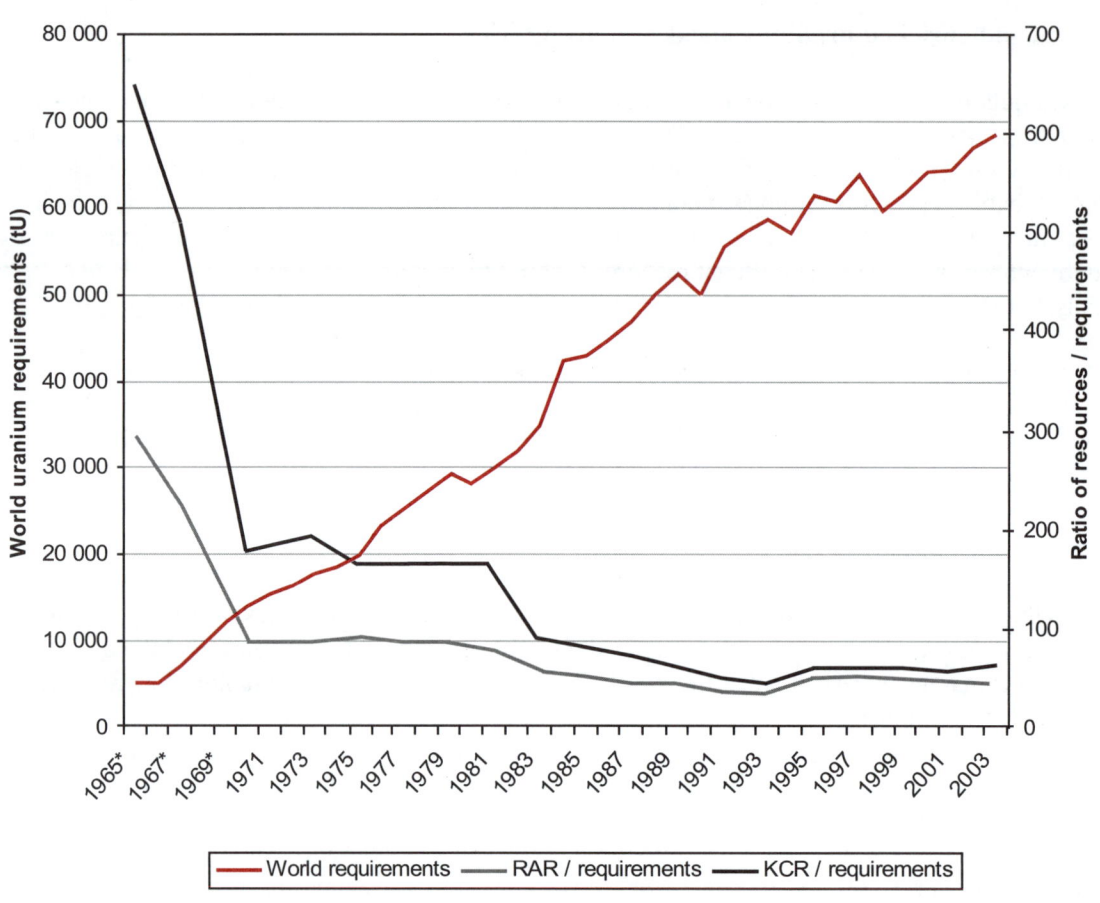

1. HISTORICAL PERSPECTIVE

Red Book history

The history of the Red Book dates back to January 1965, when the European Nuclear Energy Agency (ENEA) of the Organisation for Economic Cooperation and Development (OECD) established a Study Group on the long-term role of nuclear energy in Western Europe. A "Working Party", the precursor of the Uranium Group was formed by the ENEA to compile worldwide uranium and thorium resource estimates. The Working Party, which became a joint effort between the ENEA and the International Atomic Energy Agency (IAEA) about a year later, was responsible for preparing a series of 10 assessments of worldwide uranium supply between 1965 and 1982. A number of working groups were subsequently established under the direction of the NEA/IAEA to gather and publish information on a broad range of topics related to uranium exploration, resources and extraction.

In 1983, efforts to consolidate the various working groups that had been formed either jointly or individually by the NEA and the IAEA to provide information on various areas of uranium technology were unsuccessful. Accordingly, the NEA Uranium Group, which was formed in 1984 to plan and prepare the Red Book, included representatives from NEA member countries and a representative from the IAEA Secretariat. In 1992, non-OECD member countries were invited to attend Uranium Group meetings as part of the IAEA delegation. This broad participation has ensured greater availability to worldwide information on a broad range of uranium supply and demand issues. The Uranium Group has grown from 11 experts from six countries in 1965 to 47 experts from 22 countries and three international organisations in 2003. In the process the Red Book has expanded into an authoritative source of government-sponsored information on countries that have produced or used uranium for civilian purposes. Additional details regarding the history of the Uranium Group is available in Reference 1-1.

The first assessment of world uranium resources, which was titled *World Uranium and Thorium Resources*, was published in 1965 under the auspices of the ENEA. That first publication had a red cover, as have all subsequent editions; hence the informal name "the Red Book". Eleven experts from six countries – Canada, France, Spain, Sweden, the United Kingdom and the United States – and the ENEA contributed to the first Red Book. In that first edition uranium resources were listed for 16 countries along with a category "Others", which included Germany, Italy, Turkey and Yugoslavia. Uranium resources were listed for two confidence categories – Reasonably Assured Resources and Possible Additional Resources – and for three cost categories: USD 5-10/lb U_3O_8, USD 10-15/lb U_3O_8 and USD 15-50/lb U_3O_8.

Thorium resources (RAR and Possible Additional Resources) recoverable at between USD 5/lb and 10/lb ThO_2 were also listed for six countries and one region. In addition, the first Red Book included a separate section on "Uranium from the Sea".

Since that first edition of the Red Book the world has changed and so has the Red Book. By 1967 the ENEA and the IAEA had agreed to jointly publish the Red Book. Because of the near absence of a commercial market for thorium, the 1967 Red Book was re-titled *Uranium Resources Revised Estimates*; thorium occupied only a short section of the report. During its 38-year history to 2003, 19 Red Books were published.

Figure 1.1 shows the history of participation in the Red Book by showing the number of countries that submitted country reports for each edition. The number of country reports grew from 11 in the 1965 edition to a maximum of 59 reports in 1998. In total, 107 countries have provided data for inclusion in the Red Book through 2003. Appendix 1.1 lists countries that have participated in the Red Books by their geographical region and gives the number of Red Book editions that each country has contributed to. Appendix 1.2 lists the countries that participated in each edition the Red Book. As countries drop out of Red Book participation, information on their past activities has not always been carried forward to subsequent Red Book editions. Therefore, Appendices 1.1 and 1.2 serve as a reminder of past exploration activities that have contributed to the worldwide uranium geologic database and for which Red Books contain information on those activities.

Figure 1.1. History of participation in the Red Book process as measured by numbers of country reports that were submitted over time

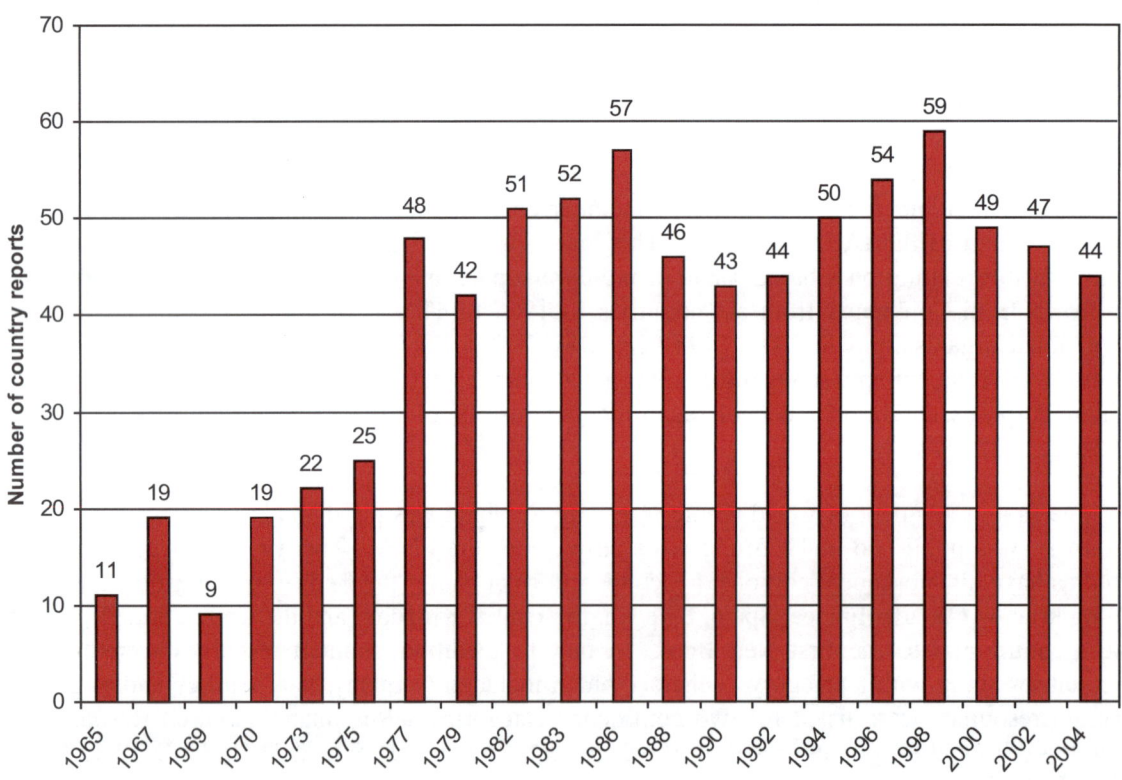

A comparison of information from the 1965, 1983 and 2003 Red Books highlights some of the significant changes in the uranium industry and in the Red Book from its beginning, at the midpoint of its 38-year history and in the 2003 edition (Table 1.1).

Table 1.1. Comparison of selected Red Book statistics

	1965	1983	2003
Number of pages in Red Book	22	348	288
Number of countries that reported RAR	16	30	42
Number of reactors operating	40	301	435
Reasonably Assured Resources (highest cost category, 1 000 tU)	1 343	2 043	3 169
Total RAR reported by Australia (1 000 tU)	15	336	735

Prior to 1989, information contained in the Red Books was almost entirely restricted to what was referred to as the World Outside Centrally Planned Economy Area (WOCA). Since then, however, membership in the Uranium Group and information contained in the Red Book has steadily grown and become more inclusive and globally representative.

The history of the Red Book has paralleled the growth of nuclear energy but has also been influenced by world events. Foremost among these was the Cold War, during which military require-ments for uranium played a major role in shaping the uranium market. Other significant world and industry events that helped shape the uranium industry and which are reflected in the Red Book over time include:

- First commercial nuclear power plant (1957).
- Oil crisis and emergence of nuclear power as a viable alternative to fossil fuel (1973).
- Discovery of unconformity-related deposits in Australia (1969) and Canada (1968).
- Discovery of Olympic Dam; the world's largest uranium deposit (1976).
- Three Mile Island accident (1979).
- Chernobyl accident (1986).
- End of the Cold War and dissolution of "Eastern Bloc" (1990) and the USSR (1991).
- Emergence of the importance of secondary supply including down blending of weapons grade uranium for civilian use (1993).
- Realisation of declining secondary sources and rising expectations of new nuclear construction driving prices upward (2003).

In response to these and other events, the scope of the Red Book has changed over time to keep up with the increasing complexity of the uranium industry. The first Red Book in 1965 was exclusively devoted to uranium and thorium resources. By 1969, the scope of the Red Book began to expand. The 1969 Red Book, titled *Uranium Production and Short Term Demand* recognised the link between supply and demand. It included discussions of uranium production and nuclear generating capacities and projections of generating capacity and uranium demand through 1980, including low and high demand cases. By 1970, only five years from publication of the first edition, the Red Book had evolved to cover a broad range of subjects at the front end of the nuclear fuel cycle. The title of the 1970 Red Book, *Uranium Resources, Production and Demand* is essentially the same as that of the 2003 edition, *Uranium 2003: Resources, Production and Demand*. The 1970 edition covered uranium supply and demand, including projections through 1985; it also included brief discussions of enrichment capability and UF_6 conversion capacity. The Red Book has continued to grow since 1970, with the inclusion of more countries and a broadening of scope to include continually emerging issues such as secondary supply and environmental aspects of uranium mining and production. In 1982, except for summary information thorium was dropped from the Red Book because of the absence of a commercial market for the commodity. In keeping with the growing importance of environmental awareness in uranium production, a section was added to the Red Book in 1995 on radiation safety and environmental aspects of uranium mining and production. The 2003 report saw inclusion for the first time of data on secondary sources of uranium including mixed-oxide fuels, former weapons materials and re-enriched tails.

The Red Book process regularly brings together representatives from throughout the nuclear community. The Uranium Group meetings during which Red Book data are reviewed and sanctioned for publication provide a forum for representatives from participating countries to exchange information on their respective nuclear industries. Data presented in the Red Book are official information submitted by the governments of participating countries. It is relatively free of influence from commercial interests and as such represents the most authoritative publicly available data source for uranium resources, exploration, production and production capacity.

Red Book Retrospective

The goals in developing this publication were to:

- Compile a complete and concise data set of uranium exploration, resources and production statistics that were published in Red Book editions between 1965 and 2003.

- Solicit new information on uranium exploration, resources, production and stocks/inventories to fill gaps in the data record which was available in previous Red Book editions to improve and broaden historical perspective.

- Provide an historical perspective of uranium supply and demand relationships over time; and

- Determine lessons to be learned from the past that may have applicability to present conditions.

Just as the Red Book has always been a collaborative effort with results depending on contributions of information by participating countries, so too has the production of this report been a collaborative effort. This collaboration went well beyond the immediate team assembled to compile and write it and involved contributions from colleagues worldwide that gave considerable time and energy to fill in gaps in the data and thus provide a more complete perspective on issues relating to uranium supply and demand.

Collectively, the 20 Red Books published from 1965 through 2004 represent arguably the most comprehensive publicly available set of data on uranium resources, exploration, production and demand. They include statistics at various points in the industry's history on uranium resources and production, the two key components of uranium supply. The Red Books also contain demand statistics and projections, which are a measure of how experts viewed the future of the industry, including estimates of nuclear generating capacity and reactor-related uranium requirements in rolling 10- to 20-year "windows" of time. This Red Book Retrospective was conceived as a way to capture the information contained in the past Red Books, thus providing an historical perspective of the uranium industry on topics ranging from resources and production capacity (supply) to nuclear generating capacity and reactor-related uranium requirements (demand).

At the time of its publication, every effort was made to ensure that each Red Book was as accurate and complete as possible. At the same time, it is recognised that new information has come to light in recent years that was not readily available when past Red Books were being prepared. Therefore, in addition to capturing information included in past Red Books, every effort has been made to fill in gaps in past information to ensure that this retrospective offers the most complete perspective on the history of uranium supply and demand possible. In particular, information previously unavailable on exploration and production activities in the former Soviet Union has been incorporated into this publication. Though every effort has been made to gather supplementary information from knowledgeable experts in their respective countries, the authors have also made estimates (duly acknowledged) to ensure the most complete data set possible.

This report includes analyses and interpretation of the Red Book data by key subjects (installed generating capacity and uranium requirements, uranium market price, resources, etc.) as well as appendices of the consolidated raw data in spreadsheet format that will allow readers to perform additional analysis, if so desired. Most chapters in the report have associated appendices with spreadsheets that include the data on which charts and graphs in the body of the text are based.

2. INSTALLED NUCLEAR CAPACITY

Civilian use of nuclear power was initiated in the United Kingdom in 1957 with the opening of the Calder Hall 1 nuclear reactor, with generating capacity during the first year of 50 megawatts electric (MWe). From that modest beginning the industry grew to 435 operating reactors with a generating capacity of over 359 400 MWe in 2003, when nuclear power accounted for 16% of the world's electricity output (Figure 2.1). In addition, in 2003 there were 33 reactors under construction in 11 countries. Generating capacity grew at an average annual rate of about 20% between 1956 and 2003. Though masked by the scale in Figure 2.1, generating capacity expanded at an annual rate of about 55% between 1957 and 1973. By comparison the rate of growth between 1973 and 1990, the peak building years, was about 13% annually. Between 1990 and 2003, the average annual rate of growth in generating capacity was less than 1% per year. The leading countries in terms of nuclear generating capacity in 2003 are listed in Table 2.1.

Table 2.1. Worldwide leadership in nuclear generating capacity (as of 1 January 2004)

Country	Generating capacity (MWe net)	Number of operating reactors
United States	98 357	104
France	63 180	58
Japan	43 197	52
Germany	21 283	19
Russian Federation	20 793	30
Others (total)	112 623	172
World Total	359 433	435

The history of nuclear generating capacity can be divided into three major eras: early growth (1957 to 1973); major expansion (1973 to 1990); and slow growth (1990 to 2003). During the early growth period capacity expanded at an average of about 2 400 MWe per year and the number of operating reactors increased from two in 1957 to 109 in 1973. Between 1973 and 1990 capacity expanded at an average of 16 060 MWe per year as more plants came on line; 109 operating plants in 1973 to 413 plants operation in 1990. The rapid growth era ended abruptly, however, mainly as a result of the Three Mile Island accident in the United States in 1979 and the Chernobyl accident in Ukraine in 1986. The increase in generating capacity between 1990 and 2003 averaged only 2 315 MWe per year. Between 1990 and 2003, 59 new plants were constructed and 50 plants were closed, resulting in a net increase of only nine reactors between 2003 and 1990.

Figure 2.2 shows the buildup of generating capacity on a country-by-country basis. Appendix 2.1 provides brief histories of the buildup of nuclear energy in each of the countries shown in Figure 2.2. Detailed data on which Figures 2.1 and 2.2 are based are contained in Appendix 2.2. Three countries, France, Japan and the United States, dominated the historical buildup of nuclear generating capacity and by 2003 accounted for 58% of worldwide generating capacity (Figure 2.2). Table 2.2 compares the numbers of reactors and the percentages of worldwide nuclear generating capacity that these three countries contributed at 10-year intervals between 1970 and 2000. The relative contributions of the three countries remained virtually unchanged during the past 30 years. Though their reactor fleets and capacity continued to grow through the 1980s, capacity elsewhere was also expanding at a comparable rate.

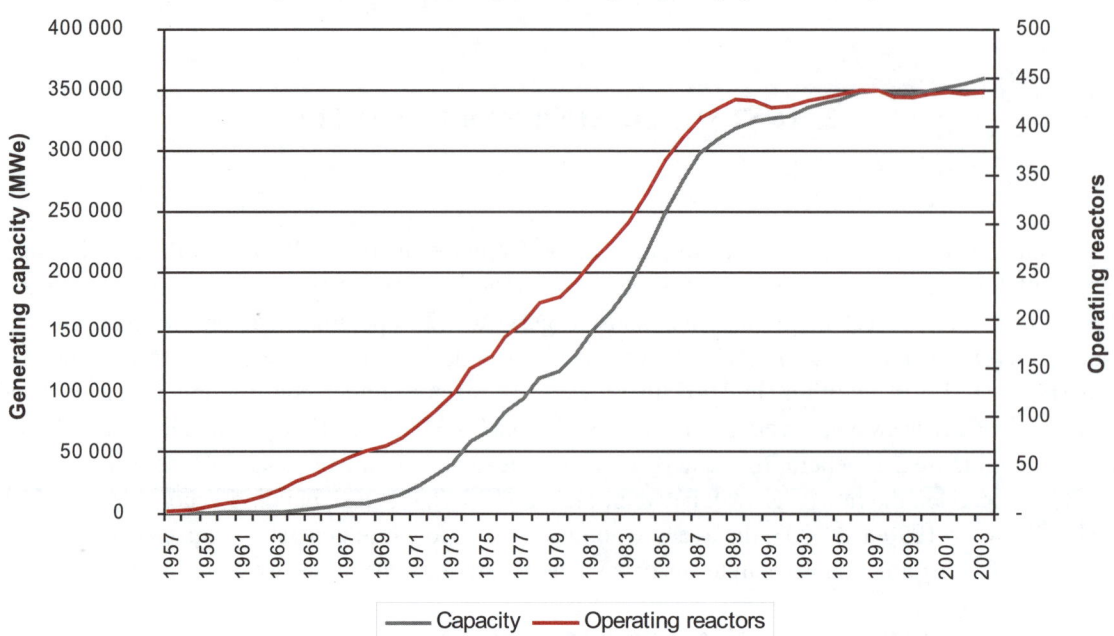

Figure 2.1. Nuclear generating capacity (1957-2003)

Capacity —— Operating reactors

Table 2.2. Growth history of nuclear power in France, Japan and the United States

	First year of nuclear power	1970		1980		1990		2000	
		Reactors	% world capacity	Reactors	% world capacity	Reactors	% world capacity	Reactors	% world capacity
France	1959	7	10	21	11	56	17	58	18
Japan	1965	4	8	22	11	40	10	51	12
United States	1957	18	39	69	38	111	31	104	28
Total		29	57	112	60	207	58	213	58

The historical growth in nuclear generating capacity resulted from a combination of increased numbers of operating plants and a corresponding increase in average plant capacity. Figure 2.3 compares the historical growth in plant capacity with the addition of operating reactors. The average generating capacity of all reactors operating in 1973 was 348 MWe. The average capacity of the reactors added between 1973 and 1990, the period of rapid growth for nuclear energy, was 797 MWe. Between 1973 and 1990, the number of reactors and the average generating capacity of operating reactors increased by 278% and 223%, respectively. Since 1990, the average capacity of new reactors has been 898 MWe.

The overall history of nuclear power has been one of the steady additions of new reactors and increased capacity. Figure 2.4, however, shows that though the overall trend has been to increase the number of reactors, there were also plant closures that slowed the growth of nuclear power. The first plant closures took place in 1968 and closures continued intermittently through 2003. Though the trend has historically favoured plant additions over plant closures, closures exceeded additions in 1990, 1991, 1998 and 2001. The closure of 20 plants (net loss of eight plants) in 1990-91 marked the end of the era of rapid growth of nuclear power.

22

Figure 2.2. Nuclear generating capacities (1957-2003)

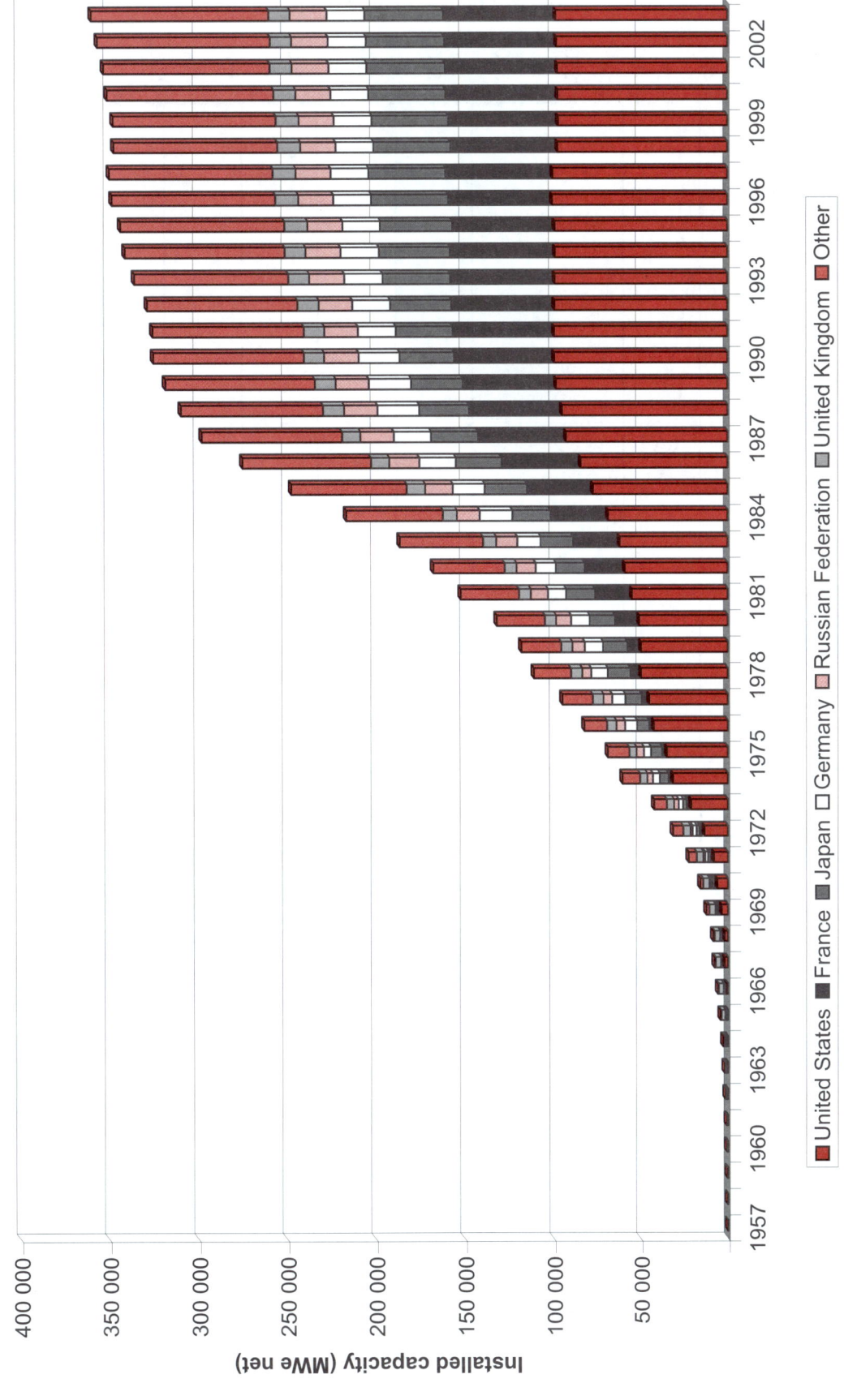

Figure 2.3. Average generating capacity compared with the number of operating reactors

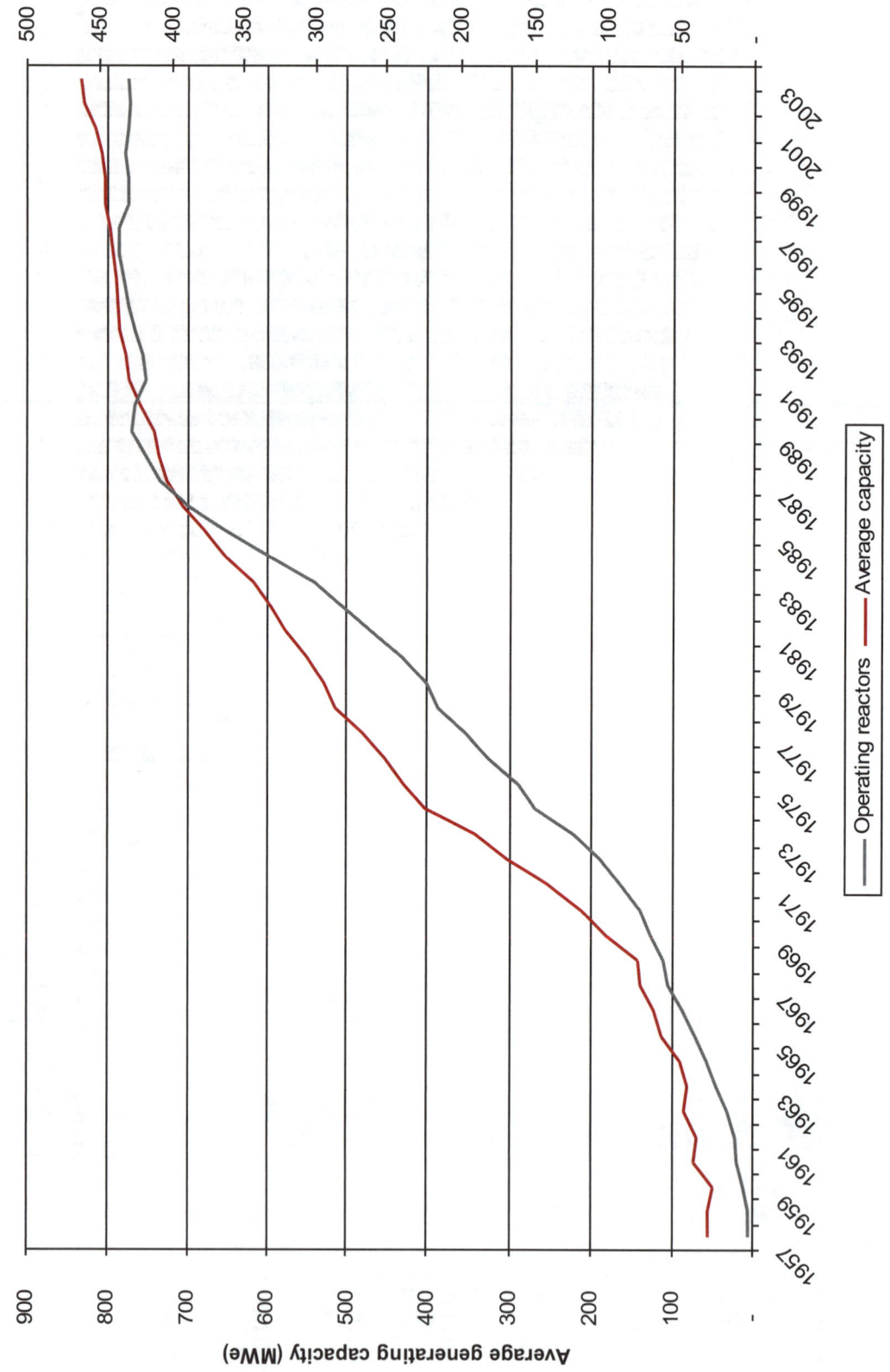

— Operating reactors   — Average capacity

Figure 2.4. History of reactor additions and closures

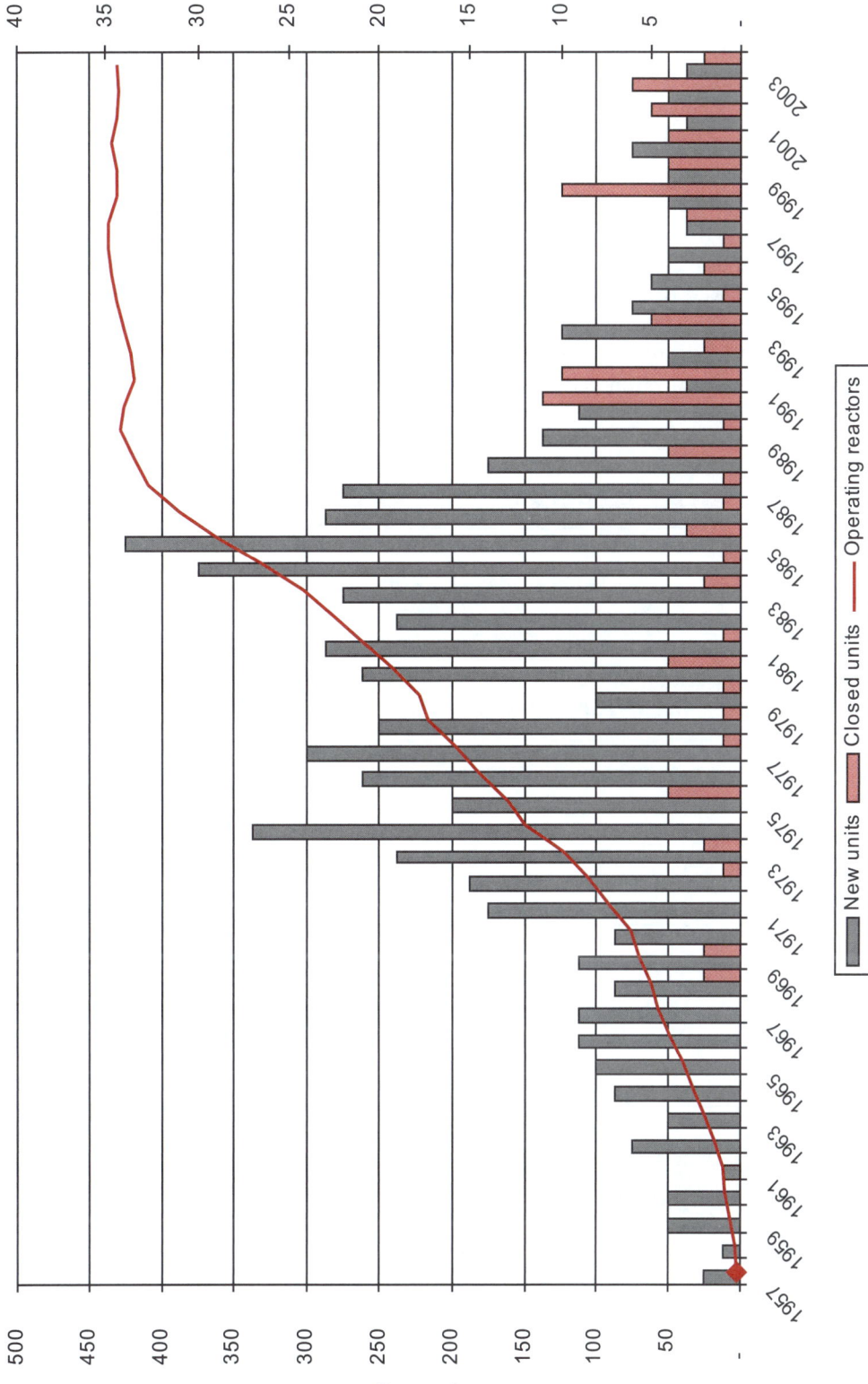

Accuracy of Red Book projections

The 1969 Red Book was the first edition to make projections of the growth of nuclear power, a tradition that has continued through the 2003 edition. One of the objectives of Red Book Retrospective is to assess the accuracy of the Red Book's projections in nuclear generating capacity, thus providing a yardstick for measuring future reactor uranium requirements. Table 2.3 compares projections of nuclear generating capacities for WOCA countries from the 1977 and 1982 Red Books for the years 1980, 1985 and 1990 with actual capacities for those same years. Figure 2.5 portrays the same information graphically for the low capacity projection. These two Red Book editions were selected because they span the time period immediately before and after the Three Mile Island accident (1979), which had a profound impact on the industry. As shown in both portrayals, the forecasts of generating capacity in the 1982 Red Book were lower than those in the 1977 edition, as the impact of Three Mile Island was already being felt by the industry. As a result of this adjustment, the low case forecast in the 1982 Red Book was only 4% higher than the actual capacity for 1985; however, the difference between the low case projection and actual capacity reaches 22% for the 1990 case.

Table 2.3. Comparison of projected and actual nuclear generating capacities for WOCA countries (GWe)

Red Book edition	1980			1985			1990		
	Low	High	Actual	Low	High	Actual	Low	High	Actual
1977	146	146	144	278	368	222	504	700	280
1982				232	258	222	361	401	280

Figure 2.5. Comparison of nuclear generating capacity projections with actual capacities for WOCA countries – 1977 and 1982 Red Books

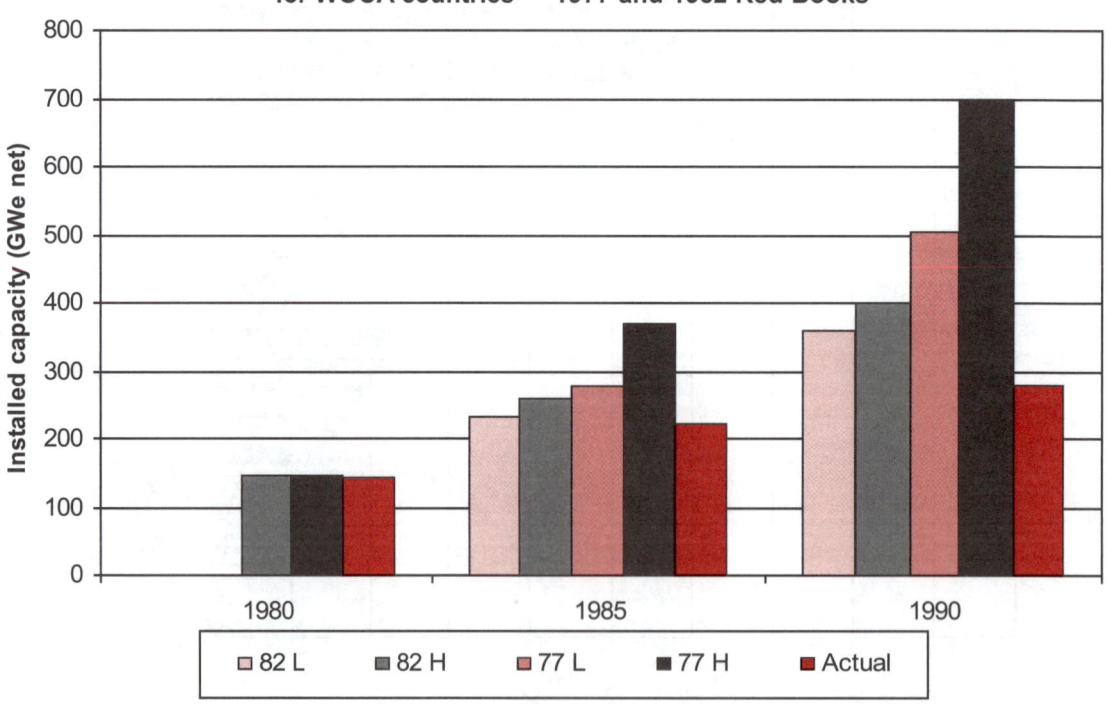

Table 2.4 compares generating capacity projections from the 1982, 1983, 1986 and 1989 Red Books. The downward trend in capacity forecasts noted in Table 2.2 between 1977 and 1982 continued through the 1989 edition. Each forecast is lower in the subsequent three Red Books than forecasts in comparable timeframes in the 1982 Red Book. The low forecast for 1995 in the 1989 Red Book is 67% of the 1982 forecast. Similarly, the high forecast for 2025 in the 1989 edition is only 28% of the 1982 forecast. Similar comparisons with Red Books after 1989 cannot be easily made because "world" statistics are presented for generating capacities rather than being limited to WOCA countries.

Table 2.4. Comparison of projected nuclear generating capacities (GWe)

Red Book edition	1995		2000		2005		2010		2015		2020		2025	
	Low	High	Low	High	Low	High	Low	High	Low	High	Low	High	Low	High
1982	451	562	585	804	725	1 120	880	1 503	1 034	1 928	1 180	2 366	1 311	2 794
1983	412	425	504	558	608	773	736	1 091	888	1 483	1 049	1 917	1 209	2 370
1986	348	360	402	500	466	677	523	875	579	1 105	630	1 325	675	1 555
1989	304	309	328	329	368	412	415	510	460	595	510	685	560	780

Table 2.5 compares actual generating capacities in 1995, 2000 and 2005 with the projections made in the 1982 through 1989 Red Books. Even the low projections for these three years significantly overestimate what capacities would be, though the projections got progressively more accurate with each successive Red Book edition. For example the 1982 low forecast for 2005 was 1.9 times actual, while the 1986 forecast for 2005 was only 1.2 times actual.

Table 2.5. Comparison of projected and actual WOCA generating capacities (GWe)

Red Book Edition	1995			2000			2005		
	Low	High	Actual*	Low	High	Actual*	Low	High	Actual*
1982	451	562	293	585	804	303	725	1 120	385
1983	412	425	293	504	558	303	608	773	385
1986	348	360	293	402	500	303	466	677	385
1989	304	309	293	328	329	303	368	412	385

* Capacity estimates for 1982-1989 Red Books were for WOCA countries. Every effort has been made to equate the actual capacities to WOCA countries to ensure a direct comparison, though later Red Books from which the actual totals were taken do not use the WOCA designation.

Conclusions relating to installed nuclear capacity

Nuclear generating capacity has grown from 50 MWe in 1965 to about 359 400 MWe in 2003. This growth occurred in three phases: "early growth" between 1957 and 1973 when annual capacity increased an average of 2 400 MWe per year; "major expansion" of nuclear power between 1973 and 1990, during which annual capacity grew at an average annual rate of 16 060 MWe, and "slow growth" between 1990 and 2003 when growth slowed dramatically to an average of only 2 315 MWe per year.

As a measure of the adequacy of uranium resources and production capacity, Red Books beginning with the 1969 edition have included forecasts of future generating capacity. Early estimates of future generating capacities consistently exceeded actual capacities, but forecasting has become more accurate with time. One lesson from this history is that long-term estimates reflect the perceptions of the time they are made as much as they do about the future, and estimates of installed capacity made today should be viewed in terms of this tendency to over-estimate during periods of rapid expansion and optimism.

3. REACTOR-RELATED URANIUM REQUIREMENTS (DEMAND)

As used in the Red Book, annual reactor-related uranium requirements *"refers to natural uranium acquisitions not necessarily consumption during a calendar year"*. Between 1956 and 2003, 33 different countries have used commercial nuclear reactors and have had reactor-related requirements that are estimated to have totalled 1 513 327 tU. Detailed annual data are available in Appendix 3.1. Table 3.1 shows the reactor-related requirements for the five major users of uranium over that period.

Table 3.1. Cumulated reactor-related uranium requirements (1956-2003)

Country	Reactor-related requirements	Percentage of world total
United States	364 180	24.1
France	173 837	11.5
Japan	163 520	10.8
Russian Federation	94 925	6.3
Germany	73 842	4.9
Others (total)	643 023	42.4
World total	**1 513 327**	**100.0**

Figure 3.1 compares worldwide historical requirements and generating capacity between 1957 and 2003. Though the two curves have the same approximate trend, they are not a perfect match, nor would they be expected to be. While both are controlled by the mix of reactor types, requirements also vary depending on load factors, fuel loading cycles, burn up, fuel management policies, strategic stockpile requirements and the extent of use of MOX fuel.

Figure 3.1. Comparison of historical generating capacity and uranium requirements

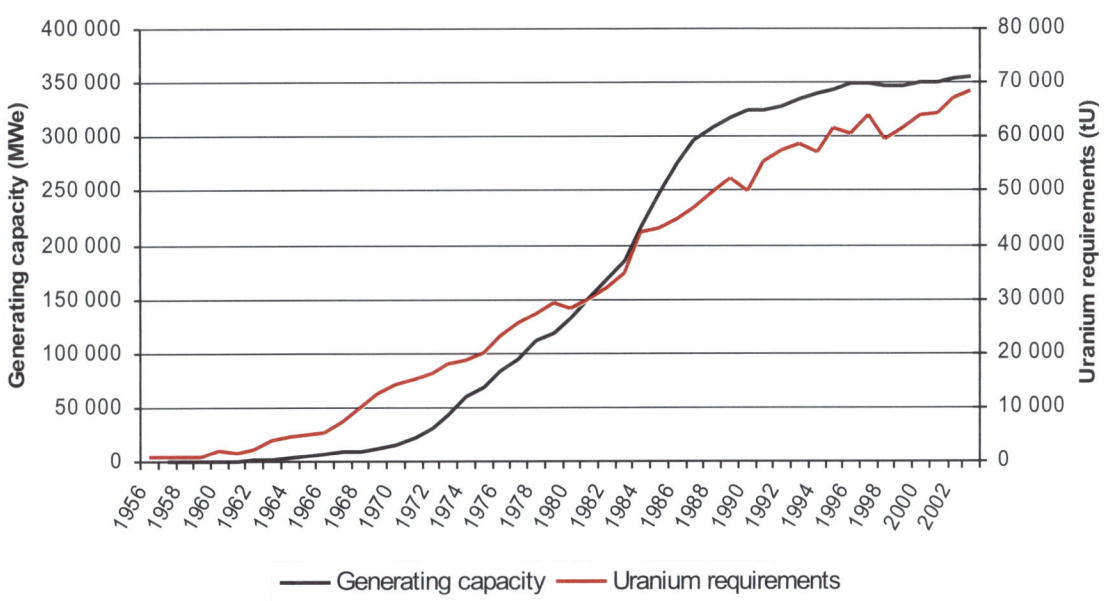

The 1969 Red Book was the first edition to make projections of uranium requirements, a tradition that has persisted through the 2003 edition. The data on which tables and figures in Chapter 3 are based are included in Appendices 3.1 and 3.2. Appendix 3.1 lists annual uranium requirements on a country by country basis; Appendix 3.2 compares forecasts of requirements with actual requirements.

Table 3.2 shows that the accuracy of forecasting uranium requirements improved between 1969 and 1973. Actual uranium requirements in 1975 were significantly lower than the forecast, at 55% of the low forecast for 1975 made in 1969 and 41% of the high forecast. Forecasting had improved by 1973; actual requirements in 1975 were 82% and 72% of the low and high projections of 1973, respectively. It should be remembered that these forecasts were made before the Three Mile Island accident and were made when nuclear power was considered the "energy source of the future". Table 3.3 summarises projections made after the Three Mile Island accident and shows how Red Book forecasting changed. In comparing the two tables it is evident that forecasting improved with the disparity between actual and forecasted requirements decreasing in successive Red Books. From 1986 to 1993 only one projection was made (i.e. no high-low range) in the Red Book.

Table 3.2. Comparison of projected and actual WOCA uranium requirements (tU)

Red Book edition	1975			1980		
	Low	High	Actual	Low	High	Actual
1969	33 840	45 370	18 750	56 140	81 538	26 000
1973	23 000	26 000	18 750	51 000	66 000	26 000

Figure 3.2 depicts the data in Table 3.3 in graphic format. The trend in forecasting as shown in Table 3.3 and Figure 3.2 has become more conservative. Forecasts published in 1982 and 1986 significantly exceeded actual requirements for the three years analysed in Figure 3.2. As shown in Figure 3.2, however, there is a very close grouping of actual and forecast lines for the remaining Red Book editions, with forecasts understating actual demand in the 1989 and 1993 Red Books, reversing the trend of over optimism of previous years. The slow growth in installed capacity beginning in the 1980s no doubt played a role in improving the accuracy of projections.

Table 3.3. Comparison of projected and actual uranium requirements for WOCA (tU)

Red Book edition	1985			1990		
	Low	High	Actual	Low	High	Actual
1982	44 000	45 000	36 100	53 000	65 000	41 919
1986				48 000		41 919
1988				44 800		41 919
1989				41 913		41 919
1993						

Red Book edition	1995			2000		
	Low	High	Actual	Low	High	Actual
1982	64 000	93 000	47 700	81 000	129 000	49 240
1986	55 000		47 700	62 000		49 240
1988	47 200		47 700	52 400		49 240
1989	43 567		47 700	48 954		49 240
1993	46 502		47 700	45 353		49 240

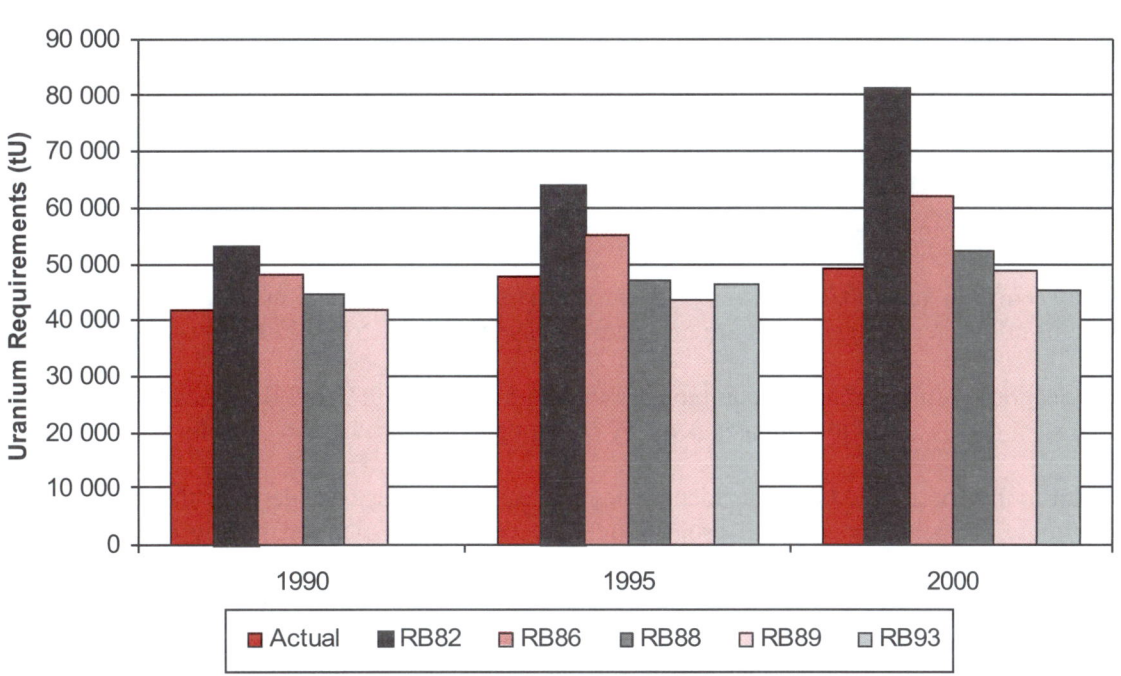

Figure 3.2. Comparison of projected and actual uranium
requirements for WOCA

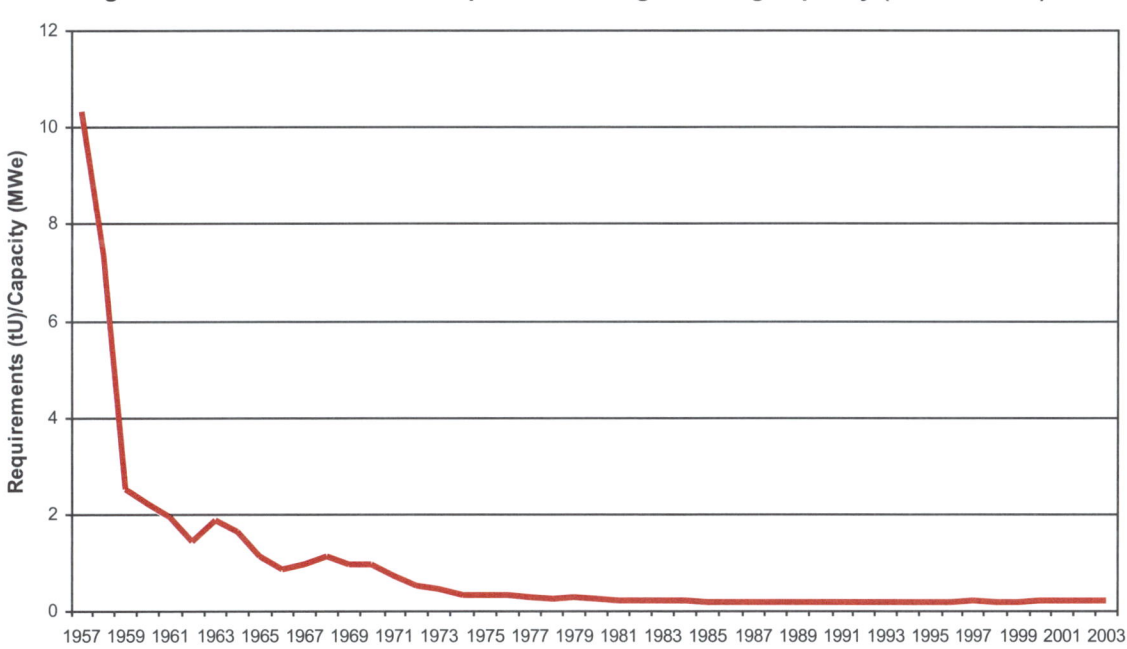

Figure 3.3. Ratio of uranium requirements to generating capacity (1957 to 2003)

There is a direct relationship between projected uranium requirements and projected generating capacity. Capacity drives requirements – if capacity projections miss the mark, so too will requirements. Figure 3.3 compares actual capacity with actual reactor-related uranium requirements between 1957 and 2003 and shows that the ratio of requirements to capacity (tU/MWe) varied dramatically between 10.3 in 1957 and 0.22 in 1980. Between 1981 and 2003, the ratio varied between 0.16 and 0.20 and averaged 0.18. This stability reflects constancy in the relationship between capacity and requirements, which may have allowed for improved forecasting beginning in the late 1980s.

Conclusions relating to reactor-related uranium requirements

Between 1956 and 2003, 33 different countries have used commercial nuclear reactors and have had reactor-related requirements that are estimated to have totalled 1 513 327 tU.

Projections of future uranium requirements were started in the 1969 Red Book and have continued to the present day as a way to evaluate adequacy of resources and production capacity.

Prior to the Three Mile Island accident, when nuclear power was considered as the energy supply for the future, capacity and demand projections were very optimistic, and greatly exceeded actual demand. Today's projections of future installed capacity and uranium requirements should be viewed in light of the tendency to over-estimate during times of optimism and rapid expansion.

Once the uranium industry had adjusted to the reality of the impact of Three Mile Island, however, Red Book projections for future uranium requirements became more conservative and were better reflections of actual demand; made easier by the slow rates of change experienced since the 1980s.

4. URANIUM MARKET PRICE

The market price of uranium is an important barometer of the perceived balance between uranium supply and demand. Sharp increases in the market price suggest potential or perceived supply shortages, thus the need for increased prices to stimulate new supply. Conversely, declining prices indicate a real or perceived supply surplus. The net effect of falling prices is to force a scaling back of the production industry including possible closure of marginal producers to bring supply into balance with demand. In both cases perception of the supply/demand balance is highlighted to emphasise that the uranium market, like all commodity markets, is in part controlled by emotion and in part by hard facts. At any point in time the utility industry's perception of the adequacy of supply translates into the reality of the market.

While the foregoing describes the uranium market of today, historically uranium prices have been influenced by a number of external factors, the most important of which is the fact that for much of its modern history (i.e. beginning in 1945) uranium demand was to a significant degree driven by military requirements tied to the build up of nuclear weapons. Military demand acted to distort uranium's market behaviour as a commodity due to national security requirements and secrecy. Since the end of the Cold War in 1989, however, uranium has gradually begun to behave more and more like a typical commodity, though availability of uranium from dismantling nuclear warheads for civilian use continues to have a significant impact on the uranium market and the balance between supply and demand.

Figure 4.1 shows uranium market price history between 1970 when civilian nuclear power was still in its infancy and 2005 when most of the research for this publication was completed. Market prices are shown both in current US dollars (USD) and in inflation adjusted constant dollars (2003 dollars). Detailed information on uranium market prices (current and constant dollars) is provided in Appendix 4.1. In both current and constant dollars the uranium market price peaked between 1976 and 1978. The 2005 market price of about USD 85/kgU is just under one-third the inflation adjusted peak price of USD 243/kgU in 1976 and 1977.

Figure 4.1. Historical uranium spot market prices

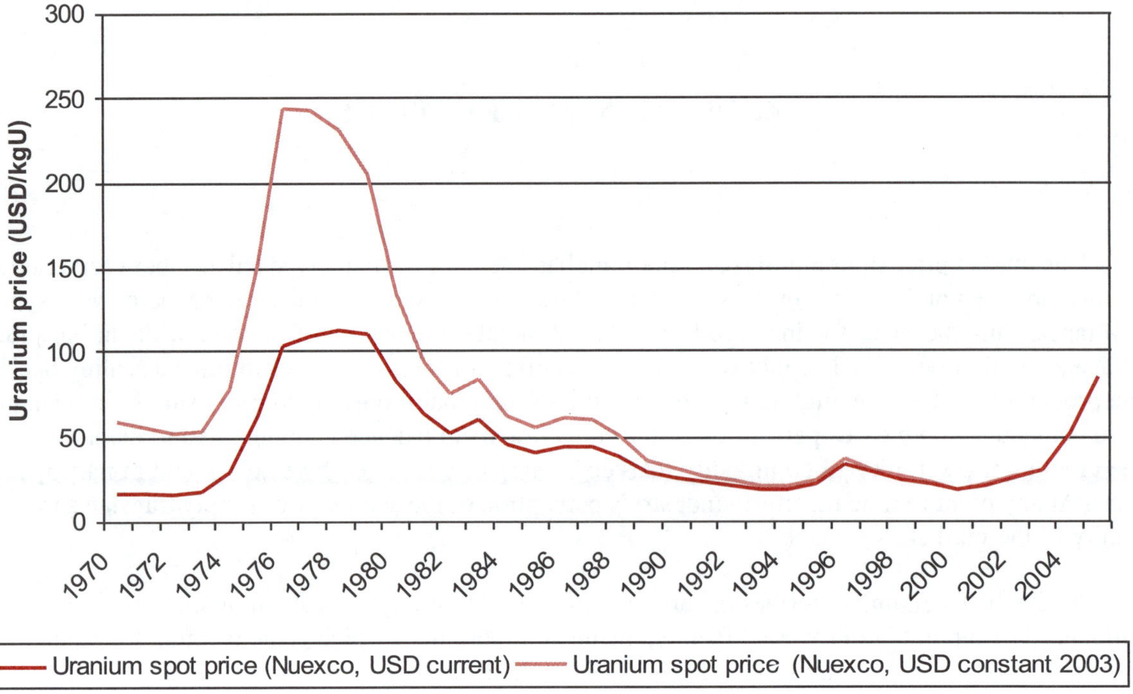

The tripling of the uranium price between 1973 and 1975 was brought about by general concerns about a possible uranium supply shortfall related to growing reactor orders and was accelerated by abrogation of supply contracts by the reactor manufacturer Westinghouse as well as ongoing military requirements. The Three Mile Island accident in 1979 and the resulting cancellation of new reactor orders precipitated the sharp price decline between 1979 and 1981. To put the price of uranium into a broader context of global economics, Figure 4.2 compares uranium price history with that of gold and oil. The information on which Figure 4.2 is based is provided in Appendix 4.2. There are broad parallels among the price histories of the three commodities, with uranium price increases and declines preceding those of oil and gold by two to four years. The parallels are, however, only very general. Uranium is not a fungible commodity; it cannot be "designed out" of nuclear reactors and replaced by a lower priced commodity. Therefore, uranium prices are somewhat detached from broader world economics and tend to move more in response to specific industry events such as:

- the Three Mile Island and Chernobyl accidents that altered the growth of nuclear power;

- accidents at mines and processing facilities that curtailed mine output;

- the influx of secondary supply, which altered the balance between supply and demand.

Though uranium prices do not parallel those of gold and oil, uranium nevertheless increasingly behaved like other commodities. As shown in subsequent chapters, price expectations tend to make supply reactions cyclical: higher prices encourage new supply; lower prices have the reverse effect.

Figure 4.2. Comparison of uranium, gold and oil market price trends

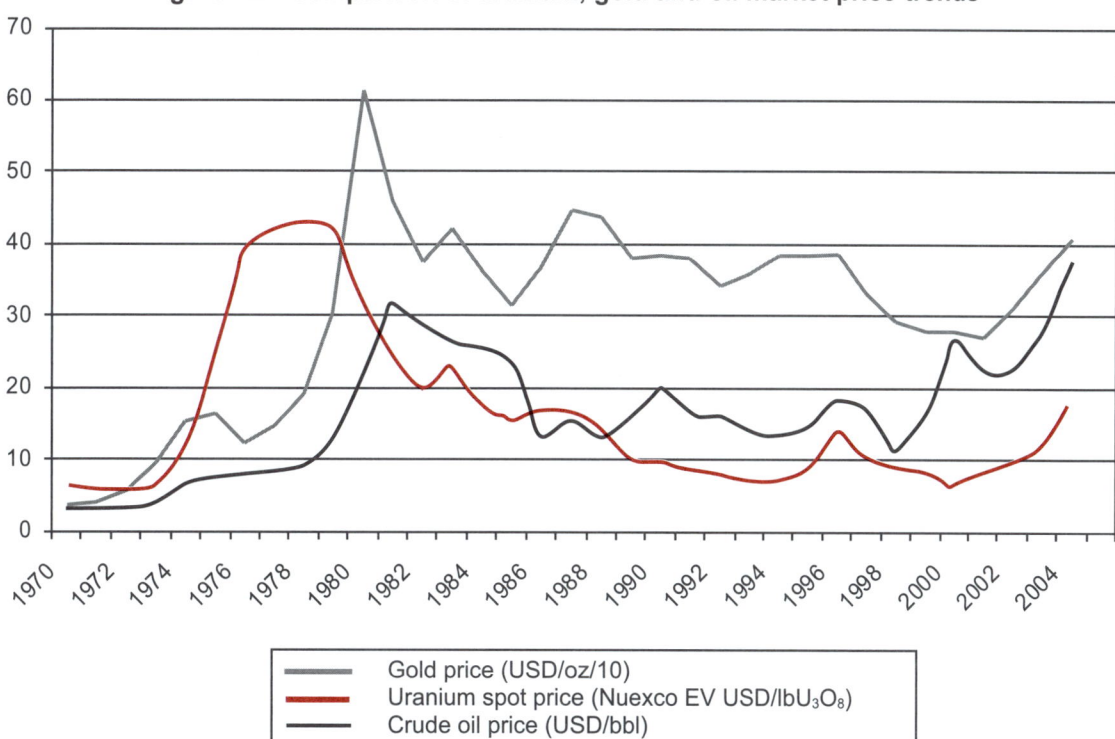

Gold price (USD/oz/10)
Uranium spot price (Nuexco EV USD/lbU$_3$O$_8$)
Crude oil price (USD/bbl)

The most readily publicly available uranium market price information reflects the price of material for delivery within the 12 months succeeding signing of a contract – the so called "spot market price". This information has been published monthly for many years by several industry trade and consulting companies, which base their information on industry contacts and announced spot market sales. Short-term sales, however, represent a relatively small percentage of overall uranium sales, as most utilities prefer the security of supply afforded by longer term contracts to meet a majority of their reactor-related uranium requirements. Therefore, in addition to spot market data, price information on longer term transactions should also be taken into account. Beginning with the 1986 edition, several governments have submitted information on uranium prices that has developed into a picture of the long-term trends in uranium prices worldwide. Figure 4.3, which is based on information submitted by governments for publication in past Red Books, compares short- and long-term contract price trends in Australia, Canada, Euratom, Niger and the United States.

As shown in Figure 4.3, prices, regardless of the time span of contracts or their geographic distribution, have followed broadly parallel trends. At the same time, however, prices have varied considerably on a regional basis over time. There is typically a premium paid for longer term contracts to help protect suppliers from the vicissitudes of the market. For example, average prices for multi-year and spot-price contracts reported by Euratom in 1990 were USD 76.20/kgU (USD 29.31/lb U$_3$O$_8$)* and USD 25.08/kgU (USD 9.65/lb U$_3$O$_8$), respectively. Overall the range of prices narrowed significantly beginning in 1997, however, multi-year and spot prices now vary in a relatively narrow range. For example, average prices for multi-year and spot-price contracts reported by Euratom in 2002 were USD 32.30/kgU (USD 12.42/lb U$_3$O$_8$) and USD 24.24/kgU (USD 9.32/lb U$_3$O$_8$), respectively, a differential of only USD 8.06/kgU (USD 3.10/lb U$_3$O$_8$) compared to USD 51.12/kgU (USD 19.66/lb U$_3$O$_8$) in 1990. Part of the reason for the narrowing of the range of market prices is market volatility. Suppliers are less willing to offer sharp discounts for material sold on the spot market, because in the current tight supply situation there are increased opportunities for longer term sales.

---

* USD 1/lb U$_3$O$_8$ = USD 2.6/kgU.

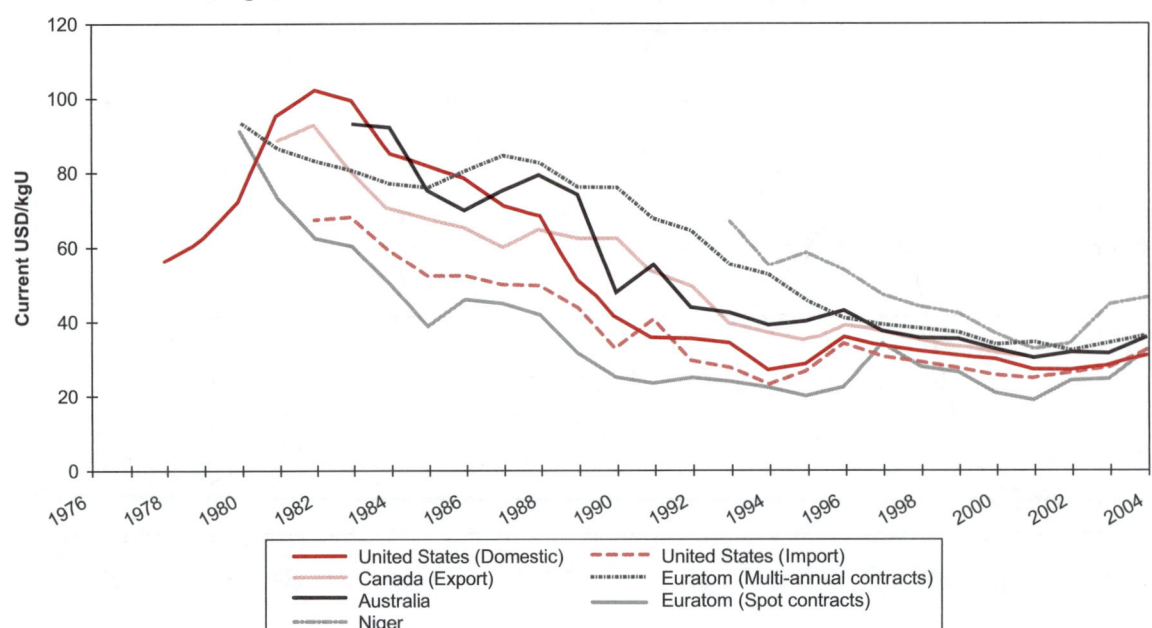

Figure 4.3. Development of uranium market prices over time

Legend:
- United States (Domestic)
- Canada (Export)
- Australia
- Niger
- United States (Import)
- Euratom (Multi-annual contracts)
- Euratom (Spot contracts)

Conclusions regarding market price

While the uranium market price is now largely controlled by perceptions of the balance between supply and demand, this has not always been the case. Prior to the end of the Cold War, military demand acted to distort the treatment of uranium as a commodity due to national security requirements and secrecy.

Even after the end of the Cold War this legacy of military utilisation has lingered, as stocks of highly enriched uranium from dismantling of nuclear weapons became available for civilian use, displacing significant quantities of newly mined and processed uranium. In addition, the Three Mile Island and Chernobyl reactor accidents slowed the growth of nuclear power. As a result, production exceeded requirements resulting in a significant inventory buildup. Drawdown of these inventories combined with the introduction of highly enriched uranium from weapons dismantlement displaced primary production and resulted in price decreases.

Uranium prices reached their historic high in the mid-1970s as a result of general concerns about a possible uranium supply shortfall related to growing reactor orders. Prices began to stagnate in 1982, drifting down to historic lows in late 2000 before beginning an upward trend that continued through 2005 when research for this publication was completed.

5. EXPLORATION

The element uranium was discovered by Martin Klaproth in 1789 in the mineral pitchblende derived from Jachymov (Joachimsthal) of the Bohemian part of the Erzgebirge (Kruzne Hory, Ore Mountains) in what is now the Czech Republic. Through the 19th century there were only limited uses for uranium, mainly ceramic glazes and pigmentation of glass (the famous green Bohemian glass) until the discovery of radioactivity by Rutherford at the end of that century. When the element radium was detected by Marie Curie at the beginning of the 20th century uranium was mined to extract radium, thus initiating the first uranium mining "boom". This early uranium mining activity did not, however, even begin to compare with the activity that took place once the fission of uranium was detected by Otto Hahn in 1938, initiating its use for military and energy purposes.

Though exploration for uranium began in the early part of the 20th century, information is not readily available as to exploration expenditure levels and specific activities until the mid-1940s. The era of exploration that spanned the 1940s to mid-1950s was largely motivated by the need to satisfy military uranium requirements. Exploration for uranium to fuel civilian power reactors gained importance starting in the mid-1950s. In all, 81 countries have reported exploration expenditures. Other countries are known to have conducted exploration for uranium but details of these activities were not provided. Worldwide exploration expenditures from 1945 through 2003 are shown in Figure 5.1. Detailed country by country information on uranium expenditures is included in Appendix 5.1. Appendix 5.2 provides a narrative of worldwide exploration activities between 1965 and 2003. Appendix 5.3 gives the definitions of the uranium deposit types used in the Red Book and Appendix 5.4 gives summary data on world and OECD exploration expenditures and drilling activities. The world leaders in total exploration expenditures during this time are listed in Table 5.1.

Table 5.1. Countries with highest exploration expenditures (1945-2003)

Country	USD million	Percentage of world total
USSR[1]	3 692	27.6
United States	2 507	18.7
Germany[2]	2 003	14.9
Canada	1 289	9.6
France	907	6.8
Others (total)	3 002	22.4
World total	13 400	100.0

1. Does not include expenditures by Kazakhstan, the Russian Federation, Ukraine and Uzbekistan since 1991.
2. Includes the GDR.

Figure 5.1 shows that exploration gradually increased from 1945 through 1970. After 1970 there was a sharp increase in exploration, which culminated in an all-time high in 1979. Information on exploration expenditures between 1945 and 1970 is, however, incomplete due to the lack of reported data for many countries. The sharp increase between 1970 and 1971, for example, is mostly an artefact of an incomplete data base. In 1970 only three countries and the Soviet Union reported exploration expenditures; in 1971, 22 countries reported expenditures, many of which represented cumulative expenditures from *all* previous years. Therefore, the increase between 1970 and 1971 was largely a matter of reporting not a true reflection of an upsurge in exploration activity. While the rapid growth of exploration expenditures through the 1970s is a true reflection of an overall industry trend, the data in Figure 5.1 should be used more in a qualitative than a quantitative way, because reporting was incomplete and/or inconsistent. For example, no annual expenditure data are consistently available for Canada until 1975, though exploration was already well underway much earlier. The same holds true for several other countries. It was only in the early to mid-1970s that reporting on exploration expenditures started to become progressively more accurate. During the 1980s exploration expenditures decreased then levelled off at a fairly low level that was maintained through late 2000.

As of 1 January 2004, cumulative worldwide exploration expenditures totalled about USD 13 383 million. Included in this total is information that became available during preparation of this publication relating to exploration in the Former Soviet Union (totalling about USD 3 692 million) and the German Democratic Republic (USD 1 900 million) that was not reflected in past Red Books. Cumulative historical expenditures in the 2003 Red Book totalled USD 8 260 million. The balance of the difference between the totals reflected in Figure 5.1 and the 2003 Red Book mostly relates to unreported expenditures in the Czecho-Slovak Socialist Republic/Czech Republic and other countries.

Figure 5.1. Historical exploration expenditures

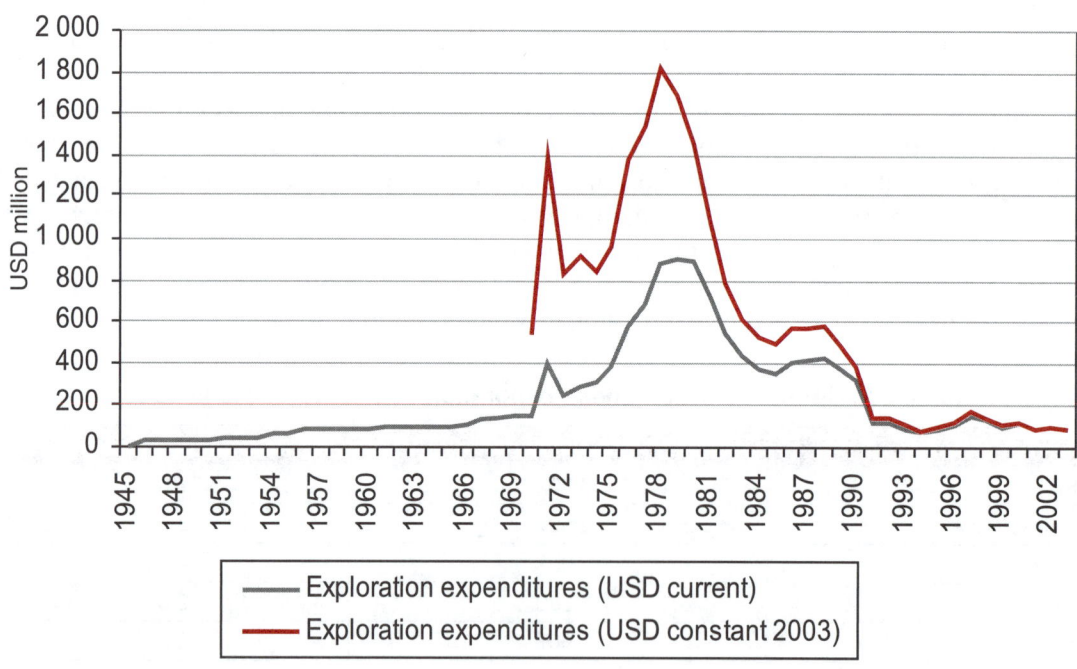

Figure 5.2 compares OECD and worldwide uranium exploration expenditures (current dollars) with uranium market prices beginning in 1970.

Figure 5.2. Comparison of annual OECD and worldwide exploration expenditures and uranium market price (USD current)

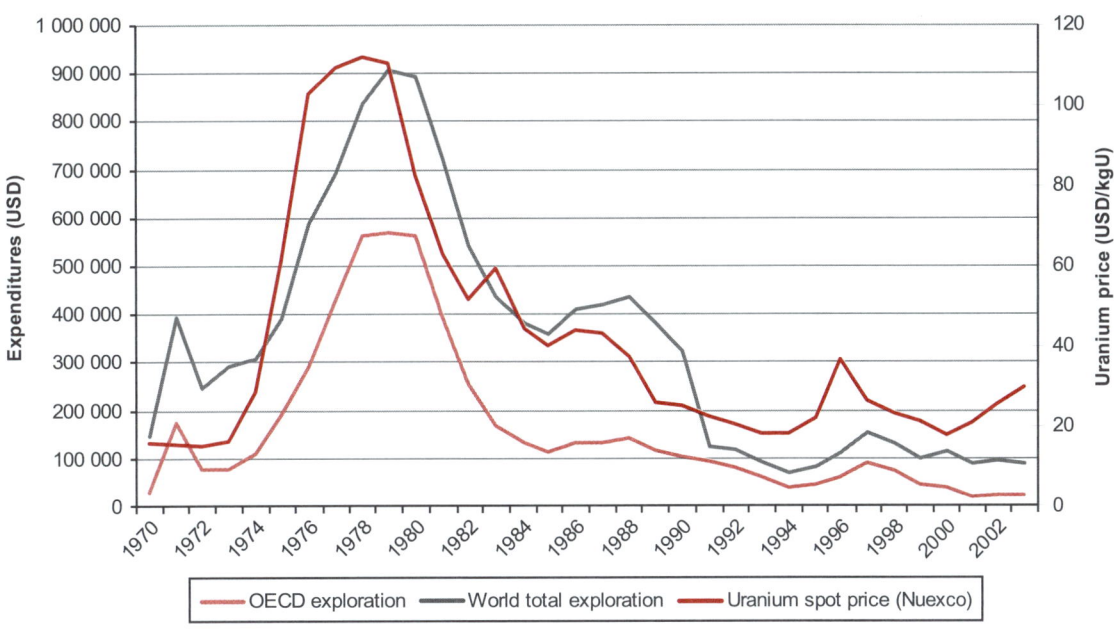

Despite being based on an incomplete data set, Figure 5.2 illustrates trends. It shows the direct parallel between uranium exploration expenditures and market price. Exploration expenditures tend to lag one to two years behind market price trends, which reflects the time the industry needs to become convinced that price increases have at least a measure of sustainability as well as the time needed to secure rights and permits and plan exploration campaigns. The anomalous spike that creates a marked separation between OECD expenditures and worldwide expenditures in 1990 resulted from the first-time inclusion of expenditures by the USSR and other former Eastern Bloc countries.

Uranium exploration expenditures typically include land acquisition and maintenance, airborne and ground geophysics, geochemical sampling, drilling and personnel costs. The Red Book collects specific information on drilling, which is typically the largest exploration expenditure. Figure 5.3 shows the close relationship between drilling and market price. From 1971 to 1981, drilling rates closely paralleled the market price. Beginning in 1982, however, drilling rates declined at a steeper rate than market price. A closer look at the numbers provides insights into what, at first glance, may seem to be a counterintuitive relationship.

Drilling rates declined sharply between 1978 and 1983 after which the trend flattened out. However, by comparing average hole depth (metres) with the number of holes drilled annually one can see that during the same time that exploration expenditures were declining (1978-1983), the number of holes being drilled annually was also decreasing, but average hole depths were increasing (Figure 5.4). One possible explanation is that fewer but deeper holes were drilled as exploration emphasis was shifting towards areas with deeper targets (e.g. the unconformity-related deposits in the Athabasca Basin at depths greater than 500 m). In this case, limited exploration budgets were being directed towards deeper drilling in targets with better economic potential. Increased drilling costs would also help explain the decrease in the number of holes drilled per year. The 1977 Red Book points out that drilling costs in the United States increased from USD 4.90/metre in 1973 to USD 10.20/metre in 1976. Therefore, even had exploration expenditures and hole depths remained at a constant level, fewer holes would have been drilled because of increased unit drilling costs.

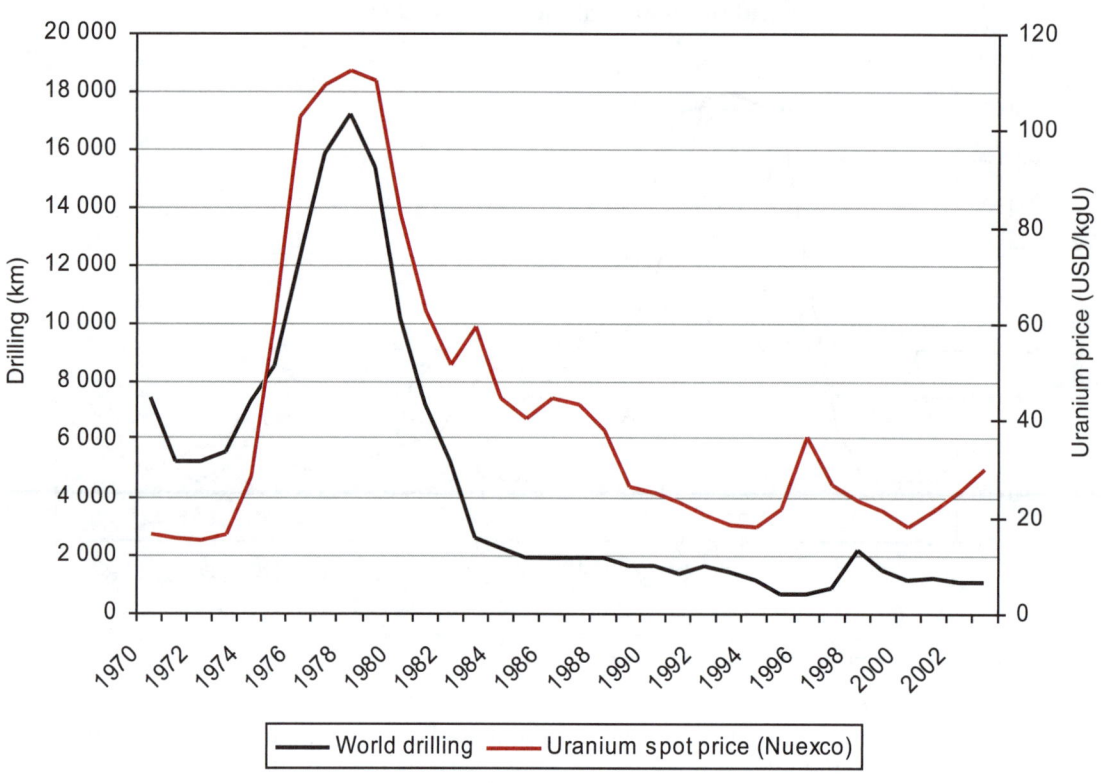

Figure 5.3. World exploration drilling and market price

Legend: World drilling — Uranium spot price (Nuexco)

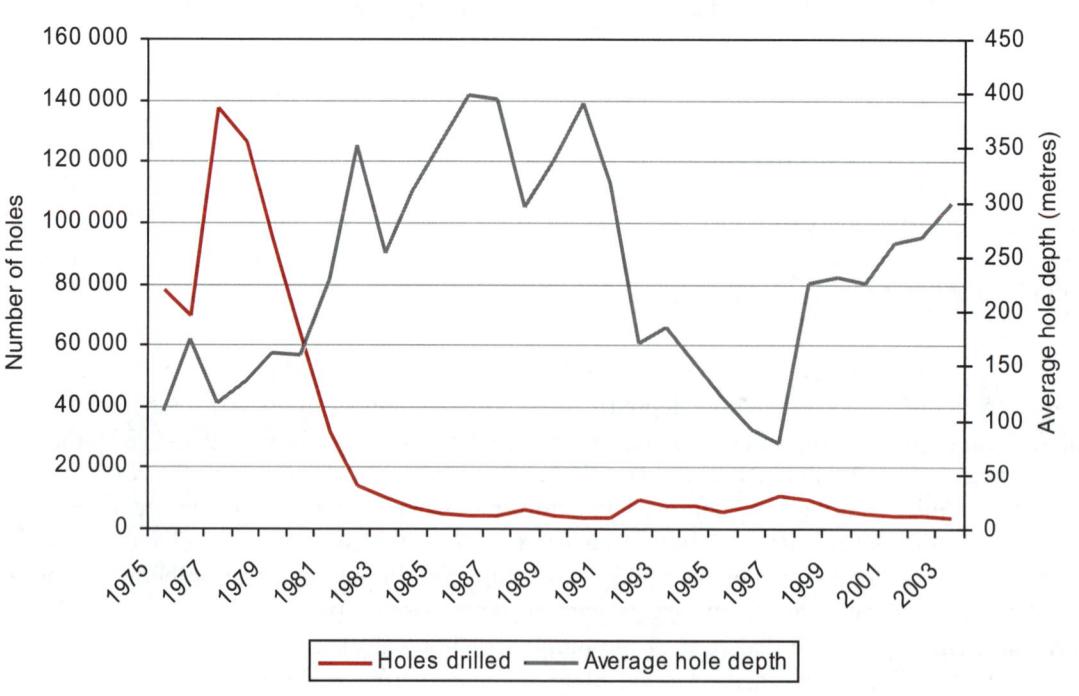

Figure 5.4. Average hole depth compared to number of holes drilled

Legend: Holes drilled — Average hole depth

Figure 5.5 compares the percentage of total exploration expenditures that three key countries and the USSR accounted for in 1978 and 1986. In the United States, where exploration emphasis was on relatively shallow ISL-amenable sandstone targets, exploration expenditures dropped from 42% of worldwide exploration expenditures in 1978 to just six percent in 1986. The other three countries all have relatively deeper target potential. The USSR doubled its share of exploration expenditures, while Australia remained about the same and Canada lost ground though its loss was much less dramatic than that of the United States. These examples show how the historical data in the appendices could provide further insight into all aspects of uranium exploration and production.

Figure 5.5. Relative percentages of exploration expenditures in 1978 and 1986

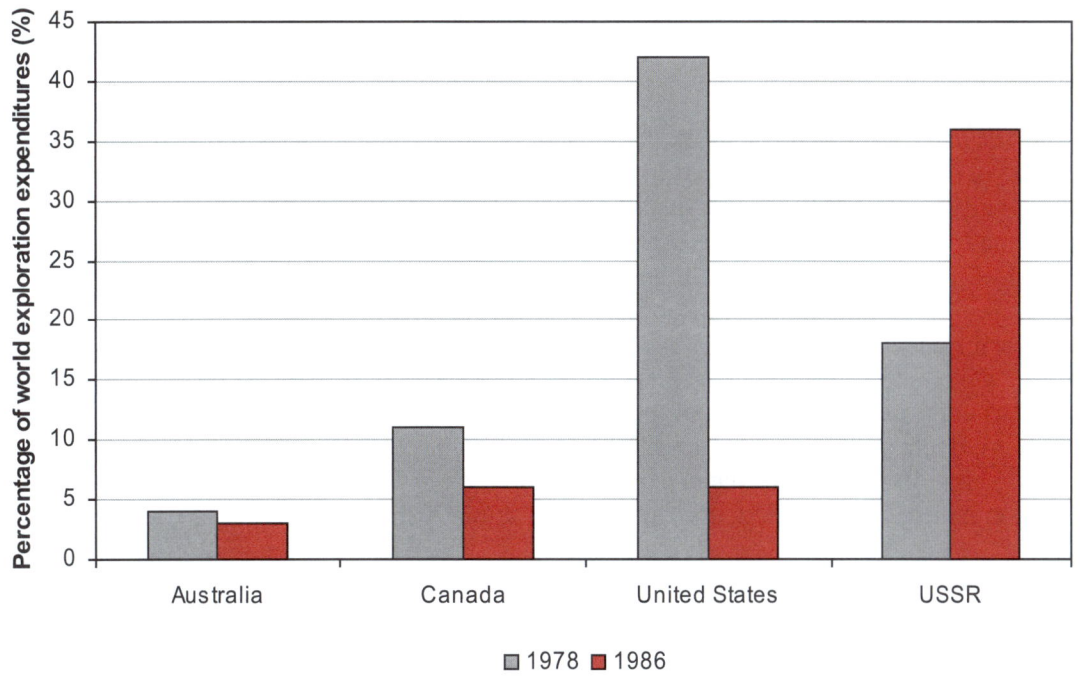

While the preceding figures and discussion are enlightening as to global trends, they do not show the very different histories of uranium exploration in different countries and how national conditions and policies, as well as market economics, played a role in influencing uranium exploration trends. These varying histories are needed to better understand the development of the uranium industry over time. Therefore, histories of exploration trends for seven major uranium producing countries or political entities follow. Australia, Canada, France and the United States were selected as examples because they are, or have historically been, leading producers and their industries are market-based. The Czech Republic, the USSR and Germany, including the German Democratic Republic are included because they too were historically important producers but by contrast their industries were operating under centrally planned economies. Annual exploration expenditures for each of these countries are compared with market price and with surface drilling (km) to show both the influence of market forces on exploration expenditures and the generally close parallel between overall exploration expenditure trends and fluctuations in drilling. In addition to the countries described below, summary information from successive Red Books on exploration histories for other countries is provided in Appendix 5.2. Detailed statistical information on exploration expenditures and drilling in the countries highlighted below is included in Appendix 5.5.

Australia

Uranium exploration in Australia can be divided into two distinct periods: 1947 to 1961, and 1966 to present. During the first period, the Australian Government introduced measures to encourage exploration, including a system of rewards for the discovery of uranium ore driven mainly be trying to supply Western defence-related demands for uranium. Active exploration was conducted throughout Australia, particularly by prospectors, that led to discovery of several deposits, many of which were subsequently mined, the largest being Mary Kathleen, Rum Jungle and Radium Hill (all of which were mined out).

Uranium requirements for defence-related purposes decreased in the early 1960s and uranium demand fell sharply. As a result, there was virtually no exploration for uranium in Australia between 1961 and 1966.

The second phase of uranium exploration in Australia commenced in 1966. The revival of exploration in Australia was encouraged by the announcement in 1967 of a new export policy designed to encourage exploration for new deposits. Most of this exploration was undertaken by companies (both domestic and foreign) with substantial exploration budgets, utilising advanced geological, geochemical and geophysical techniques. Major uranium deposits which were discovered during Australia's second phase of exploration included Ranger, Nabarlek, Koongarra, Jabiluka, Olympic Dam, Yeelirrie, Beverley, and Honeymoon.

Following the exploration boom of the late 1970s exploration expenditures declined sharply from a peak of AUD 35 million in 1980 to AUD 14 million in 1983. Surface drilling showed a similar pattern of decline. This sharp decline in exploration was due to decreases in uranium prices and energy conservation policies in response to the oil shocks of the 1970s.

In 1983 the Australian Government introduced what became known as the "three mines" policy, under which exports of uranium were permitted only from the Nabarlek, Ranger and Olympic Dam mines. This restriction effectively became a "no new mines" policy when the Nabarlek deposit was mined out and ore processing ceased in 1988. Despite the dampening effect of that policy, exploration expenditures increased from 1985 to 1988. From 1989 onwards, however, exploration declined, reaching a low of AUD 6.7 million in 1994. This decline was due to several factors including the fall in spot market prices beginning in 1976, the build up of excess uranium inventories in Western countries and the beginning of commercial uranium sales from the countries of the former Soviet Union.

From 1994 to 1997, uranium exploration expenditures increased. These increases were due to expectations for the abolition of the three mines policy and improved prices for uranium during 1996. This increase was, however, short-lived, and uranium exploration activities decreased progressively between 1998 and 2001 both in response to the decline in uranium market prices and to ongoing uncertainty as to public acceptance of uranium mining in most of Australia. Part of the decline may also have been attributable to Australia's large reserve base associated with known deposits, which assured an adequate uranium supply for short- and medium-term production. Exploration has begun to rebound, largely due to recent market price increases and partly in response to a favourable attitude toward uranium mining in South Australia.

Australian exploration expenditure data are available beginning in 1967; drilling statistics are available beginning in 1975. Figure 5.6 compares trends in exploration expenditures with market price trends and exploration drilling rates. One can see a definite parallelism among the three curves in Figure 5.6, which shows the direct relationship between exploration and market price in market-based economies.

Exploration tends to lag one to two years behind market price trends, which as previously noted reflects the time the industry needs to become convinced that price increases have at least a measure of sustainability as well as the time needed to obtain permits and financing and to plan exploration campaigns.

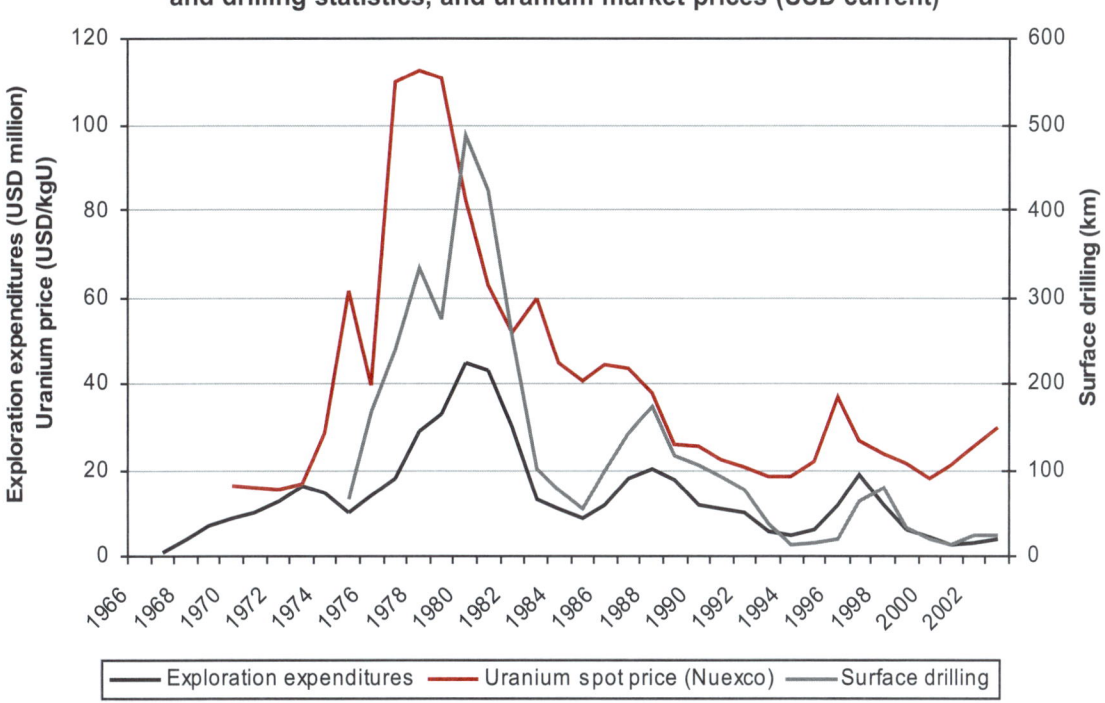

Figure 5.6. Comparison of Australia's exploration expenditures and drilling statistics, and uranium market prices (USD current)

Canada

Exploration for uranium began in Canada in 1942, with geographic emphasis shifting over time, first from the Great Bear Lake area in the Northwest Territories, to the Beaverlodge area in northern Saskatchewan (in both areas deposits are now depleted), to the Blind River/Elliot Lake area of Ontario (currently closed and decommissioned), and finally in the 1970s to Saskatchewan's Athabasca Basin, which continues to be the focus of exploration in Canada. In 1942, mining of all radioactive minerals by private persons was banned in the Northwest Territories and in a number of the provinces. Instead exploration was conducted jointly by a federal Crown company, Eldorado Nuclear Limited, and the Geological Survey of Canada. The ban on private prospecting was lifted in 1947 and various incentives were offered by the federal government in an effort to encourage exploration. By 1956, more than 10 000 radioactive occurrences had been discovered. Many proved to be viable deposits and by 1959, 23 mines were in operation in five districts.

Exploration for uranium virtually ceased in the late 1950s, and was not renewed until the mid-1960s. By 1969 exploration had returned to near record levels, but this resurgence was short-lived, and exploration again declined in response to growing over-supply and depressed prices in the world uranium market. Significant uranium deposits were discovered in the Cluff Lake area of northern Saskatchewan by the end of the 1960s. During this time both domestic and foreign companies were actively involved in exploration in Canada.

Exploration activities resumed in 1973, as consumers moved to acquire long-term uranium supplies, which led in turn to projected supply shortfalls. Discovery of the Rabbit Lake and Key Lake deposits resulted from this resurgence in exploration. Exploration levels expanded throughout 1974 and 1975, and by 1976 it was estimated that some 200 companies were spending USD 45 million for exploration annually. Though exploration was conducted in virtually all provinces and territories, most of the activity occurred in Saskatchewan, followed by Quebec, Ontario, British Columbia and the Northwest Territories. Activity reached record levels in 1979 and 1980 with expenditures totalling USD 111.9 million and USD 111.6 million, respectively.

In 1975, the Geological Survey of Canada embarked on a 10-year, CAD 30 million uranium reconnaissance programme designed to identify and delineate all areas in Canada which may be favourable for the occurrence of uranium deposits.

Between 1980 and 1981, however, exploration activity declined sharply in response to continued erosion of both the spot-market price and short-term sales prospects for uranium. Although grass-roots exploration activity in new areas continued, over 80% of total exploration expenditures and drilling activity took place in the Athabasca Basin of Saskatchewan and in the Thelon Basin in Nunavut (formerly part of the Northwest Territories). Despite the reduction in exploration activity, important discoveries were made, including the discovery of Cigar Lake in 1983 and McArthur River in 1990. In the 1990s and early 2000s, continuing low uranium prices kept exploration activity at a relatively low level, with more than half of the overall exploration expenditures attributed to advanced underground exploration, deposit appraisal activities, and care and maintenance expenditures associated with projects awaiting production approvals.

Detailed records of exploration activities in Canada are available beginning in 1975. Exploration expenditures of USD 24.9 million are also reported for the period 1971-1974, but no annual totals are available for this period. Figure 5.7 compares trends in exploration expenditures with market prices and exploration drilling rates. As with Australia, one can see that Canada's exploration expenditures are fundamentally linked to the market price for uranium, a natural result of Canada's market-based economy. However, there is less parallelism in the curves for Canada than was observed for Australia. For example, between 1986 and 1988 exploration increased even as the market price continued to decline. Similarly, there was a much steeper decrease in drilling compared to overall exploration expenditures between 1989 and 1990. It may be that during this time of limited exploration budgets, emphasis was being placed on geophysics in the Athabasca Basin at the expense of drilling.

Figure 5.7. Comparison of Canada's exploration expenditures and drilling statistics and uranium market prices (USD current)

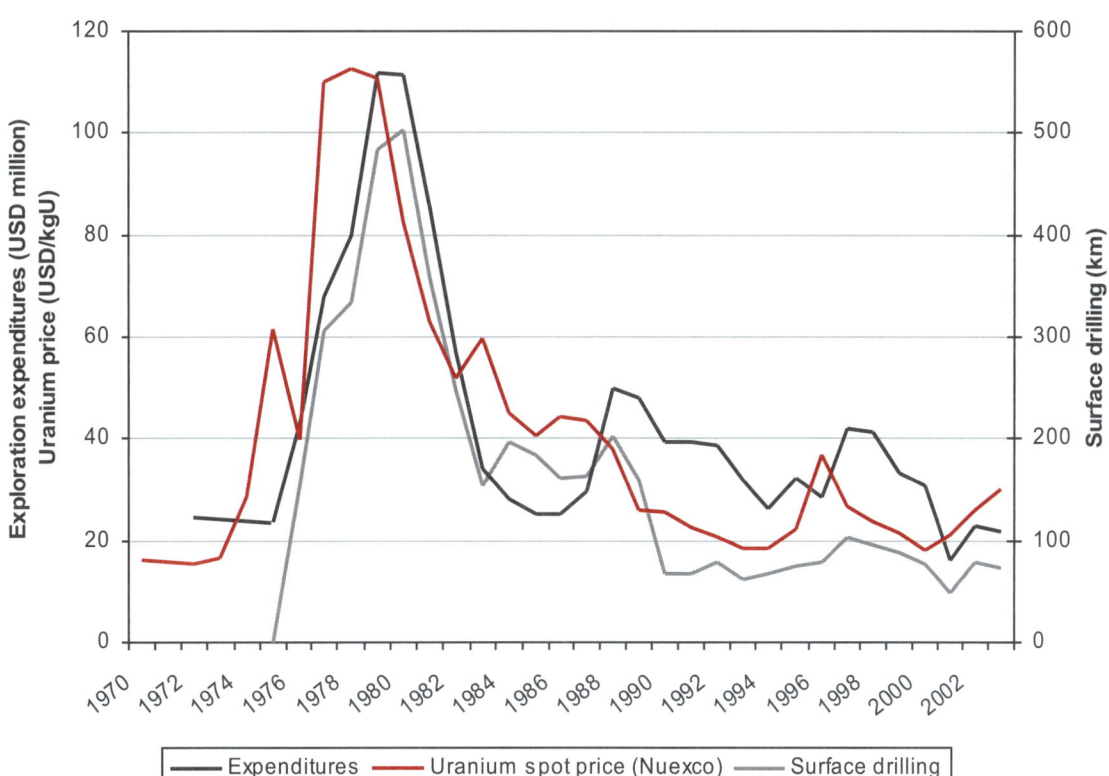

Expenditures — Uranium spot price (Nuexco) — Surface drilling

France

Uranium exploration in France began in 1946 on the basis of recorded uranium mineralisation anomalies and a few small deposits discovered in the course of radium exploration. The prospecting work, which consisted mainly of geological mapping and aerial, carborne and hand carried radiometry, led to the discovery in 1948 of the La Crouzille deposit. By 1955, deposits were discovered in Hercynian granite formations in the Limousin, Forez, Vendée and Morvan areas.

During the 1950s and 1960s, exploration was carried out in the vicinity of known deposits, which were the source of almost all production in France (Forez, Limousin and Vendée). Exploration activities were oriented towards reconnaissance and evaluation of deposits being exploited, as well as towards exploration for additional mineralisation located in proximity to existing mills. Geological studies and exploration were also conducted on sedimentary formations, primarily those of the Permian period (Basin of the Hérault, south of Massif Central).

As a result of the world energy crisis and of the consequent accelerated development of the French nuclear programme, uranium exploration increased significantly in France beginning in 1974. Exploration was conducted adjacent to deposits being mined, in areas where past prospecting had revealed the existence of occurrences which were not developed because of previous economic conditions, and in new areas, mainly within Permian and Tertiary sedimentary basins.

Exploration activities continued at a high level until the mid-1980s, despite a decrease in the uranium spot price; expenditures expressed in USD decreased, but were maintained at the same level when expressed in French Francs. Exploration was carried out in areas around active mines, but also in the Massif Central (Marche, Millevaches and Margeride), Massif Armoricain, the Aquitaine Basin and the Alps.

From 1977 to 1981, French exploration activities received government aid *(Plan d'aide à la prospection d'uranium)*, the total being USD 38 million. The purpose of this aid was to provide an incentive for prospecting in France and abroad on technically sound but high-risk projects. The ceiling of the aid was 35% of the total cost of a project, and was to be reimbursed if a recoverable deposit was found.

Exploration activities started to decline at the end of the 1980s, when they were confined to areas adjacent to deposits currently being mined. In 1995, activities were only conducted in the north-west part of the Massif Central (around the Bernardan deposit) and within the Lodève Permian basin. Exploration activities ceased in the Lodève Basin in 1996, and around the Bernardan deposit in 1998.

Figure 5.8 compares trends in France's exploration expenditures with market prices and exploration drilling rates. The curves in Figure 5.8 show similar parallelism to those for Australia and Canada.

Figure 5.8. Comparison of France's exploration expenditures and drilling statistics, and uranium prices (USD current)

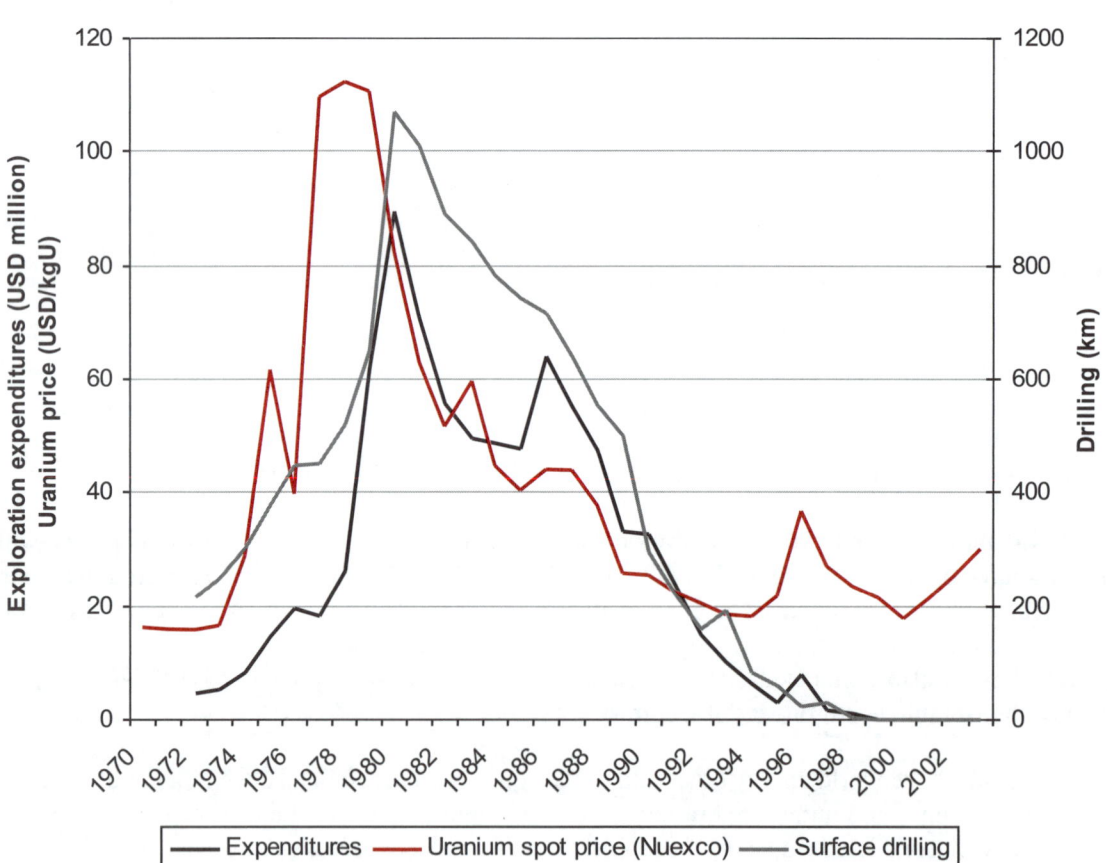

United States

From 1947 through 1958, the United States Government fostered both a Federal and a private-sector uranium exploration and production industry to procure uranium for military uses and peaceful atomic energy applications. From 1958 through 1970, the government continued support of private-sector exploration and production. This effort led to widespread exploration by prospectors throughout the western United States, who identified literally thousands of uranium occurrences. By late 1957, the number of new deposits being brought into production by private industry had increased sufficiently to meet projected requirements, and Federal subsidiaries for exploration programmes were ended.

The US government did, however, continue to support research in uranium geology and exploration. For example, it initiated the National Uranium Resource Evaluation programme (NURE). This multifaceted programme included both basic research on uranium deposit types and a broad range of field exploration work, including extensive airborne geophysical surveys, stream and groundwater sampling and compilations of past exploration activities/results. Results of the NURE programme, which were made publicly available, stimulated exploration in areas outside the traditional uranium producing districts, particularly in the eastern United States.

Uranium exploration in the United States increased throughout the 1970s in response to rising uranium prices and the projected large demand for uranium to fuel an increasing number of nuclear reactors being built or planned for electric power stations. The peak in annual surface drilling (exploration and development) was reached in 1978, when 14 700 km of borehole drilling were completed and expenditures totalled USD 374.5 million. Beginning in the late 1960s and accelerating through the 1970s, many US oil companies diversified into uranium exploration and production. The rapid growth of exploration in the 1970s can be directly tied to the entry of these well-funded oil companies into the uranium business. Their wholesale exodus from the uranium industry was a major reason for the precipitous decline in US exploration beginning in the early 1980s.

From 1966 through 1982, surface drilling totalled approximately 116 400 km, and expenditures totalled about USD 2 121 million. In response to falling uranium prices, exploration activities decreased rapidly between 1981 and 1984 before beginning a gradual downward trend reaching a low of USD 0.35 million in 2002 when government statistics indicate that there was only one person involved in uranium exploration in the United States. Expenditures increased slightly in 1997 and 1998 in response to the 1996 increase in the spot market price. This increase was short-lived and exploration quickly returned to pre-1996 levels. The recent market price increase has, however, rekindled exploration interest in the United States, with literally dozens of small companies having been formed recently to engage in uranium exploration.

Both domestic and foreign companies have been actively involved in exploration in the United States, with increased foreign participation in the industry beginning in the 1970s. Uranium exploration in the United States has primarily been directed toward discovering sandstone-type uranium deposits in districts such as the Grants Mineral Belt and the Uravan Mineral Belt of New Mexico, Colorado and Utah, the Wyoming Basins and the Texas Gulf Coastal Plain. Vein and other structurally-controlled deposits were exploration targets in the Front Range of Colorado, in Utah and in north-eastern Washington State. Several relatively high-grade deposits associated with collapse breccia pipe deposits were discovered in northern Arizona and were mined in the 1970s through early 1990s.

Experimental ISL pilot tests were conducted in the 1960s in the Shirley Basin and Powder River Basin in Wyoming. The first commercial ISL operation began production in 1964

(577 tU were produced by ISL at Shirley Basin Mine from 1964 to 1970). US companies considered ISL operations as the only potentially commercial projects in the United States in the depressed uranium market that characterised the 1980s and 1990s. Accordingly, as uranium prices continued to fall in the early 1980s exploration emphasis was concentrated almost entirely on sandstone deposits that were potentially amenable to ISL extraction. Most of the exploration activity since the early 1980s has been concentrated in the Wyoming Basins and in South Texas, both of which have hosted ISL production centres. Though New Mexico also has significant ISL amenable resources, public opposition to uranium mining has delayed development of this potential.

Annual exploration expenditures are only available for the United States from 1966 onwards. Therefore, for the period 1948 to 1965 estimates of uranium expenditures have been made based on summary figures. Annual drilling statistics are, however, available from 1948 through 2003. Figure 5.9 compares trends in exploration expenditures with market prices and exploration drilling rates. The curves in Figure 5.9 show similar parallelism to those for Australia and Canada showing that even with a strong military component uranium is primarily a market-based commodity in the United States. As shown in Figure 5.9, drilling dominates US exploration expenditures and US exploration expenditures are very sensitive to uranium market prices.

Figure 5.9. Comparison of US exploration expenditures and drilling statistics, and uranium prices (USD current)

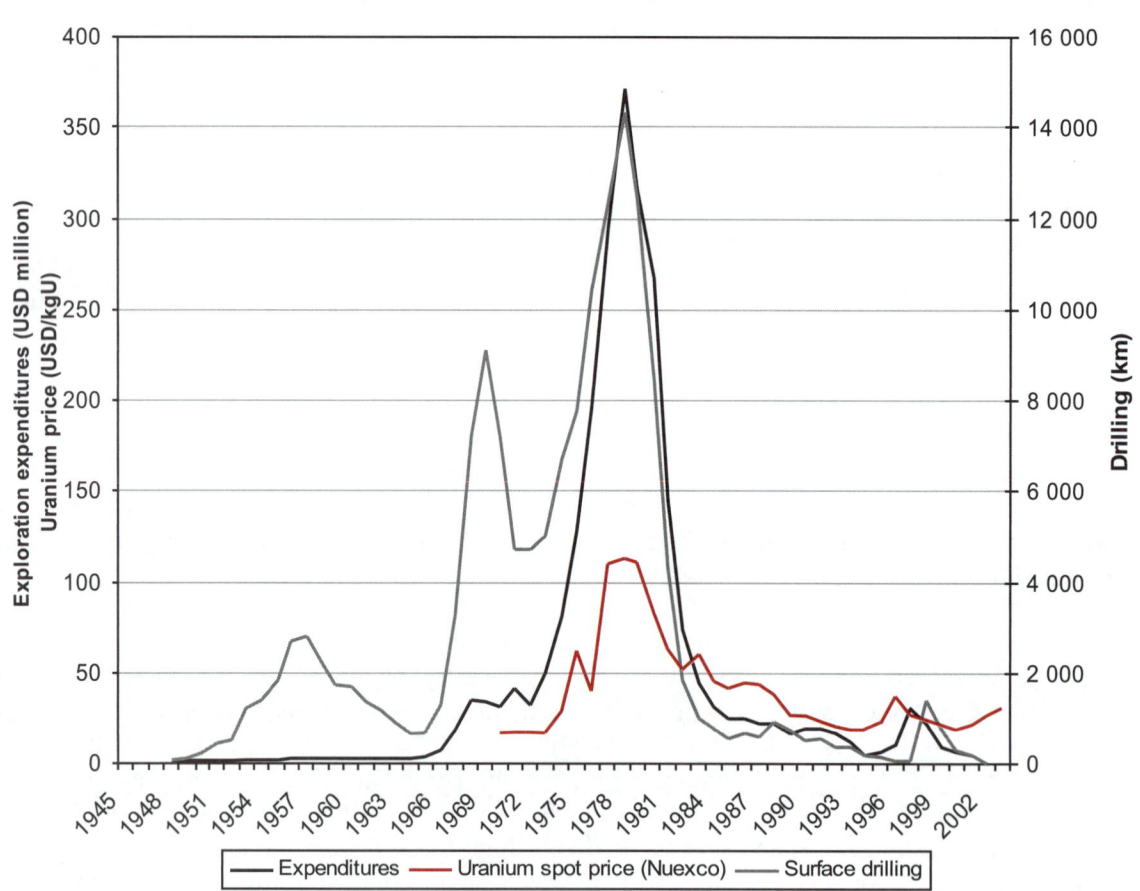

Czech Republic including the former Czechoslovakia

Following its start in 1946, uranium exploration in Czechoslovakia grew rapidly and developed into a large-scale programme in support of the country's uranium mining industry. A systematic exploration programme including geological, geophysical and geochemical surveys was carried out to assess the uranium potential of the entire country. Areas with identified potential were explored in detail using drilling and underground methods. During this period old mines with known uranium mineralisation were reactivated and new deposits in a variety of geological environments were discovered.

Exploration continued in a systematic manner until 1989 with annual exploration expenditures of USD 10-20 million and an annual drilling effort of 70-120 km. Exploration was centred around vein deposits located in metamorphic complexes (Jachymov, Horni Slavkov, Pribram, Zadni Chodov, all of which have been depleted or are dormant) and Rozna (currently producing), in granitoids of the Bohemian massif (Vitkov, depleted) and in the vicinity of the sandstone-hosted deposits in northern and north-western Bohemia (Hamr and Straz, both of which are being reclaimed/decommissioned).

Exploration activities in the Czech Republic/Czechoslovakia decreased steadily starting in 1982, except for a slight increase from 1986 to 1989. No field exploration has been carried out since the beginning of 1994. Exploration activities have been focused on conservation and processing of previously collected exploration data.

Figure 5.10 compares trends in exploration expenditures with market prices and exploration drilling rates. Information on exploration activities in Czechoslovakia for the period 1971 to 1980 is not available in detail. Expenditures totalled CSK[*] 2 475.3 million (USD 139 million) and drilling 1 215.54 km. These figures have been divided equally among the individual years during this period. Detailed annual data are available for 1985 to 2003. Because of the lack of detailed annual data before 1985, it is not possible to determine what influence, if any, market price had on exploration activities, though a direct parallel would be less likely because the Czech industry operated within a centrally planned economy until the late 1980s. The abrupt declines in exploration expenditures and drilling between 1989 and 1991 and the near absence of exploration thereafter were more the result of deteriorating conditions throughout the Czech economy than purely the result of market forces.

---

[*] Czech Koruna

Figure 5.10. Comparison of Czech Republic/Czechoslovakia exploration expenditures and drilling statistics, and uranium prices (USD current)

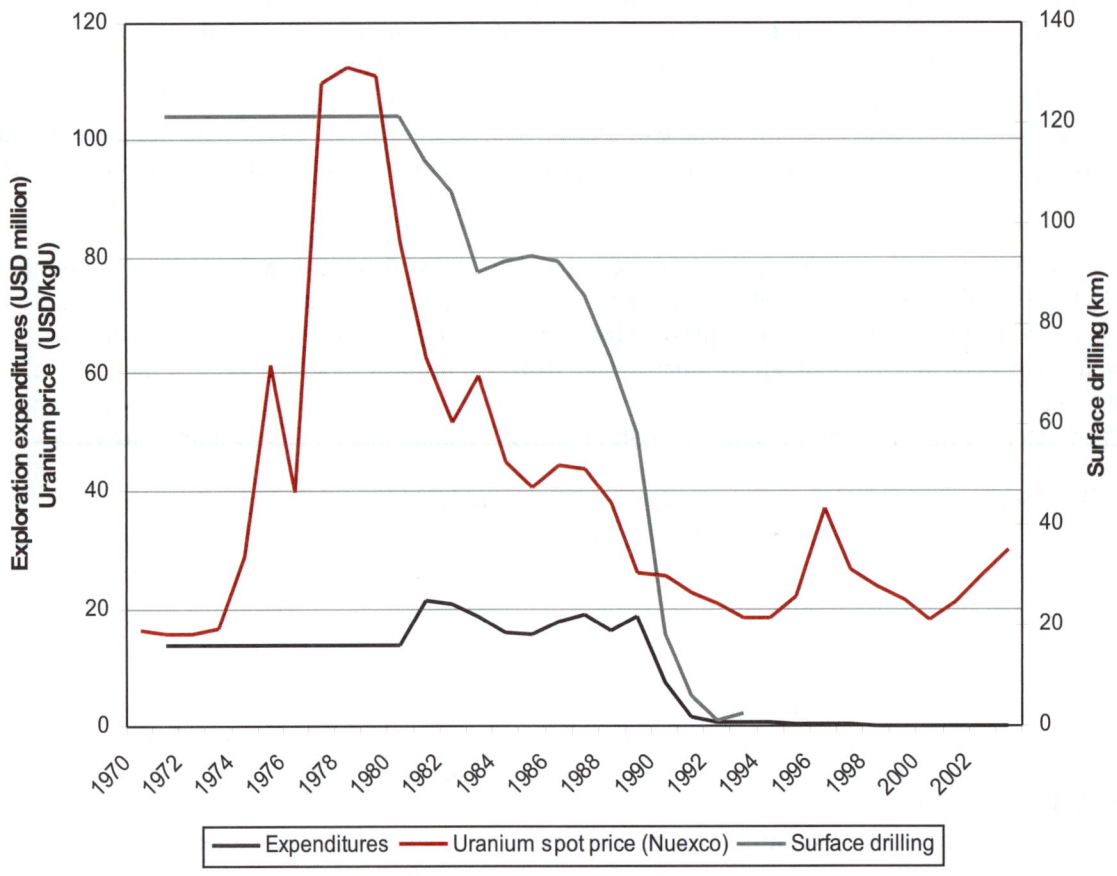

Germany including the former German Democratic Republic

Uranium exploration history in Germany basically parallels the history of the country during the second half of the 20th century. Exploration in the eastern and western parts of the country was as independent from one another as were their political systems. Information regarding uranium exploration and production in the German Democratic Republic (GDR) was a state secret under the communist regime, so no exchange of uranium statistics or technology (or product) took place between the two parts of the country. The status of the uranium industry of the GDR only became publicly known after unification in 1990. As a consequence, it is appropriate that separate sections be included for the GDR (East Germany) and the Federal Republic of Germany (West Germany).

Exploration history in the German Democratic Republic (GDR)

In the GDR, uranium exploration and mining were undertaken from 1946 to 1953 by the Soviet stock company, SAG Wismut. Activities during this time were centred in the vicinity of old silver, cobalt, nickel and other metal mines in the Erzgebirge and in Vogtland, Saxony, where uranium had first been discovered in 1789. Mining of uranium first began at the cobalt and bismuth mines near Schneeberg and Oberschlema (both now depleted). During this early period more than 100 000 people were engaged in exploration and mining activities. The richer pitchblende ore from the vein deposits was hand-sorted and shipped to the USSR for further processing; the lower grade ore was locally treated in small processing plants.

In 1954, a new joint Soviet-German stock company was created, Sowjetisch-Deutsche Aktiengesellschaft Wismut (SDAG Wismut). The joint company was owned equally by both governments. All uranium produced, whether hand-sorted concentrate, gravity concentrate, or chemical concentrate, was shipped to the USSR for further treatment.

Uranium exploration in the vicinity of the radium spa at Ronneburg started in 1950. By the end of the 1950s, mining was concentrated in the region of Eastern Thuringia and in the Erzgebirge. At the beginning of the 1970s, the mines in Eastern Thuringia provided about two-thirds of Wismut annual production.

The uranium deposits that have been explored in eastern Germany were subdivided into five major types according to their geological setting:

- Ronneburg type: Polygenetic lens-shaped to stockwork-type deposits in black shale, limestone and diabase of Paleozoic age.
- Erzgebirge type: Hydrothermal veins at the exocontact of Hercynian granites.
- Königstein type: Peneconcordant and stack-type deposits in sandstone of Upper Cretaceous age.
- Zechstein type: Seam-like deposits in Upper Permian calcareous sediments of fluviatile to lagoonal deposition.
- Freital type: Uraniferous hard coal in Lower Permian continental sediments.

Since the end of the 1980s, no exploration activities have been carried out in Germany. Commercial mining was terminated in 1990 due to economic and environmental reasons and all mines were subject to reclamation. GDR exploration expenditures are reported as follows:[1]

- Total exploration expenditures of Wismut from 1946 to 1990: GDR Mark 5 500 million.
- Expenditures from 1946 to 1953: GDR Mark 600 million divided in equal annual amounts for this period.
- Expenditures from 1954 to 1982: GDR Mark 3 900 million, divided in equal annual amounts for this period.
- Annual data are available from 1983 to 1990 (Appendix 5.1).
- Drilling totalled 27 814 km between 1956 and 1990. Details of annual drilling rates are included in Appendix 5.5.

Figure 5.11 compares trends in exploration expenditures with market prices and exploration drilling rates. Uranium exploration and mining in the GDR show no correlation with western market prices and exploration activities as would be expected for a centrally-planned economy isolated from the West. There is clearly some evidence for this premise; exploration costs increased by 50% between

---

1. Wismut was the only uranium exploration company in the former German Democratic Republic (GDR). Exploration data for the GDR are only available in GDR Marks. According to information from Deutsche Bundesbank, the IAEA, and the BGR databank, the exchange rate for GDR Mark and the Deutsche Mark (DM) was 1:1 for the entire period.

1985 and 1988 while the market price decreased by 10% during the corresponding period. There is no readily apparent explanation as to why drilling peaked in 1972 and began a steep decline while exploration expenditures continued an overall upward trend before peaking in 1988.

Figure 5.11. Comparison of German Democratic Republic exploration expenditures and drilling statistics, and uranium market prices (USD current)

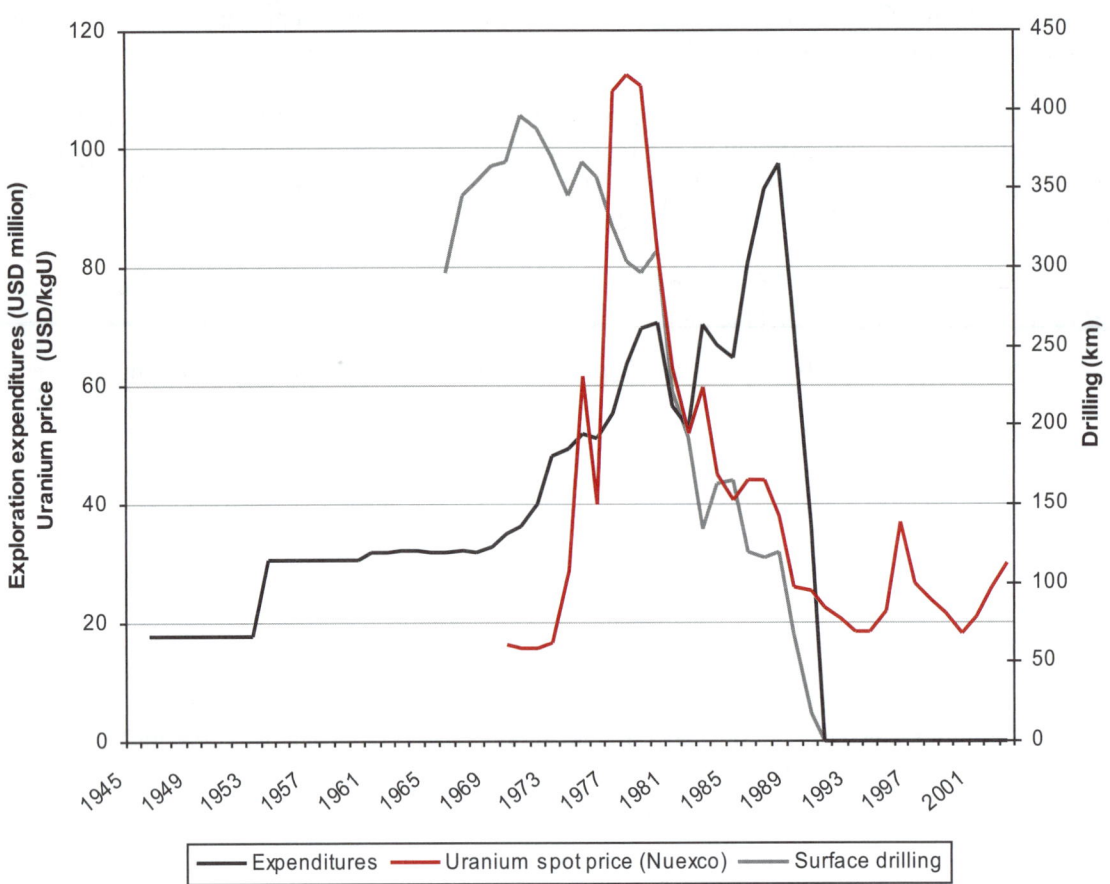

Uranium exploration in the Federal Republic of Germany (West Germany)

Uranium exploration in West Germany (Federal Republic of Germany, FRG) began in the 1950s. Mining companies and Government organisations (Geological Surveys, both Federal and some of the Länder, as well as research institutes) were involved in FRG exploration. University institutes also contributed basic research to the FRG exploration effort. Since the FRG has by law refused to use uranium for military purposes, exploration emphasis was placed on discovering indigenous uranium resources for the expected future civilian demand for uranium.

Exploration in the FRG included geochemical prospecting (water samples, stream sediments), airborne and carborne radiometric surveys and detailed exploration on promising anomalies. In prospective areas (e.g. granites and other magmatic rocks and sedimentary basins) mining companies undertook detailed exploration work, including drilling, trenching and underground excavation.

Several small deposits were found in the second half of the 1950s in Hercynian granites of Weissenstadt/Fichtelgebirge, north-east Bavaria (veins and mineralisation on joints) and in acid volcanic rocks of Lower Permian age in Rheinland-Palatinate. Due to the many promising areas located in acid volcanic and adjacent sedimentary rocks in Rheinland-Palatinate, a small mill was built near Ellweiler. Underground exploration at the Weissenstadt deposit was stopped due to the limited extent of the deposit. Mineralisation in acid volcanic rocks was of limited extent and mining was terminated after limited activity.

Detailed exploration in the southern Black Forest led to discovery of the Menzenschwand vein deposit in Hercynian granites. After excavation of few hundred tonnes of uranium from small-scale underground mines, mining was terminated in the 1980s. In the northern part of the Black Forest a small sedimentary deposit in Upper Carboniferous sandstones near Baden-Baden (Müllenbach) was investigated in the late 1970s and early 1980s by underground exploration from two adits. Due to its limited size work on this deposit has been terminated.

In the late 1970s a vein deposit in Hercynian granites near Grossschloppen, Northern Bavaria, was investigated by underground exploration; however, exploration was terminated when underground exploration drilling found that the veins were of limited extent.

Declining uranium demand during the 1980s resulted in termination of all exploration activities in the FRG. All uranium prospects and mines have been reclaimed in compliance with environmental requirements.

Russian Federation including the former Soviet Union

Uranium exploration activities in the USSR can be divided into five distinct periods: 1914-1940, 1945-1950, 1950-1960, 1960-1970, and 1970-1990. Between 1914 and 1940, the Tyuya-Muyun deposit in the Fergana valley in eastern Uzbekistan, which was discovered in 1900, was investigated, and the Taboshar and Adrasman (Tajikistan), and Malysu (Kyrgyzstan) deposits were discovered (all three mines are now depleted). These discoveries led to the recognition of the first uranium district in the USSR, Fergana Valley (Uzbekistan).

In the period between 1945 and 1950, the uranium districts of Krivorozh (Pervomaiskoye and Zheltovodskoye deposits), in Ukraine and Stavropol (Beshtau and Bykogorskoye deposits) in the Russian Federation, were discovered. Additional deposits were found in the Karamazar district (Kattasay, Chauli, Maylisay), Uzbekistan. From 1945 to 1950, exploration expenditures totalled approximately USD 71.3 million and drilling totalled 403 km.

During the 1950s, new exploration methods, including airborne radiometrics, led to the discovery of the following uranium districts:

- Zacaspiysk (Kazakhstan; Melovoye, Tomakskoye, Taybagarskoye and Tasmurun deposits, all of which are now depleted or dormant).

- Pribalkhash (Kazakhstan; Kurday, Botaburum, Kyzylsay, Dzhideli deposits all of which are now depleted or dormant).

- Kyzylkum (Uzbekistan; Uchkuduk deposit, currently in operation).

- Kokchetavsk (Kazakhstan; Manybay, Ishimskoye, Zaozernoye deposits, all of which are depleted or dormant).

These discoveries formed the basis of the large uranium mining industry in the USSR. From 1951 to 1960, exploration expenditures totalled about USD 391.5 million and drilling totalled 7 784 km.

From 1960 to 1970, scientific research work on conceptual deposit models and the use of airborne radiometrics led to the discovery of a new deposit type, vein-stockworks in continental volcanic complexes. The most significant example is the Streltsovsk deposit complex (also called Priargunsky). During this same time period, the metasomatic albitite deposit type was discovered in close proximity to the Krivorozh uranium district in what became the Kirovograd district (Ukraine; Michurinskoye and Vatutinskoye deposits, both of which are producing, and Severinskoye, dormant). Practical experience gained in ISL techniques in Uzbekistan led to increased exploration for sandstone deposits, resulting in additional discoveries in the Kyzylkum district (Uzbekistan) and the Chu Sarysu and Syr-Darya districts (Kazakhstan), which became the most important uranium districts in the USSR. From 1961 to 1970, exploration expenditures totalled approximately USD 690.8 million and drilling totalled 20 277 km.

From 1970 to 1990, additional sandstone type deposits were found in the Zauralsk and Vitimsk regions (Russian Federation) which are in the initial production phase. Discoveries were also made in black shales in the Kyzylkum and Onezhsk districts (Russian Federation, dormant). Between 1971 and 1990, exploration expenditures totalled approximately USD 2 538.7 million and drilling totalled 88 151 km.

Thus, from 1945 through 1990, exploration expenditures in USSR totalled approximately USD 3 692 million and drilling totalled 116 615 km. Exploration expenditures and drilling are available from 1945 to 1990 in 5-year intervals. Drilling has been equally divided within each 5-year period to establish annual rates. Similarly, exploration expenditures have been equally divided within each 5-year period using an exchange rate of 1 USD = 0.75 Rouble. After the dissolution of the Soviet Union in 1990, exploration activities have been conducted and are reported by the individual countries which were part of the USSR (i.e. Kazakhstan, Russian Federation, Ukraine and Uzbekistan). Uranium exploration throughout the former Soviet Union declined significantly after its dissolution.

Though the lack of annual data precludes detailed comparisons between market prices and exploration expenditures, it is apparent from Figure 5.12 that there was no relationship between exploration expenditures and Western uranium market prices. USSR drilling and expenditures moved together at an ever increasing rate between 1945 and 1990.

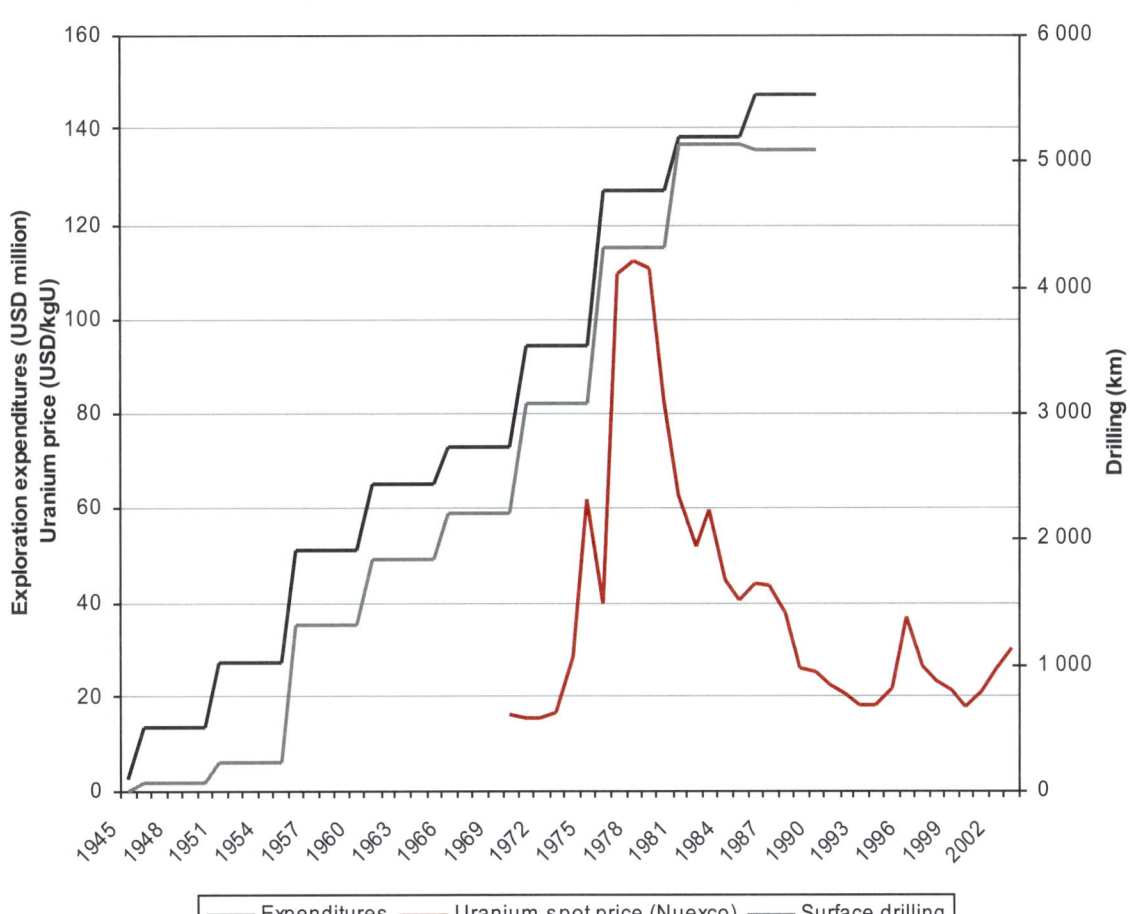

Figure 5.12. Comparison of USSR exploration expenditures and drilling statistics, and uranium market prices (USD current)

Legend: Expenditures — Uranium spot price (Nuexco) — Surface drilling

Exploration abroad

Exploration statistics have been reported in the Red Book in two categories: "Industry and Government Uranium Exploration Expenditures – Domestic" and non-domestic called "Uranium Exploration Expenditures – Abroad". Expenditures by private or government-sponsored organisations for exploration outside the borders of the funding entity (abroad) are reported as a separate category, but are also included as part of domestic expenditures, with the latter representing *total* exploration expenditures. There may, however, be exceptions to the inclusion of abroad expenditures in the domestic expenditure totals because of incomplete information in some countries. This exception is not, however, thought to be significant.

Exploration expenditures abroad were reported for the first time in the 1977 edition of the Red Book. In this first report, annual expenditures were listed for 1972 through 1977. In addition pre-1972 expenditures totalling USD 117.7 million were reported by Australia, France, Germany, Japan and the United States but without annual expenditure details. Figure 5.13 shows historical exploration abroad expenditures between 1972 and 2003. Figure 5.14 shows the relative contributions to the worldwide totals by the four countries that were most active in exploration abroad. Appendix 5.6 includes detailed information of exploration expenditures abroad.

Figure 5.13. Exploration expenditures abroad (1972 to 2003) (USD current)

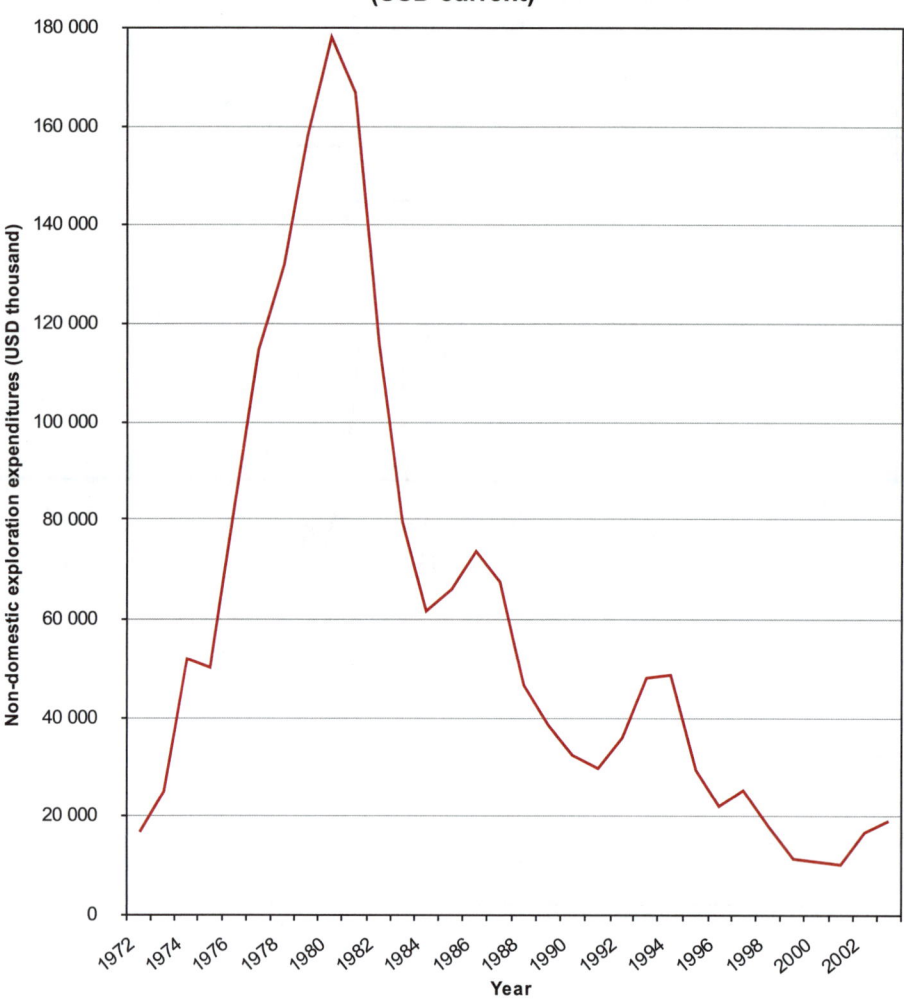

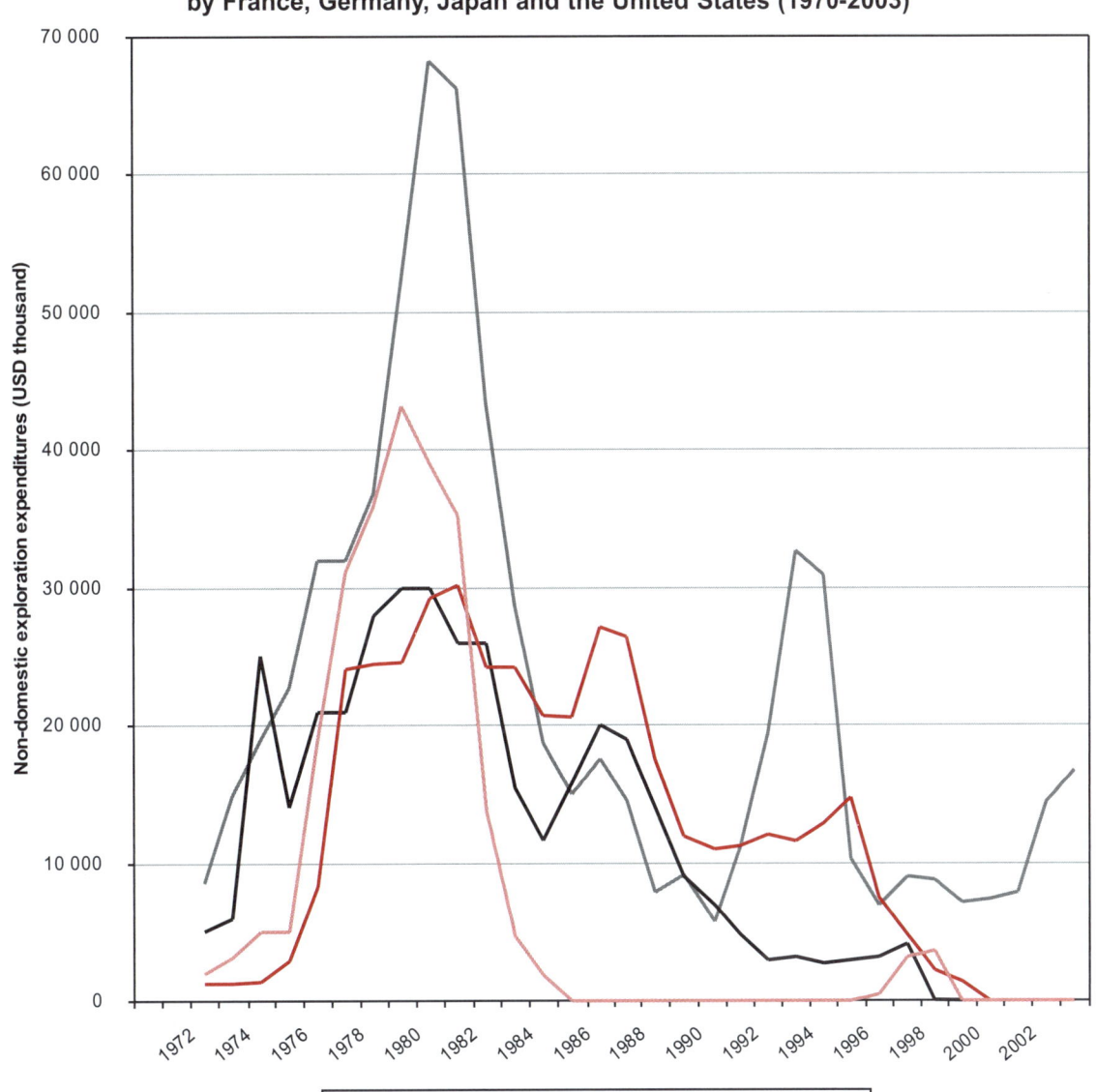

Figure 5.14. Exploration expenditures abroad by France, Germany, Japan and the United States (1970-2003)

Increased uranium prices starting in 1973 in combination with projected increases in uranium demand stimulated resurgence in uranium exploration abroad by international uranium exploration groups primarily in those countries that had major nuclear programmes and needed to identify adequate resources to meet growing demand. This interest was pursued in both established uranium provinces and in developing countries, with the latter stimulating interest in their respective countries through co-operative ventures or by granting exploration and exploitation concessions. Exploration expenditures increased from USD 16.8 million in 1972 (France, Germany, Japan and United States) to USD 82.4 million in 1976 (same countries as in 1972 plus Spain and Switzerland).

From 1977 to 1980, exploration programmes beyond domestic borders continued to increase with total expenditures of USD 178.4 million in 1980, which as it turned out was the peak year for exploration abroad. In 1980, major exploration programmes were financed by the following countries:

- France in Canada, Gabon, Indonesia, Niger, United States and South America (exploration in a total of 19 different countries in 1979).

- Federal Republic of Germany in Australia, Austria, Brazil, Canada, Gabon, Indonesia, Niger, Tanzania, United States, Zambia and Zimbabwe by companies and in Cameroon, Indonesia, Mali, Sudan and Thailand by the Geological Survey.

- Italy in Australia, Canada, Niger, United States, Zambia.

- Japan in Australia, Canada, Gabon, Niger, Zambia, Zimbabwe.

- United States in Australia, Africa, Canada, South America.

Beginning in 1981, in response to the uranium over-supply situation and declining uranium price, exploration activities by companies or governments outside of their own countries declined steadily (except for a small increase in 1986), reaching a low in 1991 of USD 29.9 million when exploration abroad was still being reported by France, Germany, Italy, Japan, Korea, Switzerland and the United Kingdom. During the period from 1981 to 1991, expenditures abroad continued to be quite important for France, Germany, Japan and the United Kingdom, but were significantly reduced by Korea, Spain, Switzerland and the United States.

From 1992 to 1994, exploration expenditures abroad again increased (mostly attributable to expenditures by France), before beginning a steady decline to a low of USD 10.3 million in 2001. Since 2000, only two countries, Canada (in Australia, Kazakhstan and the United States) and France (in Australia, Canada, Kazakhstan, Niger and Mongolia) report exploration expenditures outside of their own countries. Exploration expenditures abroad increased in 2002 and 2003, to USD 16.9 million and 18.5 million, respectively.

Discovery costs

The preceding sections have dealt with exploration history – where has exploration been conducted and how much it has cost. Among all of the data presented, perhaps the single most important statistic that has emerged in the above discussion is the fact that approximately USD 13 400 million has been spent on worldwide uranium exploration between 1945 and 2003. A logical question to ask is what did this effort accomplish? In purely qualitative terms, the wealth of exploration data generated from geologic mapping and sampling, geophysical surveys and drilling have identified geologic environments worldwide that have the potential to host economically viable uranium deposits.

Quantitatively, exploration between 1945 and 2003 has discovered the uranium resources on which past and present uranium production have been based. We should also not overlook the benefits of uranium exploration that preceded 1945 as it provided the early foundation for the start of the "modern era" of uranium exploration. As we look to the future, the past nearly 60 years of exploration have established the resource base on which almost all near- and mid-term production will be based. More will be said about the uranium resource base in subsequent sections, but it is important to put exploration in an economic framework to highlight its accomplishments.

Two different criteria have been selected to measure the effectiveness of past uranium exploration – how much was discovered and at what cost as measured by USD/kgU discovered (the "discovery" or "finding" cost). As is the case in many aspects of the Red Book, an incomplete database precludes a precise analysis of discovery costs. Nevertheless, sufficient data are available to provide a very good summation of the effectiveness of past exploration.

The first measure of the effectiveness of past exploration is what was discovered. Table 5.2 summarises resources in two confidence categories (RAR and EAR-I), which are recoverable at <USD 130/kgU and a third confidence category (Undiscovered Resources) with unspecified costs. Uranium exploration has been successful, with an estimated 20.9 million tU of resources identified between 1945 and 2003, of which 2 204 732 tU were mined (about 10% of the total). It must be emphasised, however, that undiscovered resources (EAR-II and Speculative Resources) are based on indirect evidence and/or geological extrapolation. Both are speculative and will require substantial additional exploration before they can be considered to be truly reliable resources. To provide historical perspective, in 1982, when information on worldwide uranium resources was much more geographically limited, WOCA known conventional resources at <USD 130/kgU totalled ~3.5 million tU; in 2003 the world total in the same category equalled ~4.6 million tU. Between 1982 and 2003, ~1 million tU were produced, resulting in a net increase of ~2.1 million tU. The increase is also due to the inclusion of the former Soviet Union in the totals for 2003, but not in the 1982 total.

Table 5.2 Uranium resources and historical production (as of 1 January 2003) (tU)

RAR	EAR-I	Undiscovered resources	Historical production	Total resources + production
3 149 736	1 419 450	14 222 290	2 204 732	20 996 208

For the analysis of discovery costs, expenditures are reported on an annual basis in current US dollars (USD). Exploration and development costs include the following activities: geological, geophysical and geochemical surveys, surface and/or underground drilling, geophysical logging, chemical analysis, test mining, land acquisition and maintenance costs and overhead and administration charges. Expenditures are included up to the time that sufficient reserves/resources have been identified to justify a commercial mining-processing operation including costs of completing feasibility studies.

Analyses of discovery costs have been completed for the following resource categories. In both cases, past production has been added to the resources to establish the effectiveness of past exploration:

- Known Conventional Resources (Reasonably Assured Resources (RAR) plus Estimated Additional Resources – Category I (EAR-I)) plus production.

- Total Resources (Known Conventional Resources plus Undiscovered Resources, which consist of Estimated Additional Resources – Category II and Speculative Resources) plus production.

The data set available for comparing uranium discovery costs contains a number of inconsistencies and is far from complete. It includes countries that report resources but no exploration expenditures and countries in which exploration resulted in discovery of no (or very limited) resources. Table 5.3 is a country-by-country summary of Known Conventional Resources (KCR) recoverable at <USD 130/kgU, past production, exploration expenditures and discovery costs (resources plus expenditures/production) for the 13 leading resource countries, which account for 87% of KCR plus cumulative production and 78% of historical exploration costs. A total of 40 countries have reported KCR and production. In addition, data were made available from the former USSR and from Kazakhstan, Russia, Ukraine and Uzbekistan. The average discovery cost for all countries listed in Table 5.3 was USD 1.82/kgU. Discovery costs of non-USSR countries listed Table 5.3 averaged USD 1.62; discovery costs in the former USSR countries averaged USD 2.25/kgU. Discovery costs listed in Table 5.3 range from a low of USD 0.25 for South Africa (low due to exploration being

coincident with exploration for gold) to a high of USD 10.61/kgU for France. A complete listing of all countries reporting KCR and production is included in Appendix 5.7. The average discovery cost for all the countries for which data are available was USD 1.95/kgU.

A similar analysis for 42 countries that reported KCR plus Undiscovered Resources is included in Appendix 5.8. In addition, data were made available from the former USSR and from Kazakhstan, Russia, Ukraine and Uzbekistan. With the inclusion of Undiscovered Resources, the global discovery cost was USD 0.63/kgU while in the former USSR and CIS countries it was USD 1.10/kgU. Appendix 5.9 lists that countries that reported exploration expenditures but no recoverable resources at <USD 130/kgU.

**Table 5.3. Uranium resources, production, exploration expenditures
and discovery costs in selected countries (1945-2003)
(as of 1 January 2003)**

Country	RAR (tU)	EAR-I (tU)	RAR + EAR-I (tU)	Production (tU)	KCR + Production (tU)	Exploration expenditures (USD 1000)	Discovery cost (USD/kgU)
Australia	735 000	323 000	1 058 000	113 304	1 171 304	508 949	0.43
Canada	333 834	104 710	438 544	374 548	813 092	1 288 500	1.58
Czech Rep.[1]	830	90	920	108 649	109 569	314 013	2.87
France		9 510	9 510	75 965	85 475	907 240	10.61
India	40 980	18 935	59 915	7 963	67 878	315 228	4.64
Niger	102 227	125 377	227 604	91 186	318 790	216 121	0.68
South Africa	315 330	80 340	395 670	157 618	553 288	140 919	0.25
United States	345 000	NA	345 000	356 485	701 485	2 507 113	3.57
Subtotal non-USSR	1 873 201	661 962	2 535 163	1 285 718	3 820 881	6 198 083	1.62
USSR and CIS States[2]	817 760	488 630	1 306 390	467 482	1 773 872	4 002 235	2.25
Total	**2 690 961**	**1 150 592**	**3 841 553**	**1 753 200**	**5 594 753**	**10 200 318**	**1.82**

1. Exploration expenditures from 1971 to 2003.
2. Includes totals from Kazakhstan, Russian Federation, Ukraine and Uzbekistan along with USD 3 692 550 000 in exploration expenditures in the USSR from 1945-1990.

Relationship between exploration emphasis and geological deposit types

The distribution of uranium in the world is a function of geology and uranium exploration is guided by what is known about the geology of a country or specific area. Initially, exploration may cast a broad net through the use of regional exploration techniques such as satellite imagery and airborne radio-metrics to identify favourable terrain and locate uranium anomalies. As a country's industry matures, exploration tends to become more focused on specific deposit types known to already exist in that country or on deposit types found in other countries with similar geological conditions until there is reason to alter that tendency. For example, there has been a successful cross-fertilisation between Australia and Canada regarding the knowledge of unconformity-related deposits that has benefited not only exploration in those two countries, but also that for unconformity-related deposits elsewhere in the world. Discoveries of new deposit types anywhere in the world have tended to be the catalyst for altering the pattern of exploration by broadening geologic thinking. Exploration geologists immediately begin to question where else such deposits might exist. Geological deposit models formulated in one area are applied to similar geology elsewhere, in some cases dramatically altering exploration emphasis.

Discoveries of new deposit types can set off new exploration "booms" in areas with similar geology. They can also entirely reshape a country's uranium industry. Perhaps the best recent example of this phenomenon is the discovery of high-grade mineralisation in what came to be named "unconformity-related deposits" in the Athabasca Basin, Saskatchewan, Canada. Though exploration continued in other areas, the discovery of the Key Lake, Rabbit Lake and Cluff Lake deposits immediately shifted almost all exploration in Canada to the Athabasca Basin. Eventually, development of these deposits dramatically changed the Canadian uranium production industry. Production capability had for many years been entirely concentrated in the Blind River-Elliot Lake quartz-pebble conglomerate deposits in Ontario. By 1996, however, the production centres in Ontario had been shut down and were replaced entirely by Athabasca Basin-based production.

Chapter 6 includes extensive information on a broad range of subjects related to uranium resources. Because of their relationship to exploration trends, however, it is appropriate at this point to discuss uranium resource distribution by deposit type. Table 5.4 lists the distribution of resources by deposit type, confidence category and production cost category as of 1 January 2005. These data were obtained from the 2005 Red Book, which again published information that relates resources to their respective deposit types after a hiatus of several Red Book editions. Three deposit types – unconformity-related, sandstone and hematite breccia complex – dominate the low-cost resources in both the RAR and Inferred Resources[2] categories. Unconformity-related resources form the basis for the uranium production industries of Canada and northern Australia, which together accounted for 40% of worldwide uranium output in 2004. Sandstone deposits formed the basis for the production industries in Kazakhstan, Niger, the United States and Uzbekistan, which collectively accounted for 25% of uranium production in 2004. Hematite breccia complex resources, which are mostly attributable to the Olympic Dam deposit in Australia, contributed 9% of 2004 production. These three deposit types accounted for 74% of 2004 worldwide production and for 77% of RAR recoverable at <USD 40/kgU at the beginning of 2005.

---

2. Estimated Additional Resources – Category I (EAR-I) was renamed Inferred Resources beginning with the 2005 edition of the Red Book. Similarly, EAR-II were renamed Prognosticated Resources.

Table 5.4. Identified Resources by deposit type (as of 1 January 2005)

Geological type of deposit	Reasonably Assured Resources						Inferred Resources*					
	<USD 40/kgU		<USD 80/kgU		<USD 130/kgU		<USD 40/kgU		<USD 80/kgU		<USD 130/kgU	
	10³ tU	%	10³ tU	%	10³ tU	%	10³ tU	%	10³ tU	%	10³ tU	%
Unconformity-related	433.2	22.2	492.2	18.6	498.5	15.1	151.6	19.0	169.6	14.6	171.3	11.8
Sandstone	552.5	28.4	716.5	27.1	986.6	29.9	172.9	21.6	256.3	22.1	301.6	20.9
Hematite breccia complex	513.3	26.4	513.3	19.4	522.4	15.8	281.9	35.3	286.9	24.7	288.5	20.0
Quartz-pebble conglomerate	85.6	4.4	153.3	5.8	229.3	7.0	50.5	6.3	72.0	6.2	84.8	5.9
Vein	0	0	84.0	3.2	258.8	7.9	14.8	1.9	136.1	11.7	231.8	16.0
Intrusive	63.7	3.3	150.6	5.7	202.9	6.2	60.6	7.6	81.1	7.0	109.6	7.6
Volcanic and caldera-related	49.9	2.6	135.5	5.1	140.3	4.3	1.5	0.2	5.7	0.5	7.1	0.5
Metasomatite	109.3	5.6	157.6	6.0	179.8	5.5	5.6	0.7	22.5	1.9	87.2	6.0
Other **	129.2	6.6	164.7	6.2	186.2	5.6	49.6	6.2	102.8	8.9	125.4	8.7
Unspecified	10.6	0.5	75.6	2.9	91.9	2.8	9.9	1.2	28.1	2.4	38.7	2.7
Total	1 947.3	100.0	2 643.3	100.0	3 296.7	100.0	799.0	100.0	1 161.0	100.0	1 446.2	100.0

*   Formerly EAR-I with the name changed for the 2005 edition of the Red Book.

** Includes surficial, collapse breccia pipe, metamorphic, limestone and uranium coal deposits. Rock types with elevated uranium contents such as pegmatite, granites and black shale are not included.

The 1982 Red Book edition included a summary of resources recoverable at <USD 130/kgU by deposit type. However, a direct comparison of the 1982 data with the 2005 data included in Table 5.4 is not possible because resource categories between the two data sources do not directly compare. One can, however, compare changes in resources between 1982 and 2005 for the more important resource categories (Table 5.5). Total KCR decreased from 5 013 million tU in 1982 to 4 784 million tU in 2005, despite the fact that the number of countries reporting RAR increased from 31 in 1982 to 41 in 2005. Similarly, countries reporting Inferred Resources (EAR-I) increased from 31 in 1982 to 34 in 2005. It is important to emphasise that the resources of the USSR/CIS, the Central and Eastern European countries and China were included in the 2005 totals but not in the 1982 totals. Production from 1982 to 2004 totalled 1 044 353 tU, which is more than the decrease of KCR.

Total RAR increased between 1982 and 2005, while Inferred Resources/EAR-I decreased between the two reports. This difference can be partially explained by the fact that no major new deposits were discovered that could have added to the EAR-I total. Instead, reduced exploration expenditures were directed toward upgrading the confidence level of Inferred Resources/EAR-I to RAR. Though the Olympic Dam deposit was discovered in 1978, the breccia complex deposit type was not introduced until the 1989 Red Book, prior to which Olympic Dam was included in the Other Deposit Types category, "awaiting a type definition of its own".[3]

Sandstone and quartz-pebble conglomerate deposits both declined significantly as a percentage of KCR. This is partly attributable to the addition of other deposit types in the 2005 analysis that were not included in 1982 (e.g. hematite breccia complex, intrusive, vein and volcanic deposits) and in the case of quart-pebble conglomerates the closing of the Elliot Lake district in eastern Canada and the elimination of these resources from Canada's resources.

---

3. 1986 Red Book, p. 301.

Table 5.5. Comparison of 1982 and 2005 distribution of resources by deposit type

Geological type of deposit	Reasonably Assured Resources		EAR-I/Inferred Resources	
	% in 1982	% in 2005	% in 1982	% in 2005
Unconformity-related	15.0	15.1	15.0	11.8
Sandstone	40.0	29.9	40.0	20.9
Hematite breccia complex		15.8		20.0
Quartz-pebble conglomerate	15.0	7.0	20.0	5.9
Vein	5.0	7.9		16.0
Intrusive		6.2		7.6
Volcanic and caldera-related		4.3		0.5
Metasomatite		5.5	15.0	6.0
Other*	25.0	5.6	10.0	8.7
Unspecified		2.8		2.7
Total Resources (1 000 tU)	2 293	3 317	2 720	1 467
Number of countries reporting	31	41	31	34

* Includes black shale, phosphorite, surficial, limestone and collapse breccia pipe deposits as well as deposit types reported by some countries that do not conform to the Red Book deposit types.

From its very first edition, country reports have been an important part of the Red Book. They contain information on each contributing country's exploration programme, and by reading successive editions of the Red Book one can track the exploration history of contributing countries. Appendix 5.2 provides a synthesis of exploration in contributing countries beginning with the 1965 Red Book. Appendix 5.3 describes uranium deposit types.

Conclusions relating to uranium exploration

Exploration expenditures have been reported by 81 countries. Worldwide uranium exploration expenditures are estimated to have totalled about USD 13 400 million between 1945 and 2003.

Uranium exploration peaked at about USD 908 million in 1979 before declining to a low of USD 70 million in 1994.

An estimated 154 300 km of exploration drilling were completed between 1970 and 2003.

There is a close parallel between exploration expenditures and uranium market prices in countries with market-based economies. Because drilling is the largest component of exploration expenditures, drilling expenditures also closely parallel market price trends.

Exploration expenditures were much less sensitive to market price in centrally planned economies. Therefore, expenditure trends did not begin to parallel market price in the USSR and Eastern Bloc countries until 1991 following dissolution of the USSR and the correlation is tenuous.

Exploration expenditures from 1945 through 2003 among the 13 leading non-CIS producing countries that accounted for 87% of 2003 production totalled about USD 6 200 million. KCR plus production for these countries totalled about 3.8 million tU. Discovery costs are estimated to have been about USD 1.62/kgU. The average discovery cost for the 40 countries that have reported KCR and production information was USD 1.82/kgU.

6. URANIUM RESOURCES

Uranium resources are the cornerstone of the Red Book, and along with uranium production capacity, represent the basis of uranium supply and as such they are vital to the future of the nuclear energy. Reporting of uranium resources has evolved considerably during the 40-year history of the Red Book, with a progression of changes in response to the evolving uranium market and growing sophistication in resource estimation. Resources are also among the most subjective topics addressed in the Red Book. To help offset that subjectivity, beginning with the very first Red Book in 1965 to the most recent edition, uranium resources have been reported according to varying confidence levels.

The completeness of the database for uranium resources is a function of the willingness of participating countries to fully report their respective resources. Each Red Book includes as complete a statement on worldwide resources as is possible within the limits of the completeness and accuracy of the data supplied by the individual countries. The Uranium Group and the OECD/NEA-IAEA Secretariat have historically made a concerted effort to ensure as complete a resource data base as possible in each edition of the Red Book. It remains the policy of the Uranium Group to use the data reported by individual countries. To supplement where data are missing, however, the Secretariat has included previously submitted resource information or estimates as long as this information was properly identified. The countries with the largest reported uranium resources are listed in Table 6.1.

Table 6.1. Countries with largest Known Conventional Resources recoverable at <USD 130/kgU (2003)[1]

Country	tU	Percentage of world total
Australia	1 058 000	23.1
Kazakhstan	847 620	18.5
Canada	438 544	9.6
South Africa	395 670	8.6
United States[2]	345 000	7.5
Others (total)	1 503 166	32.7
World total	**4 588 000**	**100.0**

1. Includes RAR and EAR-I resources at <USD 130/kgU.

2. The United States does not report resources in the EAR-I category.

Comparisons of historical resource estimates that span the entire history of the Red Book are not statistically appropriate or significant, because reporting parameters have changed as have the number of countries that contributed resources estimates over time. Therefore, to avoid "mixing apples and oranges", care must be taken in selecting the parameters by which to measure changes in resources during a given time period. While perhaps not statistically appropriate, a comparison of resources reported in the 1965 and 2003 Red Books helps put the scope of the changes in resource reporting into context. The first Red Book (1965) reported resources totalling 3 209 000 tU projected to be recoverable at <USD 30/kgU. Resources included in the 1965 Red Book, which were reported by 16 countries, were divided into two confidence categories – Reasonably Assured Resources and Possible Additional Resources.

By way of comparison, the 2003 Red Book included resources reported in 56 countries. The number of confidence levels had grown from two in 1965 to four in 2003. Cost categories remained the same at three for the two Red Book editions, but the ranges among the cost categories evolved significantly over time, reflecting changes in the uranium market price and worldwide economics, which in turn have affected uranium production costs. Resources reported in the 2003 Red Book in all confidence and cost categories totalled 14 382 488 tU. Appendix 6.1 provides a detailed chronology of the changes that have taken place in resource category definitions since publication of the first Red Book and offers a perspective as to the care that must be taken in analysing resource changes over time.

Changes in resources over time

As has already been suggested, one of the values of the Red Book series lies in the statistical information they contain on various aspects of uranium supply and demand. While changes in resource reporting parameters make historical comparisons that cover the entire Red Book history of limited value, data contained in each Red Book have statistical importance, particularly on a country-by-country basis. In addition, it is instructive to analyse changes in resources during periods of relative consistency in resource reporting parameters. Therefore, to enhance the value of Red Book Retrospective, extensive historical data on different confidence categories of resources are included in Appendices 6.2 through 6.10. Appendices 6.2 and 6.3 show how RAR changed from Red Book to Red Book for a wide range of production cost categories. Appendices 6.4 and 6.5 do the same for Estimated Additional Resources – Category I. Appendix 6.6 provides data on Other Known Conventional Resources while Appendix 6.7 gives summary data on KCR from 1965 through 2003. Appendix 6.8 provides data for Estimated Additional Resources – Category II and Speculative Resources (together known as "Undiscovered Resources") and Appendices 6.9 and 6.10 provide data on the evolution of total resources over the history of the Red Book for selected countries.

The one resource category that has remained relatively unchanged as to its definition is Reasonably Assured Resources (RAR). Figure 6.1 shows changes in RAR between 1965 and 2003 and compares these changes with uranium market price evolution. Though RAR have been included in all of the Red Books, the price ranges have changed, which explains why curves representing the cost categories are discontinuous (e.g. RAR<USD 40/kgU).

RAR in the different cost ranges parallel one another closely. There does not, however, seem to be a close relationship between RAR and market price. In contrast to the steep price decline that began in 1979, RAR decreased only slightly before levelling off in about 1985. In 1993, RAR steadily increased while, except for the brief 1996 price increase, market prices remained relatively flat until the recent price increase, which began in late 2000. Reclassification of CIS resources (which between 1991 and 1993 were reported as "Other Known Conventional Resources") based on Red Book resource terminology accounts for most of the increase in RAR after 1993.

Figure 6.2 shows a much closer parallelism between EAR-I and market price, though the curves for the different cost categories are rather disjointed. Before drawing conclusions from this relationship, it should be noted that while RAR has been consistently used as a confidence category beginning with the first Red Book, EAR-I became a new category in 1983. Estimated Additional Resources, which had been in use since 1967, was subdivided in 1983 into two categories – EAR-I and EAR-II. In addition, the United States has never adopted the EAR-I – EAR-II separation. Instead it only recognises the EAR-II category, with the tacit understanding that EAR-II reported by the United States includes an unknown amount of EAR-I.

Despite these internal inconsistencies, there remains a parallelism between EAR-I and market price. Though this parallelism may be only a coincidence, one possible explanation for this relationship is that as is shown in Figure 5.2, declining market prices beginning in 1979 were accompanied by a parallel decrease in exploration costs. With limited exploration budgets, emphasis may have shifted from "greenfields" exploration characterised by geophysics and broadly spaced drilling, which would have tended to increase EAR-I, to converting already identified EAR-I to RAR by more closely spaced drilling.

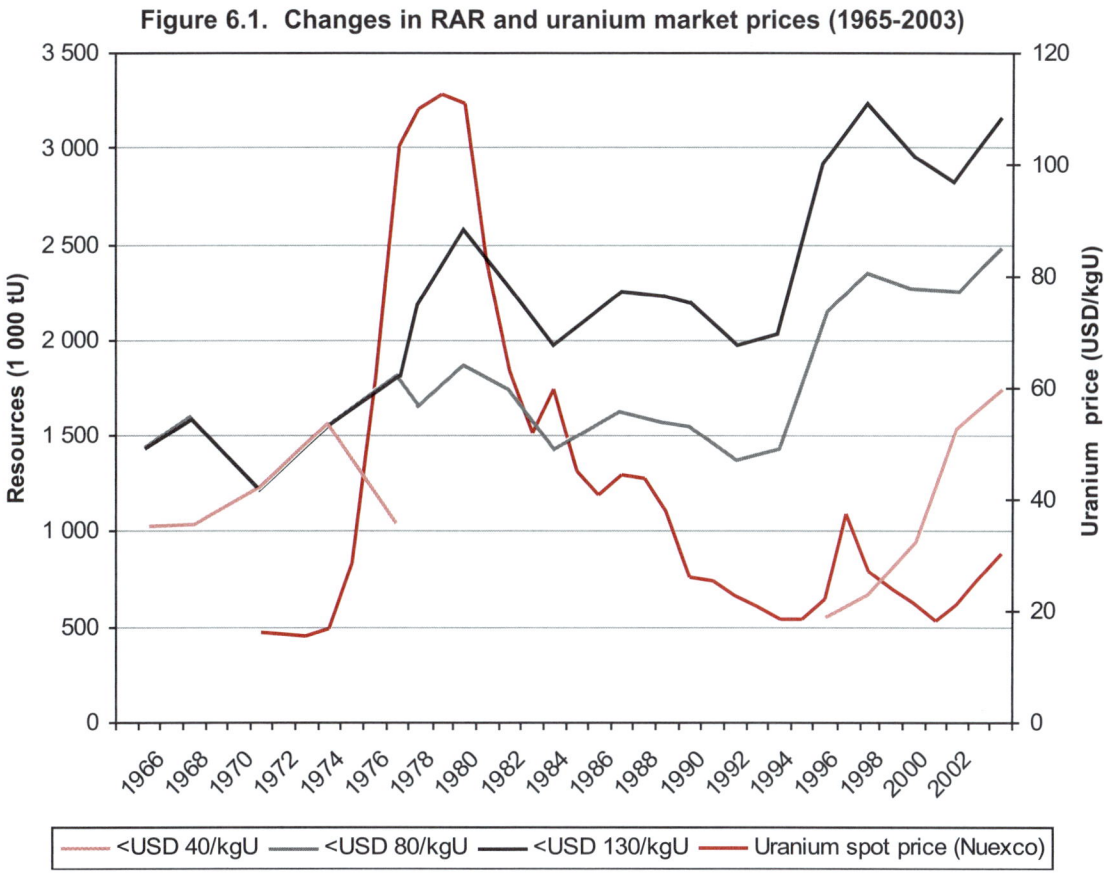

Figure 6.1. Changes in RAR and uranium market prices (1965-2003)

Figure 6.2. Changes in EAR-I and uranium market prices (1965-2003)

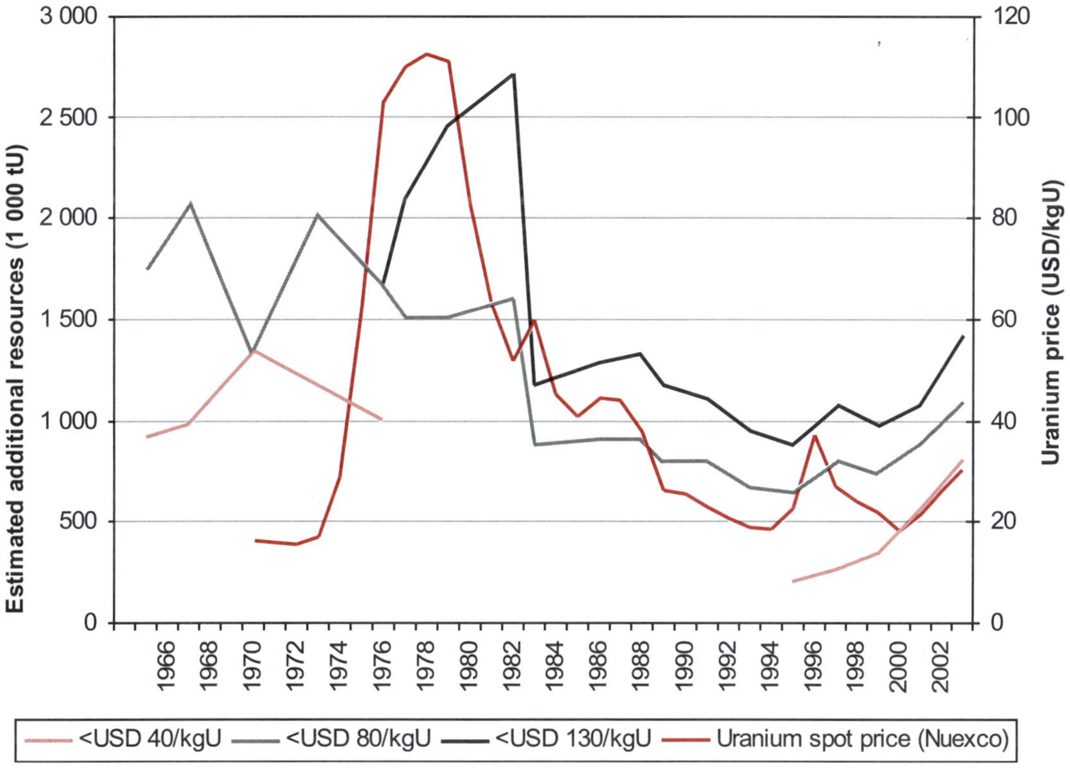

Appendix 6.8 shows changes in Undiscovered Resources (EAR-II + Speculative Resources) between 1982 and 2003. Figure 6.3 shows the evolution of Undiscovered Resources and market prices. The EAR-II curves show very gradual increases until about 1992 and then remain nearly unchanged through 2003; EAR-II data show little relationship to market price trends. Countries tend to not re-evaluate their EAR-II estimates on a regular basis, which at least in part explains the limited changes shown in Figure 6.3. In contrast with the EAR-II trends, both categories of Speculative Resources show considerably more volatility. Both parallel the decline in market prices between 1982-1983 and 1986. Speculative Resources projected to be recoverable at <USD 130/kgU continued to approximate the variations in market price, while those in the USD 130-260 cost range and in unassigned cost category depart from the price trend in 1988 and begin to increase through 1994 before starting to level out. The increase between 1988 and 1994 is largely attributable to more countries reporting data – nine countries in 1988 compared to 17 countries in 1994.

Figure 6.3. Changes in Undiscovered Resources over time compared to uranium market price (1981-2003)

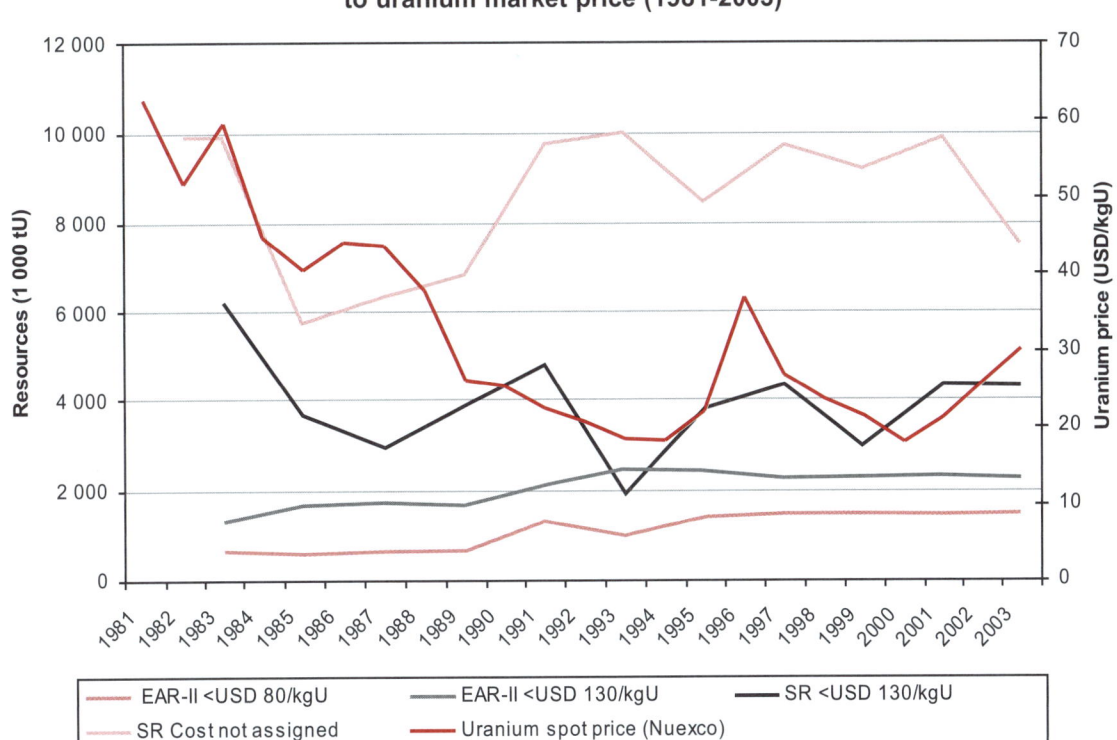

In the years immediately following dissolution of the Soviet Union, the newly independent Soviet countries (Kazakhstan, Russia, Ukraine and Uzbekistan) and most of the Central and Eastern European Countries continued to report resources using Soviet terminology, which was inconsistent with the NEA/IAEA Red Book terminology (Appendix 6.1). Therefore, resources reported by the former Soviet countries between 1991 and 1995, which were not in agreement with the Red Book classification, have been listed under "Other Known Conventional Resources" in Appendix 6.6. After 1995, following a series of meetings to harmonise resource categories, all countries reported their resources according to the Red Book terminology.

In keeping with Red Book convention, RAR plus EAR-I is referred to as Known Conventional Resources (KCR), to denote that these resource categories do not include such unconventional resources as uranium hosted in phosphate deposits, black shale, etc. Unconventional Resources are discussed in Chapter 9 of this report.

Total KCR included Reasonably Assured Resources plus Estimated Additional Resources in the period from 1965 to 1982 and Reasonably Assured Resources plus Estimated Additional Resources – Category I from 1983 to 2003. In Appendix 6.7 a summary overview is given of the development of KCR in all cost categories between 1965 and 2003 including the "Other Known Conventional Resources" described in Appendix 6.6. Total KCR recoverable at costs below USD 130/kgU are reported throughout the period from 1965 to 2003 and have continuously grown from 3.2 million tU in 1965 to nearly 4.6 million tU in 2003. In 1986 and 1989, when high growth rates of uranium requirements were anticipated, resources recoverable at a cost of between USD 130-260/kgU were reported by a few countries; however in subsequent years this high cost category was no longer reported.

Changes in resources over time for selected countries

Figures 6.1 through 6.3 show "global" trends in resource trends based on the total of all data reported for a given year. Explanations for abrupt changes in resource totals or trends are not readily apparent from the summary data. They may, however, be contained within individual country reports. There are several reasons for changes in resources in a given country:

- Production may increase faster than resources are replenished by exploration.

- Periodic analysis may move resources to a higher cost category, which is not included in analysis criteria, for example, a study limited to resources recoverable at <USD 80/kgU. This is a very important point to remember. The resources do not just "go away", at least not in a physical sense. They may be placed in a higher cost category, but they are still potentially available for the future in a different economic environment.

- Resources may be removed from country totals when associated production centres close.

- Resources may also be removed from a country's resource base because of environmental concerns. Here again, the resources are not necessarily gone forever. New technology may one day make them economically and environmentally viable.

It is beyond the scope of this report to include a comprehensive country-by-country analysis of resource trends. However, the following examples of KCR trends for a few key countries analyse changes in resources and offer possible explanations for these changes. These are examples that illustrate the kinds of information available in the Red Book series. It is important to emphasise that the Red Book does not provide detailed information on exploration results, discoveries and evaluation methods (including cut-off grade). Therefore, it is not possible to fully explain changes in resource trends, even at the country report level. Significant changes in resources in individual countries can, however, help explain changes in global resource trends. Though the CIS countries are becoming increasingly important both from a resource and production standpoint, there is not enough historical information to include them in this analysis. Table 6.2 shows the lack of variability of CIS resource since they were first reported by the individual countries beginning in the 1995 Red Book. Only Uzbekistan showed a significant change, with RAR increasing by 38% between 1999 and 2001.

Table 6.2. Comparison of CIS resources (RAR <USD 130/kgU)
reported between 1995 and 2003 (tU)

Red Book edition	1995	1997	1999	2001	2003
Kazakhstan	598.7	601.0	598.6	594.8	530.0*
Russia	NA	145.0	140.9	138.0	143.0*
Ukraine	81.0	84.0	81.0	81.0	64.6*
Uzbekistan	NA	83.7	83.1	115.0	79.6*

* Resources adjusted to reflect recoverable resources versus *in situ* resources as reported in previous Red Book editions.

Figure 6.4 shows how KCR plus production have varied over time for Australia, Canada, France and the United States. The reasons for the variations shown in Figure 6.4 are discussed in the following sections. Appendices 6.9 and 6.10 provide information on changes in KCR for select countries over time and exploration expenditures, drilling activities and production information to help analyse the changes.

In the following sections changes in resources are graphically compared with exploration expenditures and drilling (km/year). Without exploration, it is a foregone conclusion that resources will decrease. At the same time, it is important to understand that the relationship between exploration and additions to resource totals is complex. It typically takes several years from the start of exploration to discovery of a new deposit. Once a discovery is made, it can take several more years of exploratory drilling before the first preliminary resource estimate can be completed. Therefore, exploration during a given period does not necessarily translate into additions of resources during that same period. It is more likely that successful exploration will not become apparent through increases in resources for several years. Production continually depletes the resource base; replenishing the resource base requires ongoing exploration and development.

Figure 6.4. Variations in Known Conventional Resources adjusted for production in Australia, Canada and the United States

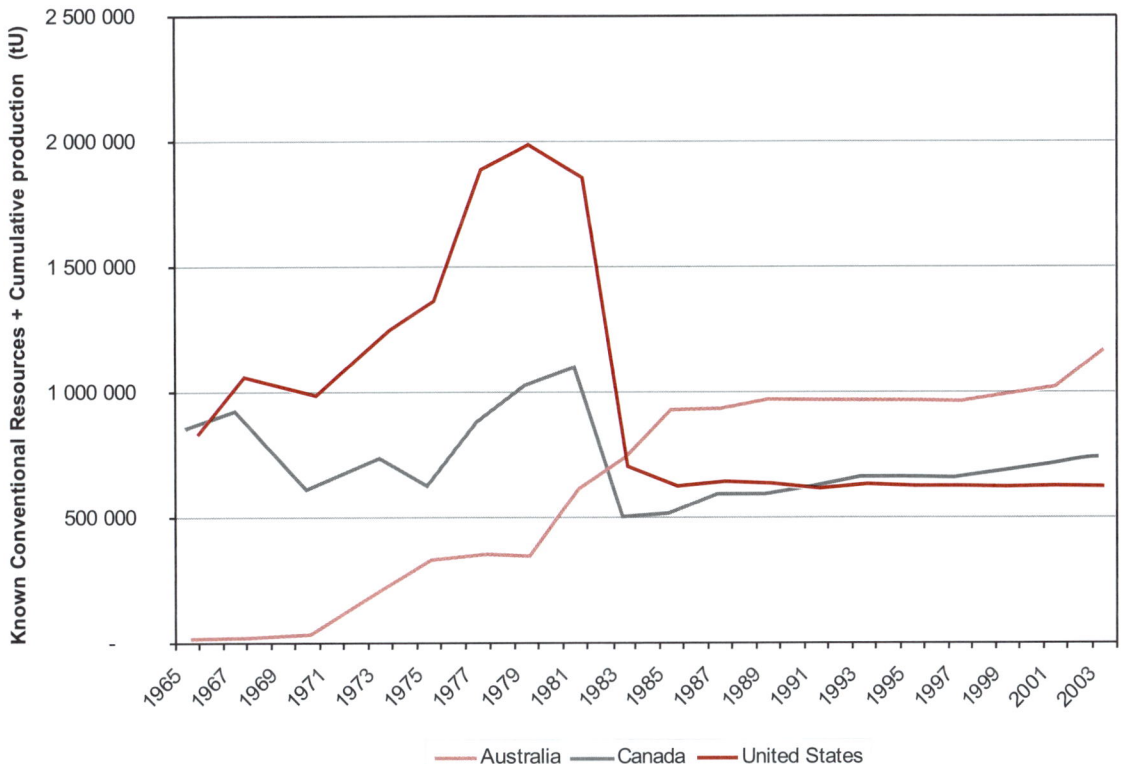

Australia

Figures 6.5 and 6.6 show that RAR in the <USD 80/kgU category has steadily increased since about 1971, as have KCR, despite the sharp drop in price that began in 1979. According to the 1982 Red Book, the steep increase in EAR-I between 1979 and 1981 is the result of an "increased effort in assessment of resources in this category". There is a decrease in EAR-I between 1983 and 1985 with a corresponding increase in RAR, which suggests that exploration (or more complete information) was successful in moving a portion of Australia's EAR-I into a higher confidence category. Figure 6.6 quantifies the magnitude of changes in KCR from one Red Book to the next, both in terms of resources and it includes drilling data to compare with the changes in resources from year to year. There is a marked coincidence between increased resources and drilling in 1981

and 1982. Whether the two are connected is very doubtful, however. If new discoveries resulted from the drilling, it is more likely that they would show up in resource additions in subsequent years.

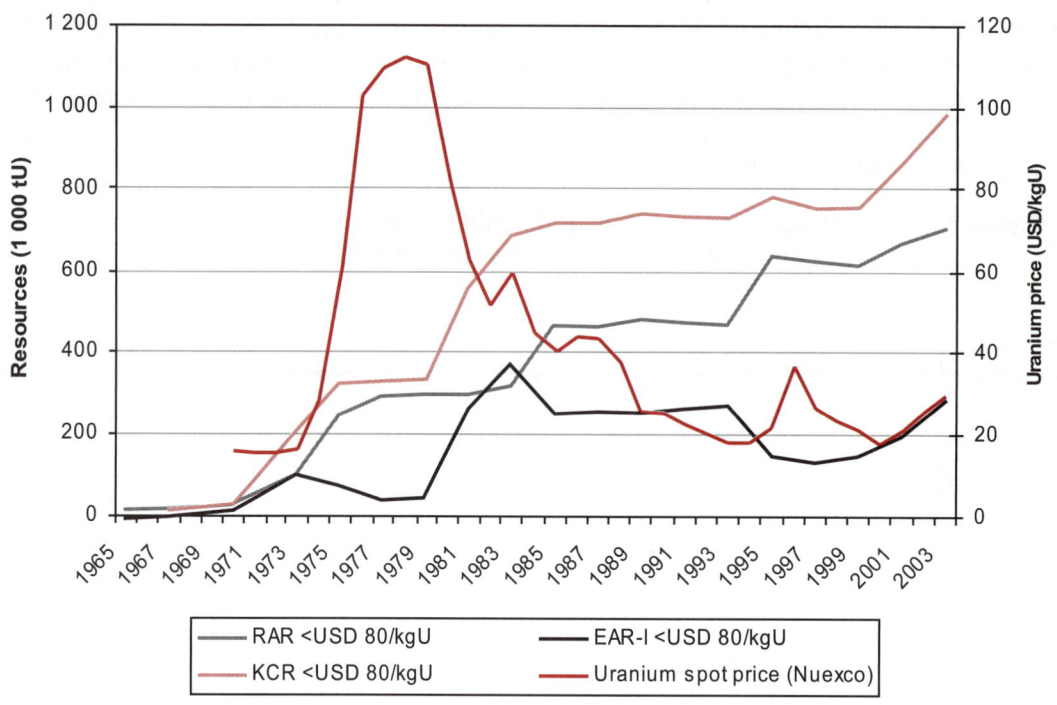

Figure 6.5. Comparison of resource trends in Australia and uranium market price

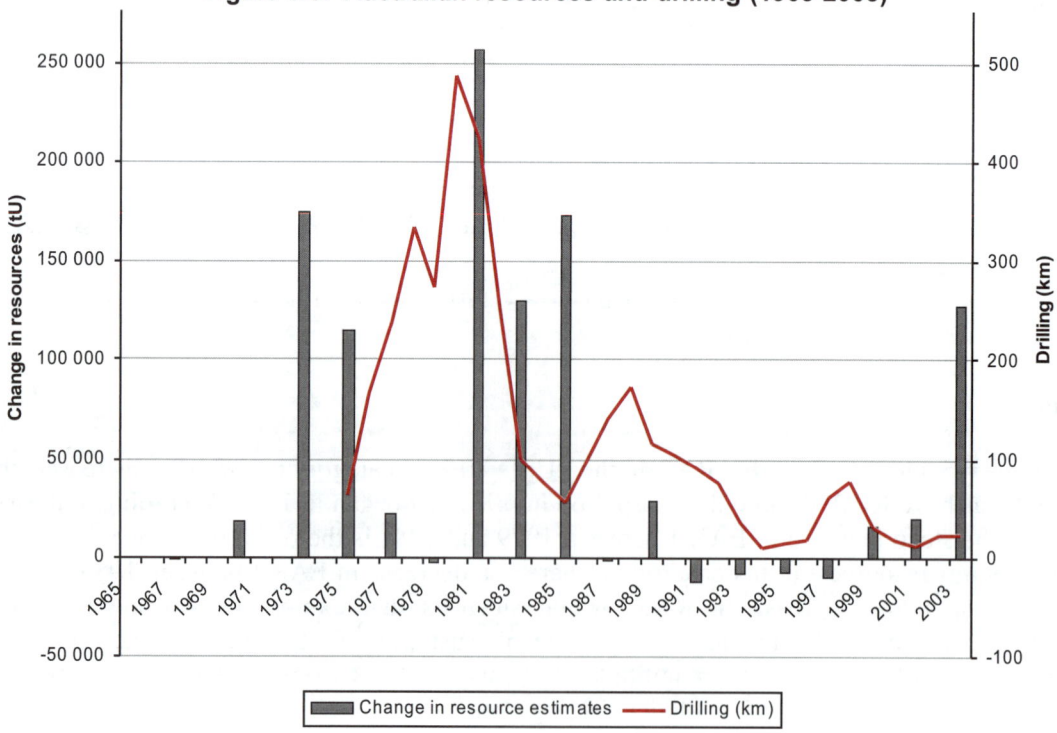

Figure 6.6. Australian resources and drilling (1965-2003)

Canada

As noted in Figure 6.7, Canada's Known Conventional Resources (KCR) steadily declined between 1967 and 1997, even during the sharp market price increase between 1973 and 1979. Figure 6.8 quantifies the magnitude of changes in KCR from one Red Book to the next and it includes drilling data to compare with the changes in resources from year to year.

Between the late 1960s and the early 1980s, more than 50% of Canada's KCR base was associated with quartz-pebble conglomerates in the Elliot Lake-Blind River districts. The steady decline in KCR that bottomed out in 1989 is in part attributable to the fact that production costs for the quartz-pebble conglomerates were increasing. As a result, more and more of the resources associated with quartz-pebble conglomerates were being pushed into higher cost categories. This is particularly true of the 604 000 tU resource reduction between 1982 and 1983 that stands out so dramatically on Figure 6.8. The increases in RAR between 1991 and 1993 resulted from exploration successes in the Athabasca and Thelon Basins, which allowed for reclassification of EAR-I to RAR.

Figure 6.7. Comparison of resource trends in Canada and uranium market prices

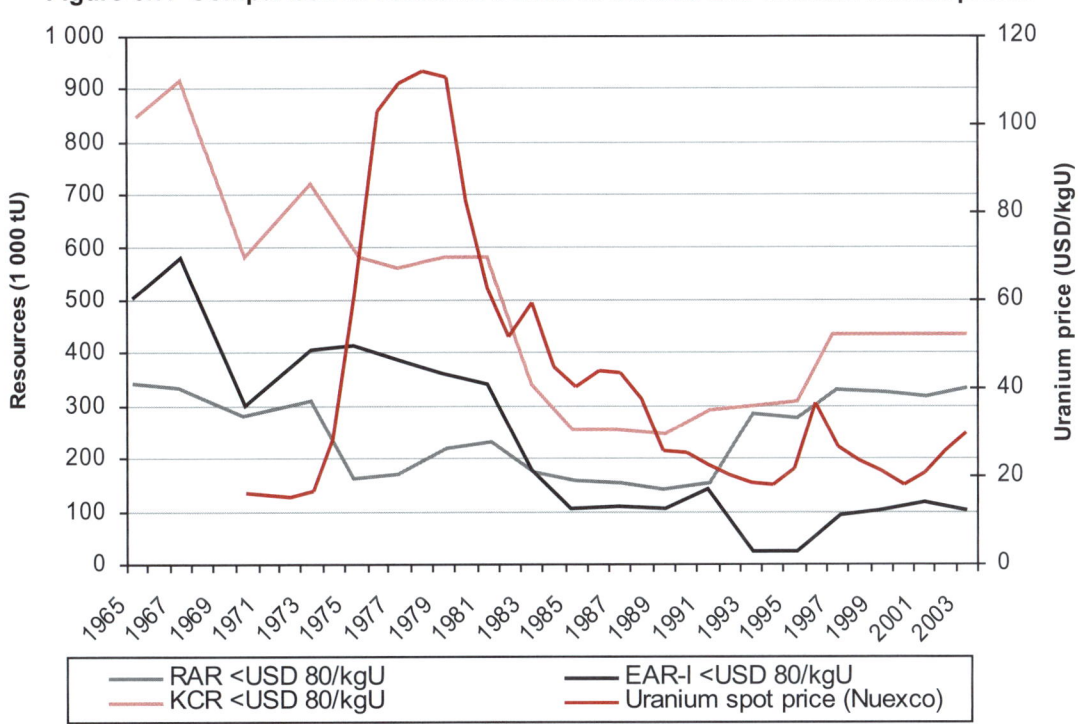

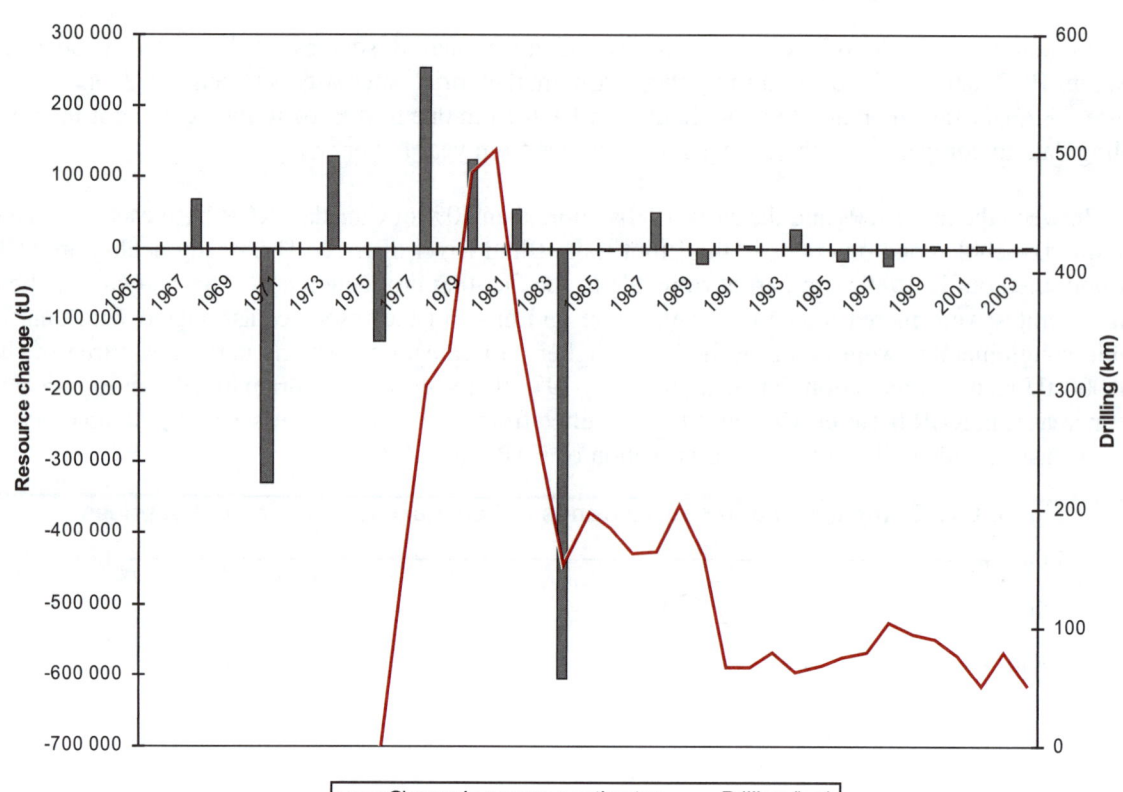

Figure 6.8. Canadian KCR and drilling (1965-2003)

United States

Known Conventional Resources (KCR) in the United States steadily increased between 1965 and 1979 (Figure 6.9). The steep decline between 1981 and 1983 resulted from changes in Red Book resources nomenclature. As previously noted, the confidence category – Estimated Additional Resources, which had been in use since 1967, was subdivided into two categories – EAR-I and EAR-II. The United States has never adopted the EAR-I – EAR-II separation. Instead, under an agreement with the Uranium Group, the United States only reported EAR-II; this accounts for the 1.21 million tU decrease in KCR between 1981 and 1983 (Figure 6.10). Though KCR had begun to decline along with the market price in 1980, the precipitous decline in 1983 was an artefact of reporting and not a true decline of resources in the United States.

Figure 6.9. Comparison of US resource trends and uranium market prices

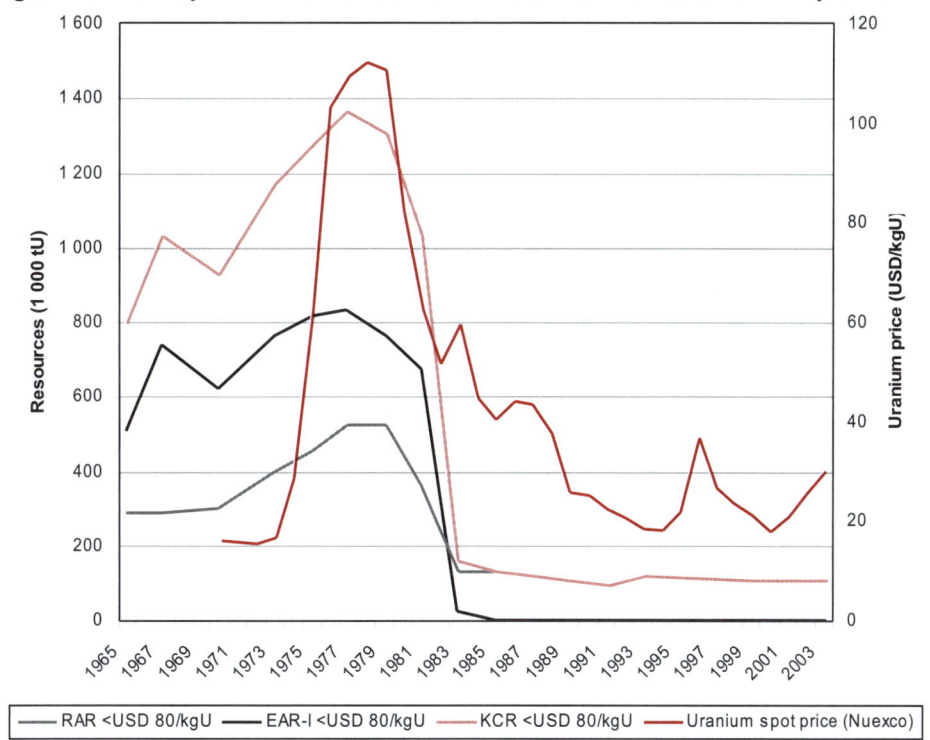

RAR <USD 80/kgU   EAR-I <USD 80/kgU   KCR <USD 80/kgU   Uranium spot price (Nuexco)

Figure 6.10. United States KCR and drilling (1965-2003)

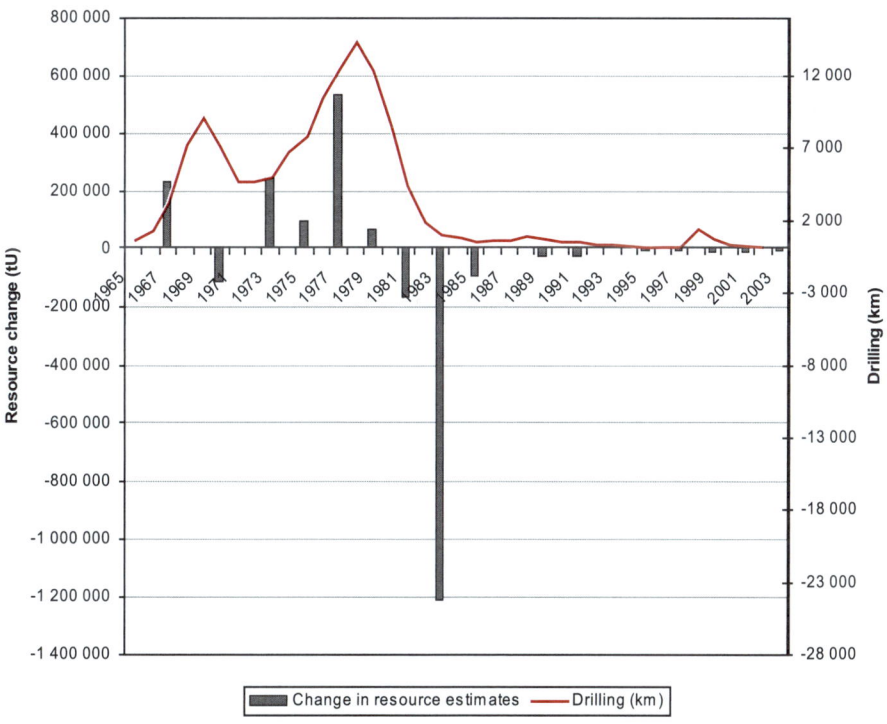

Change in resource estimates   Drilling (km)

Other selected countries

Appendices 6.9 and 6.10 contain information on other countries that have undergone substantial changes in their resources over time. Brazil's resource base steadily increased between 1977 and 1983 after which it levelled off, largely because Brazil did little exploration after 1984 (Figure 6.11). The growth in Brazil's resources in the late 1970s and early 1980s resulted from successful exploration programmes and resource additions associated with prior discoveries, namely the Lagoa Real and Itataia deposits.

In France, resources began to decline in 1989 about three years after exploration expenditures started to decline while mining continued to deplete resources (Figure 6.12). In addition to lack of replenishment of resources through exploration, the decline was also partly due to reassessment of France's resource base and reassignment of some resources to higher cost categories and to closure of some French mines.

Namibia first began reporting resources in 1979; its resource base began a steady decline in 1985 (Figure 6.13). Its resources increased dramatically between 1993, when they were based on Secretariat estimates, and 1995 when Namibia again reported official resource totals. No reason was provided for why resources increased so dramatically in 1995 compared to the Secretariat estimate in the previous Red Book. Namibia did not report exploration expenditures between 1983 and 1992 and only reported USD 2.4 million between 1992 and 2003. Therefore, new discoveries were not the reason for the sharp increase in resources between 1993 and 1995. Part of the discrepancies in estimates may be also due to political changes (transfer from South West Africa to independent Namibia in 1990).

Niger's resources increased between 1975 and 1985, largely as a result of an increase in RAR and again between 1983 and 1985, with this increase attributable to EAR-I (Figure 6.14). Niger showed another increase in resources between 2001 and 2003, with equal increases in RAR and EAR-I. No explanation is readily available for these three increases. Niger reported a sharp increase in exploration expenditures between 1978 and 1979 and expenditures remained at a relatively high level through 1981, which could explain the increase in resources between 1983 and 1985. It again reported increased exploration expenditures beginning in 2001, but these increases were modest and would not likely have contributed substantially to the increase in resources between 2001 and 2003.

The final example of changing resource totals over time in individual countries is South Africa, which is a special case because uranium was only produced as a by-product of gold mining operations and of the Palabora copper mine. Resource estimates in South Africa are a complex relationship between gold prices and the value of the Rand compared to the US dollar. Changes in uranium resources showed little direct relationship to either the price of gold or the price of uranium, though KCR did peak in 1989 following an increase in gold prices in 1987 (Figure 6.15). Except for the 55% decrease between 1989 and 1993, South Africa's KCR have not varied significantly over time. This lack of variability resulted partly from the fact that exploration virtually ceased in South Africa in 1983. The 1989 to 1993 decrease came about as a result of re-estimation of resources to reflect higher production costs at the Witwatersrand gold mines that tended to push resources into higher cost categories.

Figure 6.11. Comparison of resource trends in Brazil and uranium market price

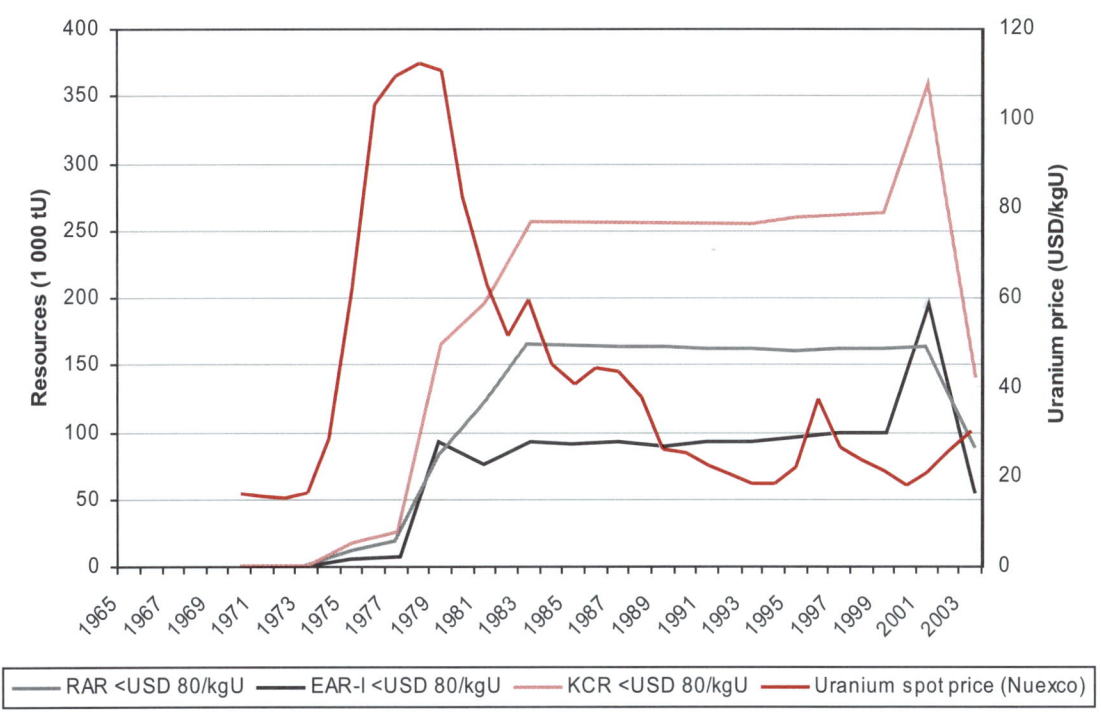

Figure 6.12. Comparison of resource trends in France and uranium market price

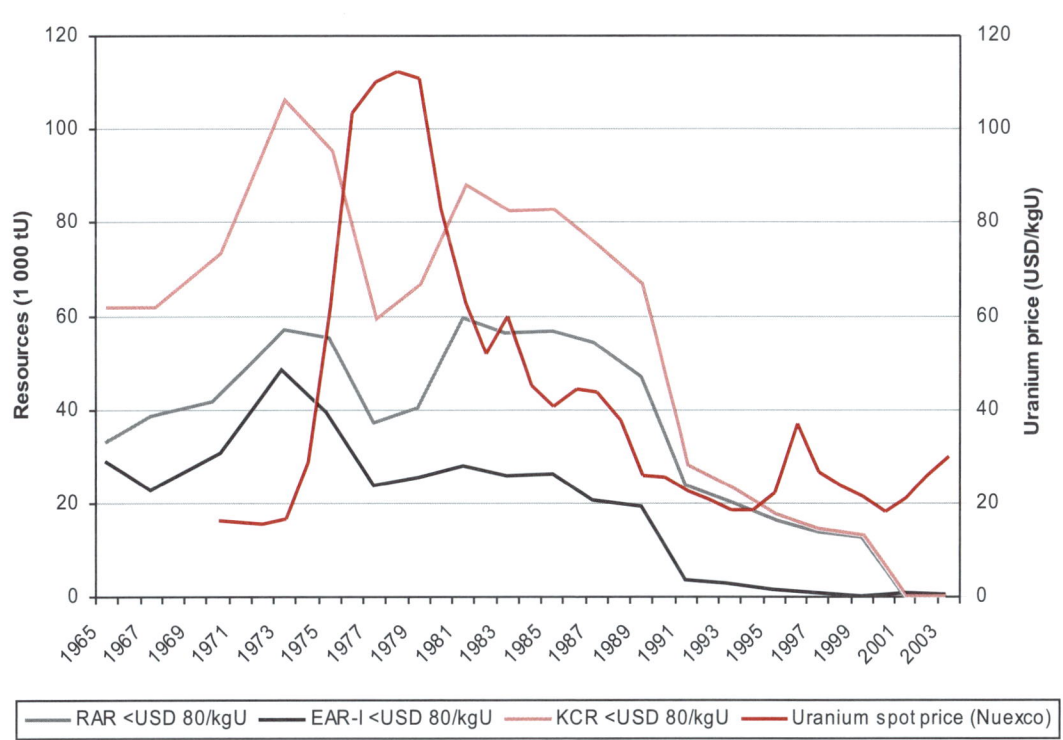

Figure 6.13. Comparison of resource trends in Namibia and uranium market price

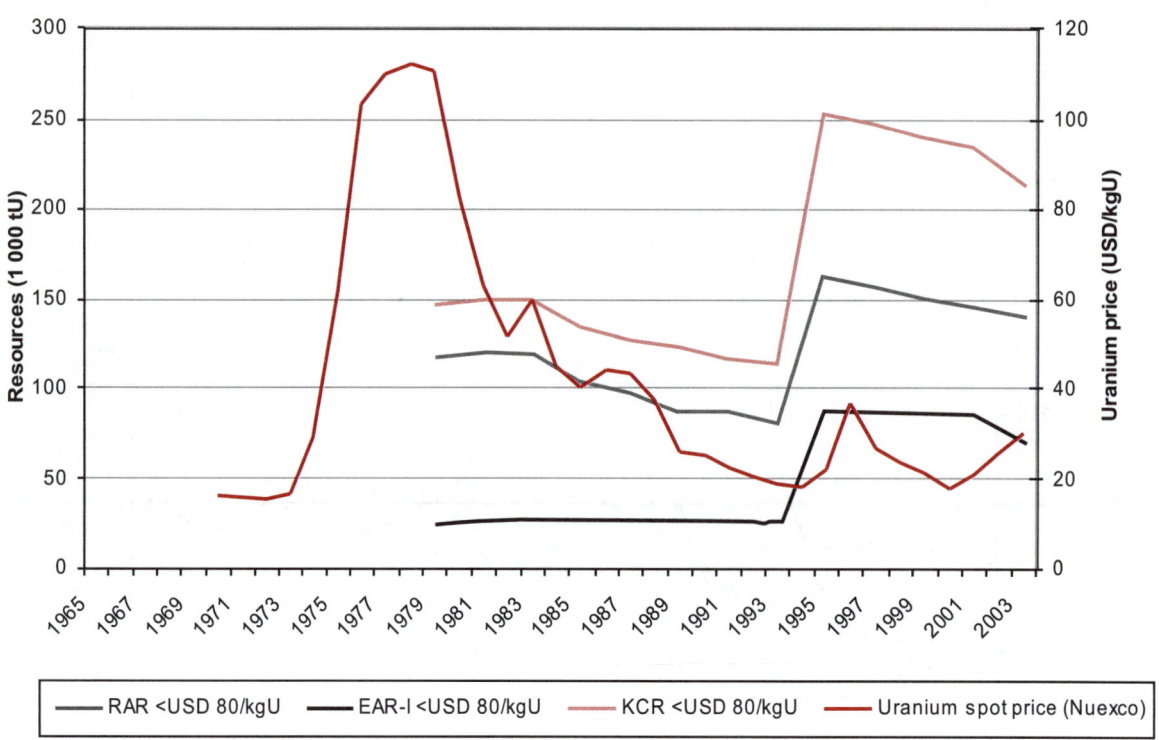

Figure 6.14. Comparison of resource trends in Niger and uranium market price

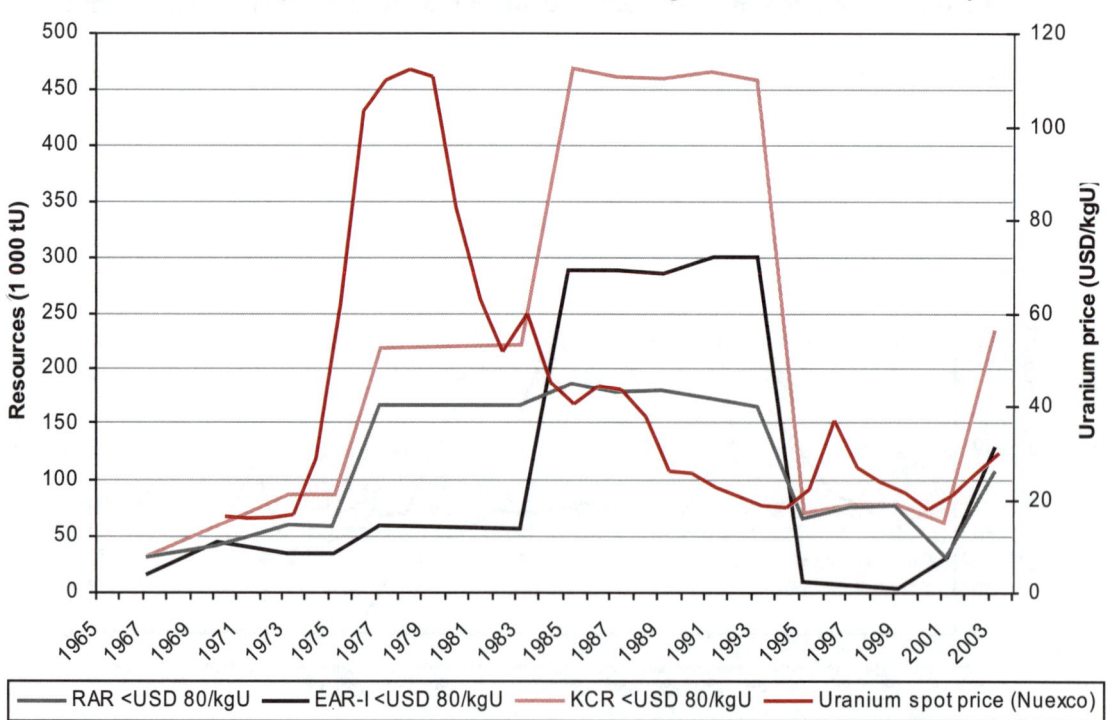

**Figure 6.15. Comparison of resource trends in South Africa
with market prices of gold and uranium**

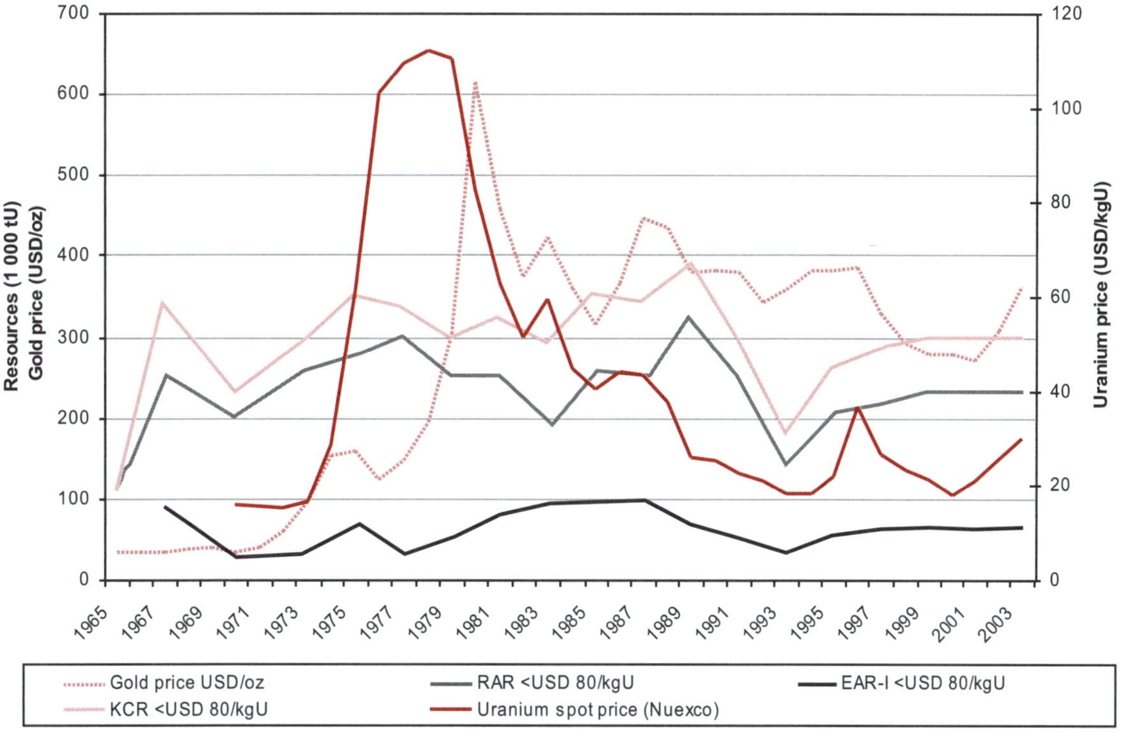

Relationship between uranium resources and reactor-related uranium requirements

A comparison between uranium resources and uranium requirements is a key measure of the future balance between supply and demand. Figure 6.16 compares the ratio between RAR and KCR and annual reactor-related uranium requirements. In 1965, uranium requirements totalled 4 920 tU compared to RAR of 1 451 000 tU and KCR totalling 3 209 000 tU. Accordingly, the ratios in 1965 were 295 and 651 for RAR and KCR, respectively. By 1970 the ratios had dropped to 88 for RAR and 183 for KCR (Figure 6.16). This decrease resulted from the combination of a 2.8 times increase in requirements and 16% and 20% decreases in RAR and KCR, respectively. In 2003, the ratios were 46 and 67 for RAR and KCR, respectively.

From a historical perspective, it is evident that uranium resources were more than adequate to meet requirements between 1965 and 2003 (Figure 6.16). It is encouraging that the ratio between resources and requirements has remained relatively unchanged since about 1990. This would suggest that additions to KCR have approximately kept pace with production. Reclassification of resources from the Commonwealth of Independent States (a one-time event) also contributed to an increase in KCR. It will be important, however, to continue to carefully monitor this balance to ensure that resources are adequate to meet future requirements. It is also important to emphasise that resources are only one part of the supply side of the supply-demand equation. Of equal importance is production capacity – getting the resources to market – which will be addressed in a subsequent chapter of this report.

Figure 6.16. Ratio between uranium resources and annual uranium requirements

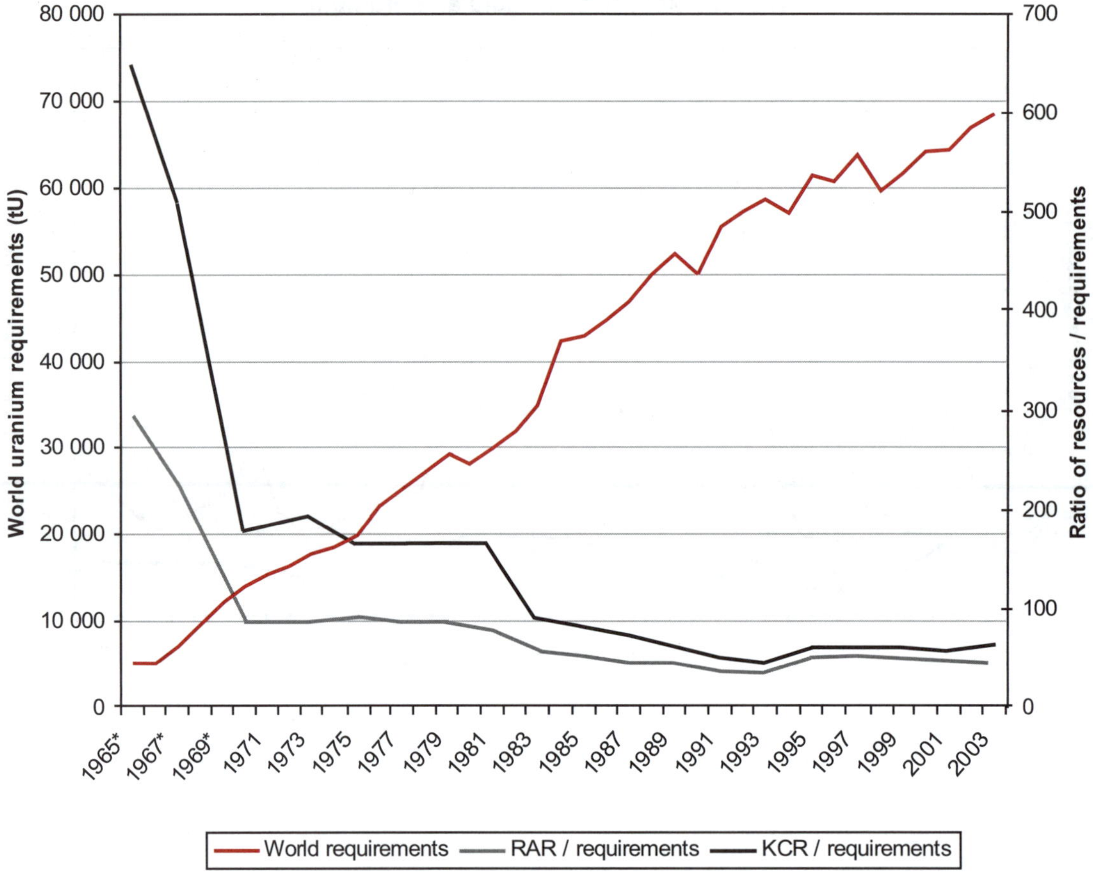

Legend: World requirements — RAR / requirements — KCR / requirements

Relationship between resources and production

Uranium resources are replenished by exploration and are depleted by production. This section will examine the relationship between production and changes in resource bases in selected countries. Appendices 6.9 and 6.10 provide the data regarding the relationship between KCR and production for the countries included in the following discussion. The conclusions that are reached in the individual country discussions are somewhat counterintuitive. Rather than being a cause of decreases in the resource base, for the most part production accounted for only small incremental changes over time. The larger changes were attributable to factors other than depletion of resources by production.

Australia

Figure 6.17 compares changes in "total uranium" (KCR + cumulative production) with production in Australia. Changes in Australia's resource base were mostly positive with depletion of resources by production more than offset by new discoveries and/or re-evaluation of the resource base that moved resources into lower cost categories. Changes in production largely reflected new mines coming on line and/or adjustments in production in response to market conditions and sales commitments.

6.17. Relationship between changes in resources and production in Australia

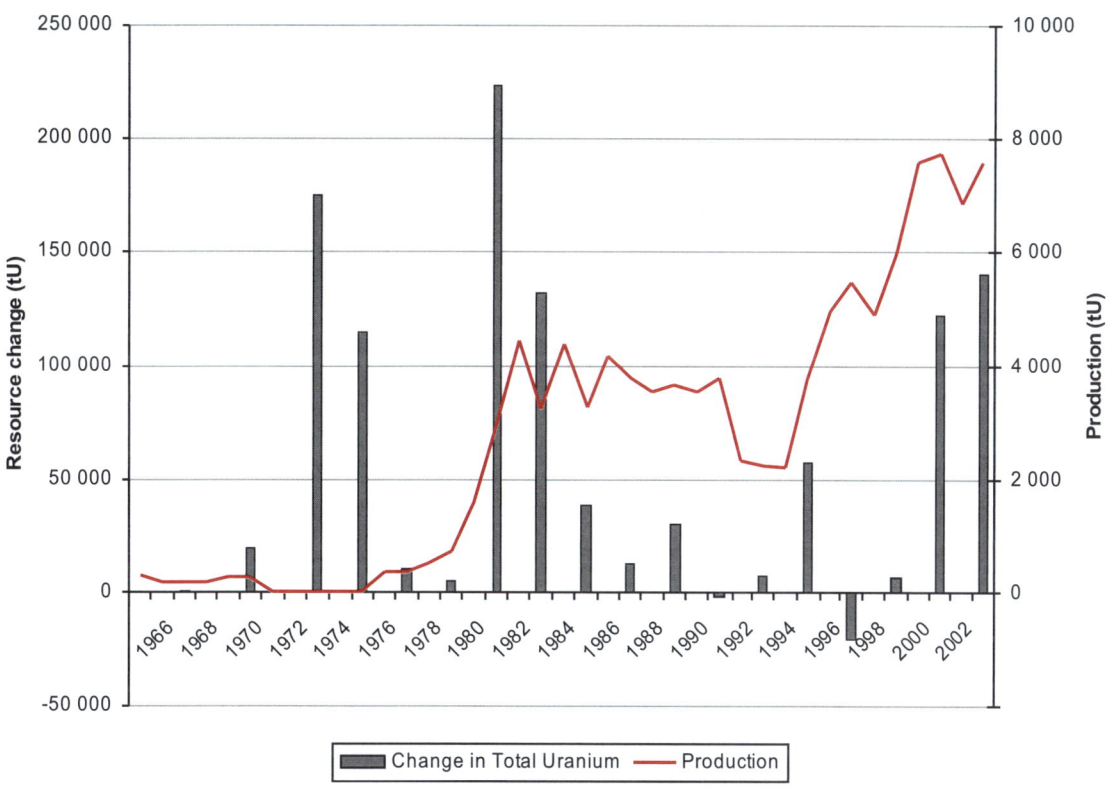

Canada

Unlike Australia, Canada experienced more significant negative changes in its resource base than positive changes (Figure 6.18). These changes, however, were independent of production. The negative change between 1969 and 1970 was largely a result of changes in nomenclature whereby Estimated Additional Resources was restricted to resources that occur in *known* uranium districts. In addition, some of the resources associated with quart-pebble conglomerate deposits were moved to higher cost categories and exploratory work in known uranium districts was not advanced enough to attribute resources to what was perceived as a more restrictive set of criteria.

Figure 6.18. Relationship between changes in resources and production in Canada

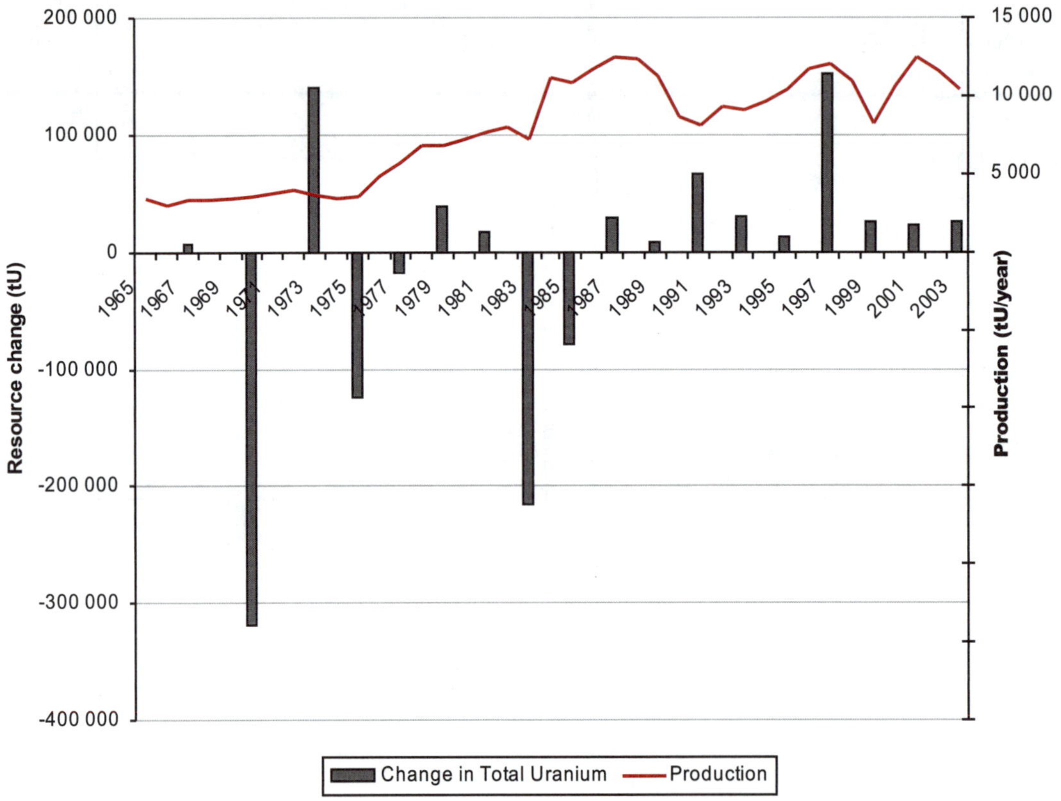

France

Production in France increased steadily between 1969 and 1988, before beginning a steady decline that concluded with closure of its uranium production facilities in 2001 (Figure 6.19). During the increase in production, France recorded several negative and positive changes in resources. The negative changes do not, however, appear to be directly attributable to depletion by production. No explanation, for example, is available for the approximately 30 000 tU decrease in resources between 1975 and 1977. Decreases in both RAR and EAR contributed to the decrease. Cumulative production between 1975 and 1977 totalled only about 5 700 tU or less than 20% of the total decrease. Therefore, production was only one contributing factor to the change between 1975 and 1977. The decline in resources between 1989 and 1991 was due to a combination of depletion through production and to reassessment of France's resource base resulting in reassigning some resources to higher cost categories.

Figure 6.19. Relationship between changes in resources and production in France

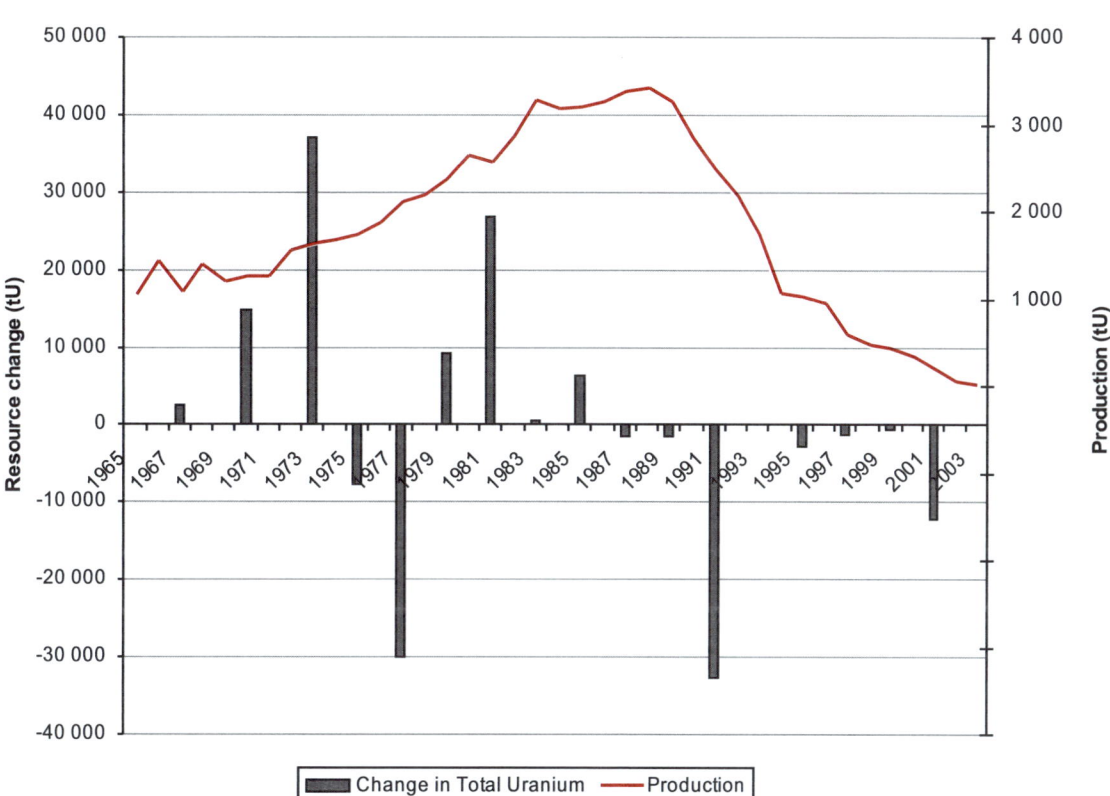

Namibia

Negative changes in Namibia's resource base between 1982 and 1995 were largely attributable to production in the intervening years. Limited exploration in Namibia during this time exacerbated depletion of resources (Figure 6.20).

Figure 6.20. Relationship between changes in resources and production in Namibia

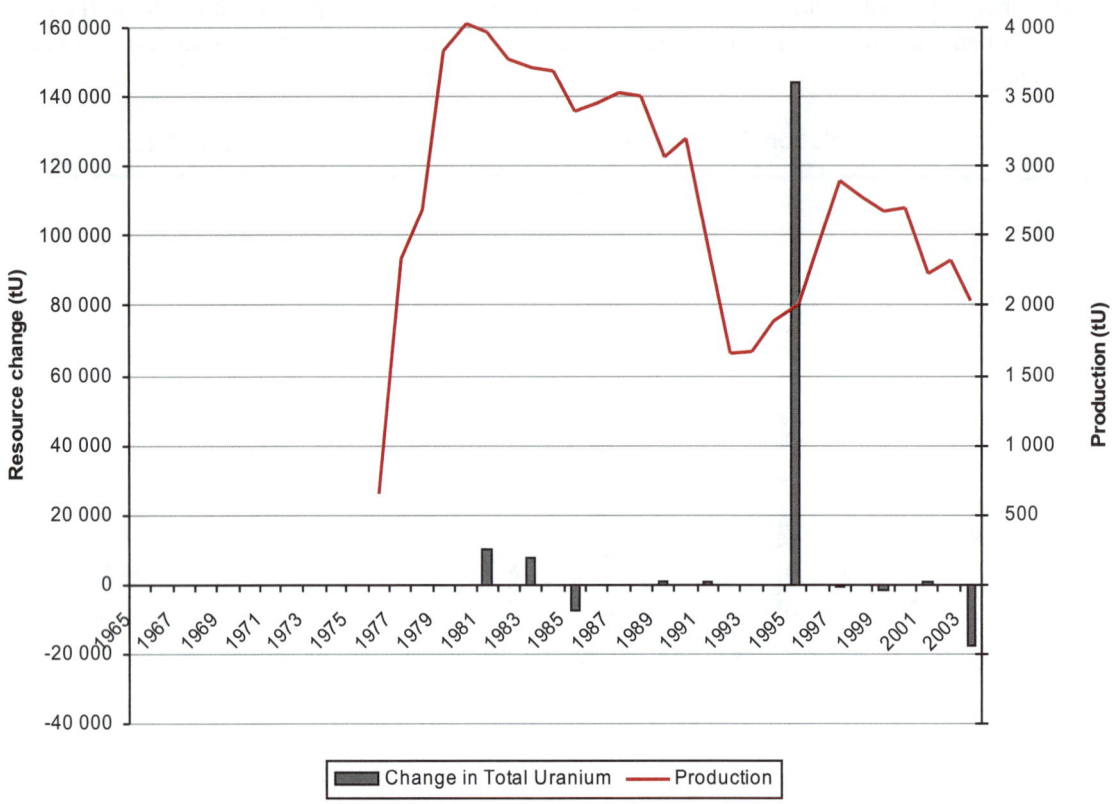

Niger

Depletion by production had very little effect on Niger's uranium resource base when the only significant decrease in KCR occurred between 1993 and 1995 (Figure 6.21). This decrease of about 386 000 tU was reportedly the result of a thorough reassessment of KCR completed by Niger's operating companies. In addition, prior to 1995, Niger had reported in situ resources; in 1995 it began to report recoverable resources, so a portion of the decrease in resources can be attributed to first time accounting for mining and processing loses.

Figure 6.21. Relationship between changes in resources and production in Niger

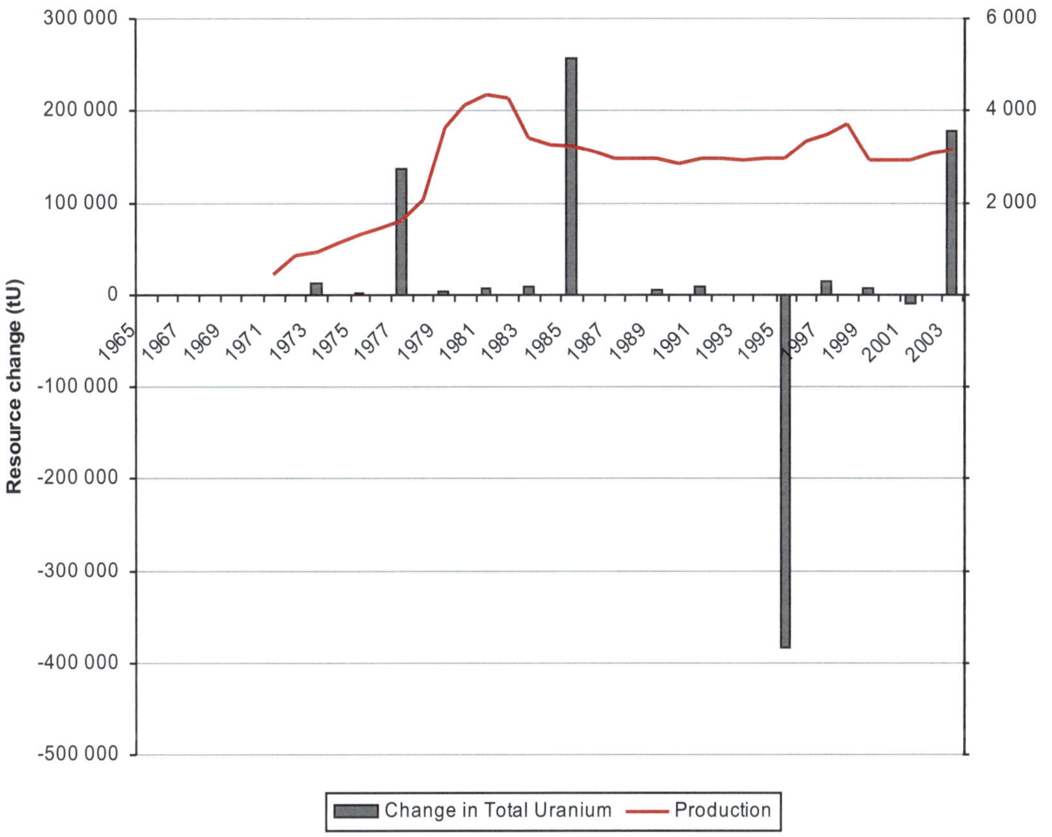

South Africa

South Africa experienced several decreases in its resource base between 1970 and 1993 (Figure 6.22). There is no ready explanation for the nearly 98 000 tU decrease in resources between 1967 and 1970. Cumulative production during that period totalled 9 232 tU or only 10% of the decrease. Uranium resources in South Africa are very closely related to gold resources so the 1967 to 1970 decrease was likely attributable to a reassessment of gold resources before prices began to increase. Uranium exploration largely ceased in South Africa in 1983. This fact also contributed to changes in the resource base, because although production steadily declined between 1983 and 2000, it nevertheless continued to be a steady drain on the resource base. As previously indicated, the 1989 to 1993 decrease came about as a result of re-estimation of resources to reflect higher production costs at the Witwatersrand gold mines that tended to push resources into higher cost categories.

Figure 6.22. Relationship between changes in resources and production in South Africa

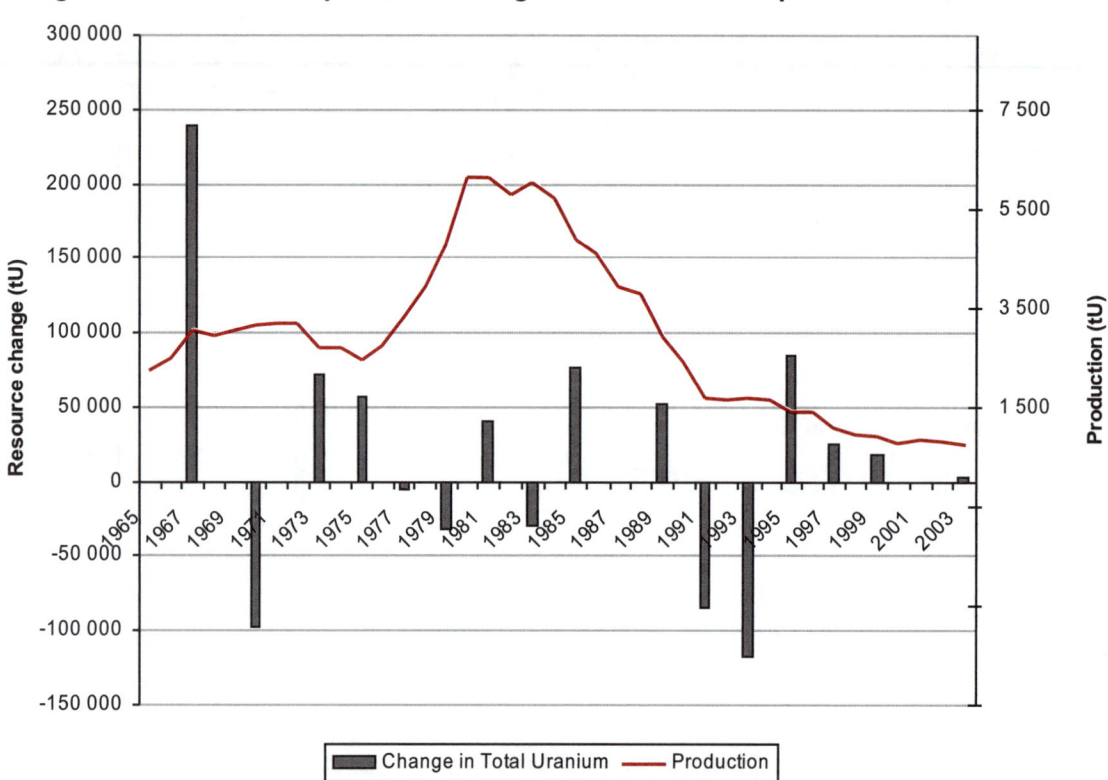

United States

The United States recorded significant negative changes in KCR in three successive Red Books – 1979, 1981 and 1983 none of which could be entirely attributable to production (Figure 6.23). The change that took place between 1977 and 1979 was, however, mostly attributable to depletion by production (cumulative production 40 000 tU compared to a decrease in resources of 57 000 tU). The decrease between 1979 and 1981 of 230 000 tU involved losses of both RAR and EAR. The losses of RAR in the <USD 80/kgU category were largely attributable to increased costs, which resulted in reallocation of some resources into higher cost categories; the same was true for EAR. "Re-delineation of resource area boundaries" was also cited as a reason for the decrease in EAR.

As has already been discussed, the steep decline between 1981 and 1983 resulted from changes in Red Book resources nomenclature. Estimated Additional Resources, which had been in use since 1967, was subdivided into two categories – Estimated Additional Resources – Category I (EAR-I) and Estimated Additional Resources – Category II (EAR-II). The United States has never adopted the EAR-I – EAR-II separation. Instead it only reported EAR, which was accounted as EAR-II in the Red Books. Thus, the 1.21 million tU decrease in KCR is due to moving resources to the EAR-II category.

Figure 6.23. Relationship between changes in resources and production in the United States

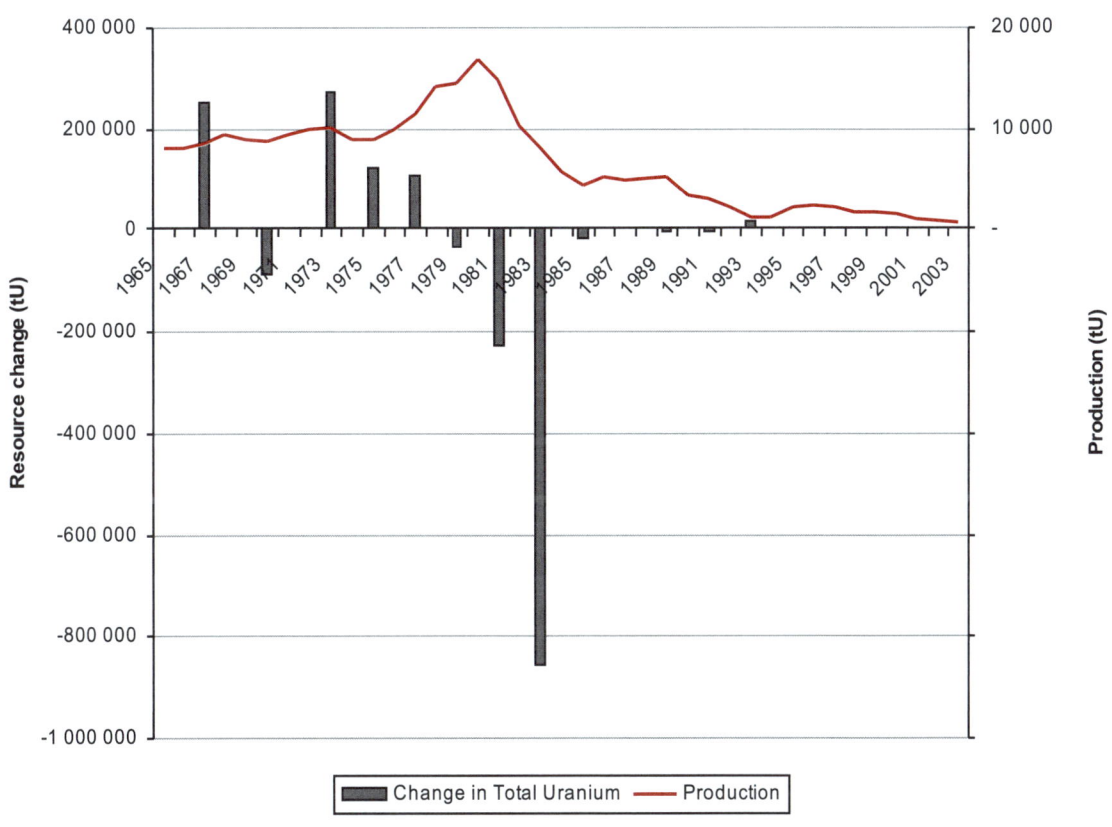

Conclusions regarding reporting on resources

Uranium resource nomenclature and cost categories have evolved over time in response to changes in the uranium market and growing sophistication in resource estimation technology. Changes in resource nomenclature and in the numbers of countries that have reported resources over time make long-term comparisons of resource totals difficult. There have, however, been relatively long periods of consistency for some resource categories for which analyses of trends are appropriate. Some conclusions can be reached as a result of the analyses described in this chapter.

- Subdividing resources by confidence levels has helped offset the subjective nature of resource estimating and reporting.

- Global trends that include all countries reporting resources in a given confidence or cost category are useful for relating resources to non-resource parameters such as market price, exploration expenditures and production. However, the answers as to why trends abruptly change are most often only apparent in individual country reports or statistics.

- Whereas the relationship between resource totals and exploration is unquestioned, it is difficult to directly relate changes in resource totals to specific trends or periods of exploration. Several years of exploration may precede a discovery. Once a discovery is made additional years are usually required before a resource estimation can be completed. Therefore, increases in resources typically reflect exploration expenditures incurred several years before.

- Market price indirectly affects resources because it affects exploration expenditures. However, because of the time lag between exploration and the reporting of related resources, this relationship is seldom readily apparent.

- Depletion of resources through production has not yet become a factor as far as the adequacy of supply is concerned. In fact, despite cumulative production of 2 204 732 tU from 1945 through 2003, the ratio between KCR and reactor-related uranium requirements in 2003 was 52. The average ratio since 1985 has been 47, so additions to resource totals have kept pace with production. A satisfactory ratio today does not, however, ensure that resources will be adequate in the future and the resources to requirements ratio should be carefully monitored.

7. URANIUM PRODUCTION

Though the element uranium was discovered in 1789 its uses were mainly limited to ceramic glazes and pigmentation of glass until the discovery of radioactivity by Ernst Rutherford at the end of the 19th century. When radium was detected by Marie Curie at the beginning of the 20th century uranium was then mined to extract radium, thus initiating the first uranium mining "boom". The modern era of uranium production began after fission of uranium was detected in 1938, followed shortly by a period of military application. In less than 25 years, from the first research for military purposes uranium production underwent a dramatic series of transformations, changing it from a commodity with minor commercial interest to major one of strategic significance before evolving to its current role as a fuel for generating electricity.

Each change in the uses of uranium brought with it increased interest in uranium exploration as well as improvements in uranium extraction technology. The military demand for uranium in the early 1940s was met mainly with rich pitchblende ore from the Shinkolobwe deposit in the Belgian Congo and the Great Bear Lake deposit in Canada. Early recovery of uranium was, by today's standards, very primitive and was based on an ether extraction technique adapted from analytical procedures that required multiple stages of selective precipitation. Processing costs were high and recovery efficiency was very low compared to modern standards.

Largely driven by early military requirements, and with the support of governments around the world, research efforts succeeded to improve uranium processing technology rapidly. In the West, process development programmes were begun almost simultaneously in Australia, Canada, France, United Kingdom, South Africa and the United States. Similar programmes were also underway in the Soviet Union. Collectively these efforts led to greater utilisation of lower grade ores than was previously possible, effectively resulting in a significant expansion of the uranium resource base.

Uranium production in 1945 is estimated to have totalled 507 tU. By 1965, when the first Red Book was published, production totalled 31 564 tU; production peaked in 1980 at 69 692 tU from 22 countries. In 2003, uranium production was reported by 19 countries with output totalling 35 492 tU. In all, 35 different countries have produced uranium since 1945. The leading countries in cumulative uranium production from 1945-2003 are listed in Table 7.1.

Table 7.1. Leading uranium producer countries based on cumulative production (1945-2003)

Country	tU	Percentage of world total
USSR[1]	377 613	17.1
Canada	374 548	17.0
United States	356 485	16.2
Germany[2]	219 239	9.9
South Africa	157 618	7.1
Others (total)	719 229	32.7
World total	2 204 732	100.0

1. Only includes production until 1991.

2. Includes production of German Democratic Republic (1946-1989) and Federal Republic of Germany (1961-2003).

Every aspect of uranium production technology has made significant advances since the early days when high-grade ore was hand sorted to provide raw material for military use. Today uranium ore is recovered by open-pit and underground mining methods and by in situ leaching. Radiation risks associated with high-grade ores encountered in the Athabasca Basin in Canada have necessitated use of sophisticated remote mining methods such as jet boring and raise boring, which eliminate the need for miners to enter the ore body. Similarly, remote handling of uranium ore in the grinding and crushing circuits significantly reduces exposure risks. Similar innovations have been made in uranium milling, such that recovery factors ranging between 90 and 99% are now the rule rather than the exception. In addition, radiation exposure risks to workers have been significantly reduced. Conventional milling of ore from underground and open-pit mines remains the dominant processing technology, though small quantities of uranium are recovered annually by surface heap leaching and underground stope or block leaching.

Similar technological improvements have been made to mitigate the environmental impact of uranium mining and processing. For example, below grade storage of uranium mill tailings substantially reduces long-term risks to the environment. Similarly, the pervious surround system of tailings management now in use in Canada reduces the potential for groundwater contamination from tailings disposal. Strategic placement of monitoring wells surrounding ISL operations ensures early detection of migration of leach solution beyond the leach area or potential contamination of overlying or underlying aquifers. Finally, improved monitoring of operating personnel through wearing exposure badges and better record keeping helps ensure the safety of mine and mill workers.

The countries producing uranium have changed significantly over time – new producing countries have been added while others have dropped out, either because their resources were exhausted or because production costs exceeded market price. Table 7.2 summarises some of the changes in the industry by showing the output from the major producing countries, which between 1965 and 2003 accounted for 94-98% of worldwide annual production. Appendix 7.1 provides comprehensive annual uranium production data on a country by country basis for the period 1945-2003.

Table 7.2. Historical production in key uranium producing countries (tU/year)

Country	1965	1970	1975	1980	1985	1990	1995	2000	2003
Australia	285	254	0	1 561	3 252	3 519	3 700	7 579	7 573
Canada	3 418	3 520	3 560	7 150	10 880	8 729	10 473	10 683	10 455
China	500	500	500	850	800	800	500	700	730
Czech Republic (includes Czechoslovakia)	2 839	2 627	2 402	2 482	2 623	2 142	600	507	452
France	1 032	1 250	1 731	2 634	3 189	2 841	1 016	296	9
Germany (GDR+FRG)	7 090	6 389	6 884	5 245	4 470	2 972	35	28	150
Namibia	0	0	0	4 042	3 400	3 211	2 016	2 715	2 037
Niger	0	0	1 306	4 120	3 181	2 839	2 974	2 914	3 157
South Africa	2 262	3 167	2 488	6 146	4 880	2 460	1 421	798	763
United States	8 033	9 900	8 900	16 800	4 300	3 420	2 324	1 522	769
USSR	4 300	8 300	13 000	15 700	15 900	14 500	x	x	x
Kazakhstan	x	x	x	x	x	x	1 630	1 870	3 327
Russian Federation	x	x	x	x	x	x	2 160	2 760	3 073
Ukraine	x	x	x	x	x	x	800	750	800
Uzbekistan	x	x	x	x	x	x	1 644	2 028	1 603
Sub-total (Key producers)	29 759	35 907	40 771	66 730	56 875	47 433	31 293	35 150	34 898
World total	31 564	37 736	43 208	69 692	60 202	50 026	32 942	35 755	35 492
% of world total	94	95	94	96	95	95	95	98	98

Trends in uranium production and relationship with market price

Uranium production in 1945 is estimated to have totalled 507 tU. Uranium production peaked in 1980 at 69 692 tU after which it began a steady decline to 31 503 tU in 1994 (Figure 7.1). Between 1994 and 2003, production varied within a narrow range of between 31 500 and 37 000 tU annually and only broadly paralleled the market price. The increase in production that began in 1975 lagged the steep price increase that began in 1973 and was much less dramatic. Production began to decline sharply in 1987; seven years after prices began their sharp descent. There are many reasons for the disparity in performance between uranium price and annual production. Producers in centrally-planned economies are less affected by market price. Producers with long-term contracts, the most common type used, that are well above falling prices could continue to operate as long as the contracts were in force. Once higher priced contracts expire, however, producers are again subject to the vicissitudes of the market. Cumulative worldwide uranium production between 1945 and 2003 totalled 2 204 732 tU.

Figure 7.1. Relationship between historical production and uranium market price

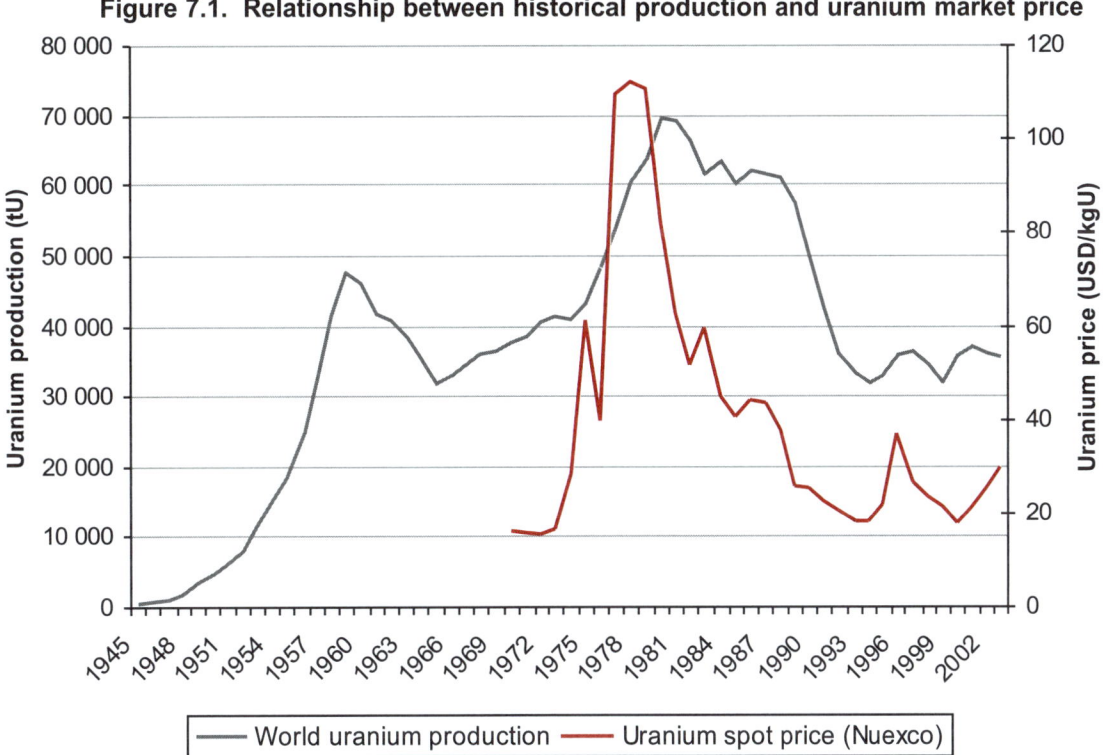

Figure 7.2 shows the relative contributions to annual production from different economic systems, with WOCA contributions showing considerably more variation than those from countries with centrally planned economies (CPE).

Figure 7.2. Contributions to worldwide uranium production by different economic systems

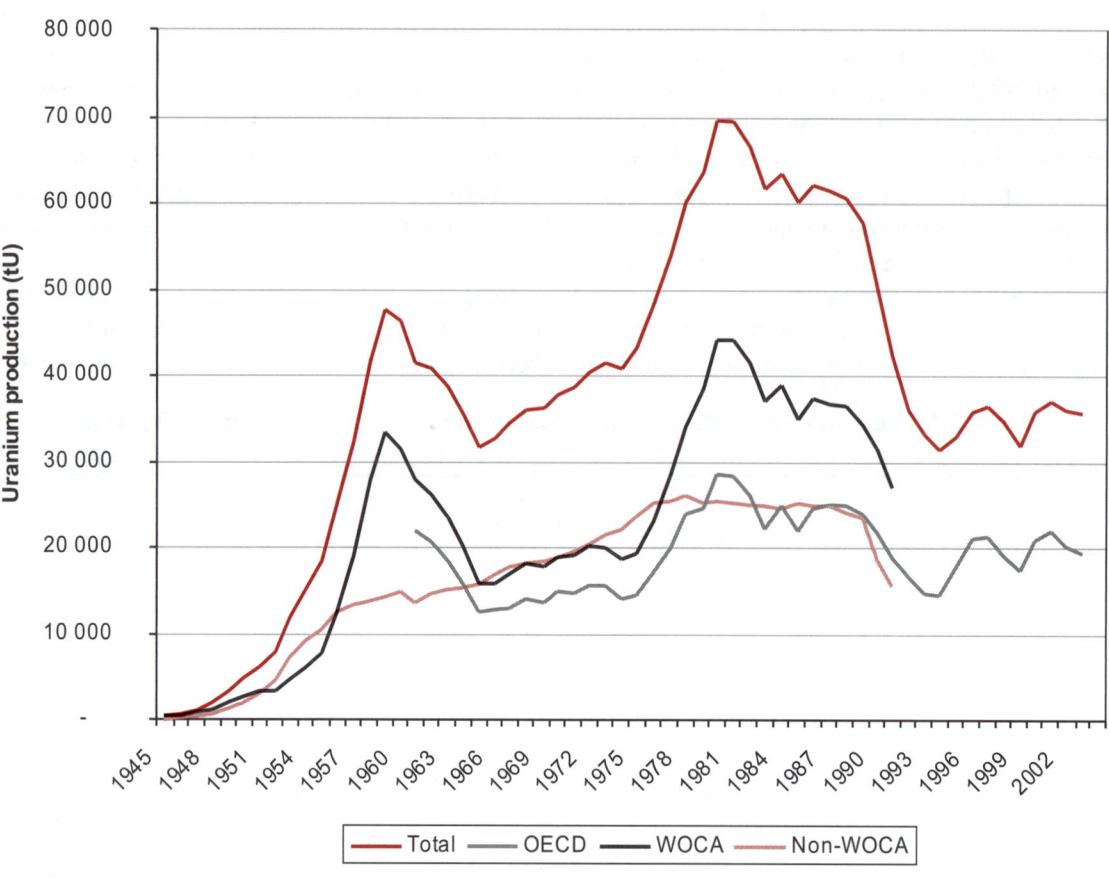

Uranium production at individual production centres may periodically be adjusted to approximately match sales commitments. In addition, mine output may not always be adequate to ensure that the related mill has sufficient feed to operate at capacity. Therefore, annual output may be less than production capacity for a given year. Production capacity, as used in this report, denotes the nominal level of output (expressed as tU/year), based on the design of the existing plant and facilities over an extended period, under normal commercial operating practices. Existing production centres are those that exist in operational condition, and include plants which are closed down but which could be readily brought back into operation. Production capacity is reported as of 1 January of each year; therefore, production from previous years can exceed capacity reported for succeeding years if adjustments have been made to plant operational capability.

Appendix 7.2 provides annual statistics on worldwide production capacity between 1968 and 2003. Figure 7.3, which compares annual production capacity with worldwide uranium requirements, shows that capacity exceeded requirements between 1968 and 1992. Except for the period between 1972 and 1984, however, the ratio between capacity and requirements declined steadily, until it flattened out beginning in 1992. The declining ratio resulted from a steady increase in requirements compared to a more volatile history for production capacity. The sharp increase in capacity between 1974 and 1982 was in response to an expected increase in requirements associated with a projected rapid growth in nuclear power. Those expectations were, however, undermined by the Three Mile Island and Chernobyl accidents after which capacity declined or remained static between 1983 and 1991. The sharp decline in capacity between 1991 and 1993 resulted from mine/mill closures in the

German Democratic Republic, Kazakhstan and Uzbekistan. Thereafter, mine/mill closures in Gabon, Western Europe and the United States were offset by development of new production centres or expansion of existing operations in Australia and Canada.

Following the end of the Cold War, increasing amounts of uranium previously held in stockpiles for military purposes were made available for commercial use. Additionally, as discussed earlier, stockpiles of commercially-held uranium were increasingly used. These "secondary sources" of already-mined uranium played a significant role in displacing primary production after 1991. They acted to drive down prices, which led to the closure of higher-priced production centres. This downward pressure lasted through 2001. As a result capacity fell below requirements beginning in 1992 (Figure 7.3). After 2001 prices began to increase leading to plans to increase production capabilities around the world though the impacts of these plans has not yet been significant as it will take some time for the full effect of these increases to be visible.

**Figure 7.3. Comparison of annual production capacity
and reactor-related uranium requirements**

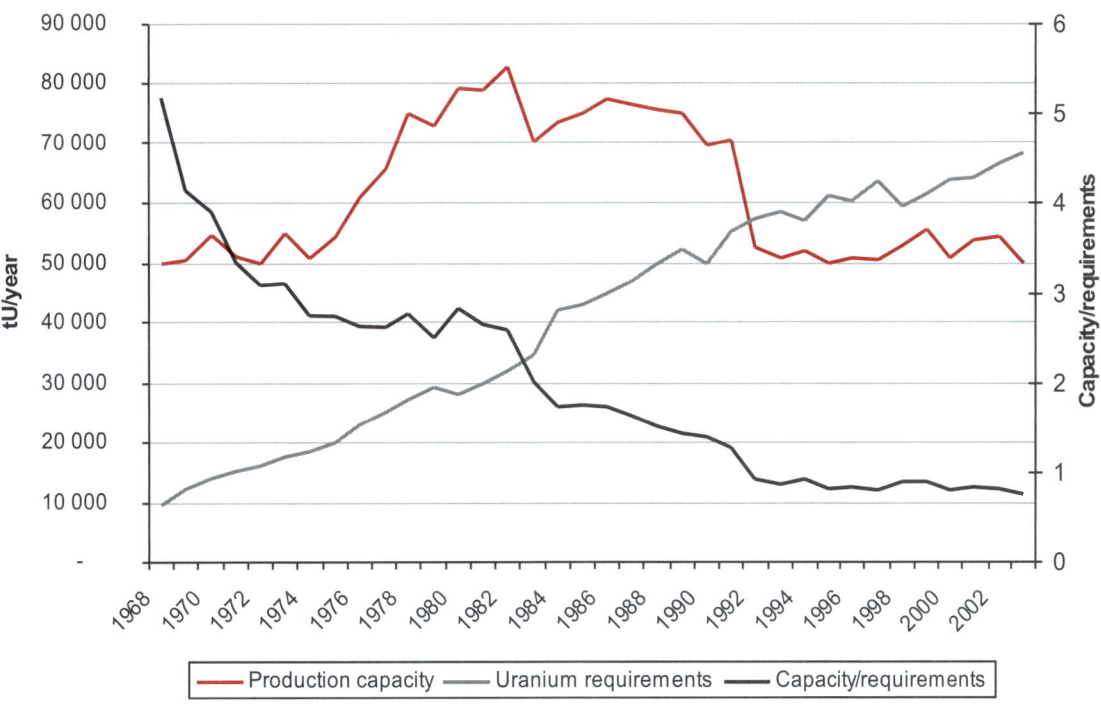

Market price is an indirect measure of demand. Subsequent to the market price adjustment beginning in 1981 that followed the Three Mile Island accident, capacity utilisation and market price were nearly coincident (Figure 7.4). Prices continued to fall through the 1980s and the early 1990s as inventory drawdown and other secondary supply sources competed with primary supply. Producers adjusted their output (or in some cases closed their operations) in response to both price and demand.

Figure 7.4. Relationship between capacity utilisation and market price

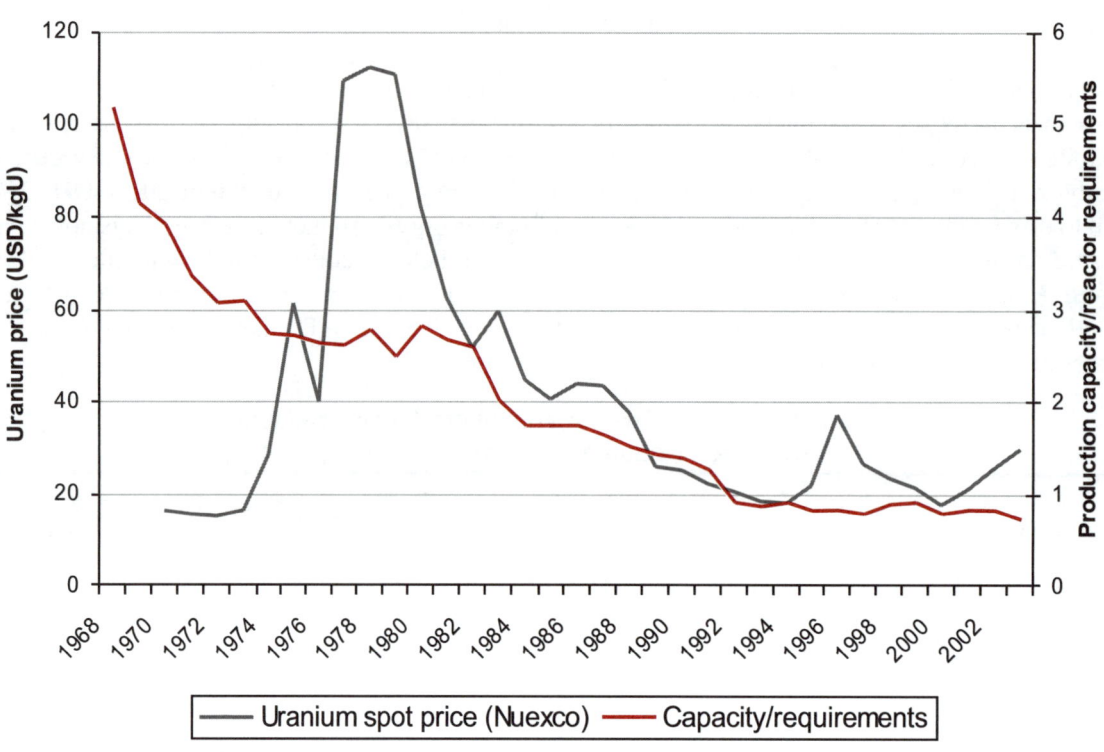

Relationship between supply and demand

Since production does not always equal capacity, it is equally important to examine the balance between reactor-related uranium requirements and annual production (Figure 7.5). In a perfect world, one would expect these two curves to approximately coincide with a slight advantage of production over requirements to ensure the availability of a strategic inventory to offset supply disruptions or other unforeseen eventualities. Instead what we see is wide disparity between supply and demand, with production exceeding demand until 1990 when that relationship was reversed.

The early years of uranium production were devoted to satisfying military requirements for nuclear material; the first civilian reactor did not start operating until 1957. Satisfying military needs continued to be the most important underlying stimulus for uranium mining at least through the mid- to late 1960s. Beginning in the early 1970s satisfying nuclear energy requirements began to gain equal footing with military requirements on the demand side of the supply-demand equation. The military component, however, continued to contribute to the disparity between production and requirements. In addition, the price of uranium tripled between 1973 and 1975 as a result of concerns about a possible uranium supply shortfall related to rapidly increasing reactor orders (Figure 7.1). A corresponding increase in production followed two to three years later, widening the gap between production and requirements.

Figure 7.5. Worldwide annual production and reactor-related requirements (1945-2003)

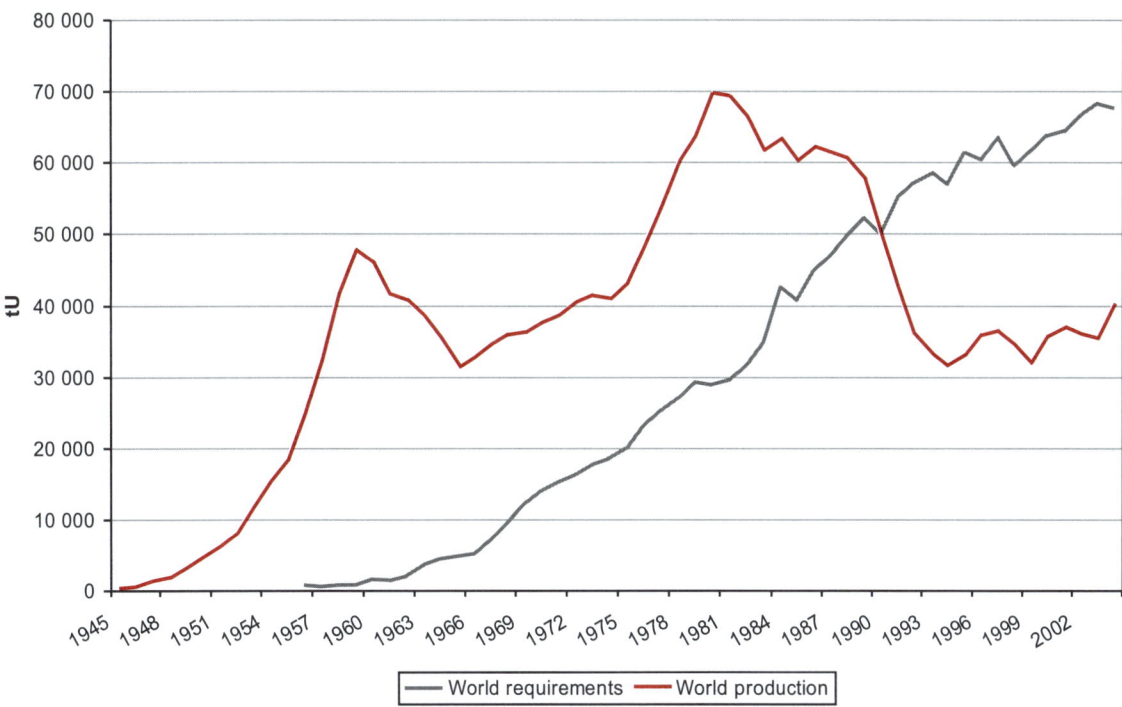

The Three Mile Island and Chernobyl accidents precipitated cancellations of new reactor orders and production began to decline in 1982 as adjustments were made to this new reality. Exactly how much of total production went into military programmes is not available. Nevertheless, it can be said with certainty that the net result of 58 years of production (2 204 732 tU) exceeding requirements (1 513 327 tU) resulted in a large inventory of already-mined uranium in both military and civilian stockpiles (Figure 7.6).

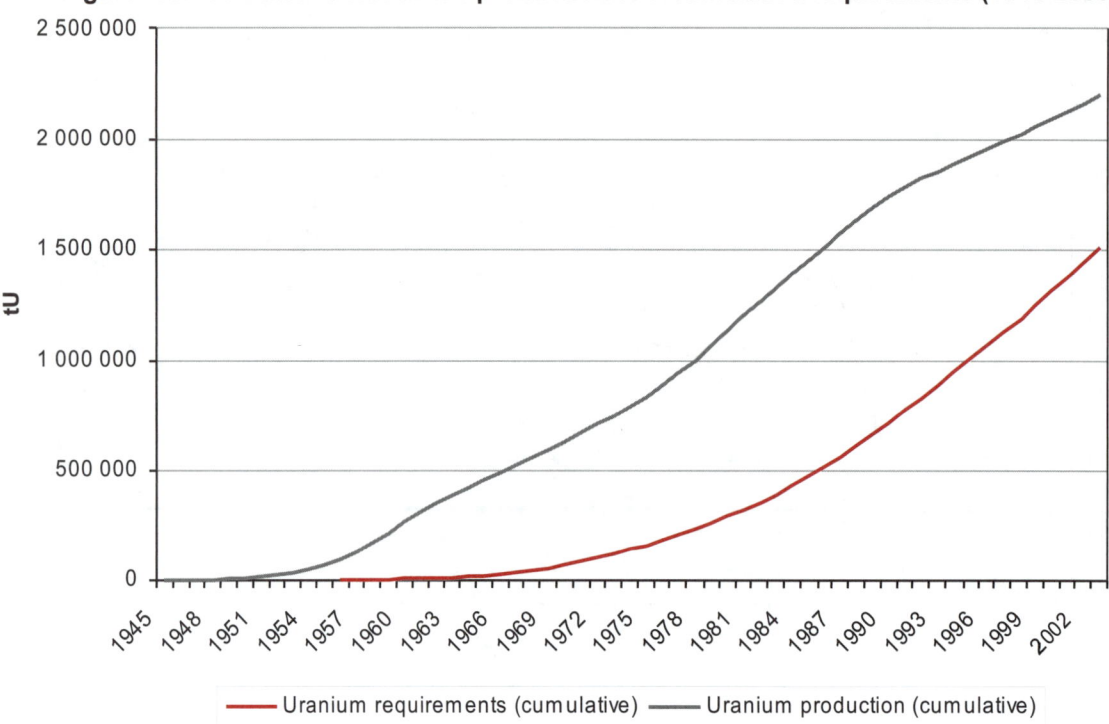

Figure 7.6. Cumulative worldwide production and cumulative requirements (1945-2003)

However, beginning in the late 1980s, nuclear utilities, which controlled most of the civilian inventory, began to draw down that inventory, which in turn reduced demand for newly mined and processed uranium or "primary supply". While production was declining, demand continued to grow, eventually resulting in the 1990 cross over from excess production to a deficit between primary supply and demand. The gap between primary supply and demand grew from about 13 000 tU in 1991 to nearly 33 000 tU in 2003. Production (primary supply) satisfied only about 52% of reactor-related demand in 2003; the balance or the gap between primary supply and demand was filled by secondary supply.

Secondary supply

Secondary supply, also known as "already-mined uranium" or "above ground resources", is material that has been held in inventory or that has been previously used but has then been reprocessed into a form suitable for further use. The 2003 Red Book provided the first substantive discussion of secondary supply, which was defined to include the following categories:

- Stocks and inventories of natural and enriched uranium, both civilian and military in origin.
- Nuclear fuel produced by reprocessing of spent reactor fuels and from surplus military plutonium.
- Uranium produced by re-enrichment of depleted uranium tails.

Secondary supply started to be an important factor in satisfying demand in the early 1980s. The ratio between production and requirements declined from nearly 2.5 in 1980 to 1.0 in 1990 (Figure 7.7). Inventory drawdown was largely responsible for this decline. After 1990, requirements increasingly exceeded primary production, with secondary supply filling the gap between supply and demand (Figure 7.5).

Figure 7.7. Ratio of production to reactor-related uranium requirements (1980 to 2003)

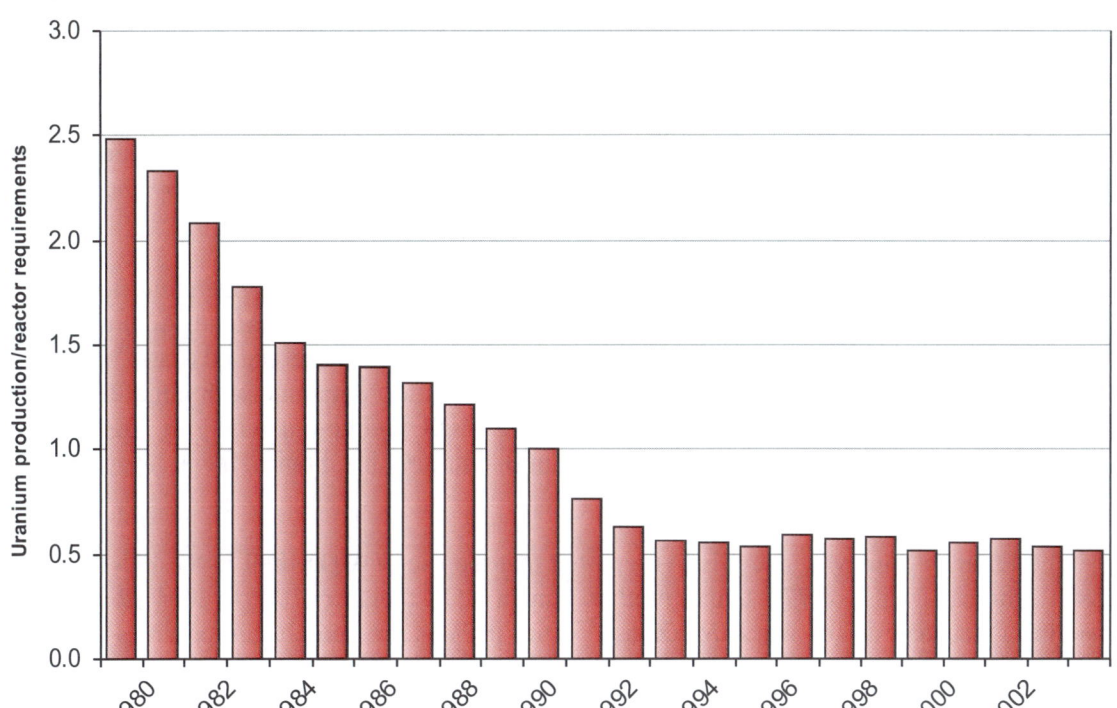

Stocks and inventories of natural and enriched uranium, both civilian and military in origin

Stockpiles of uranium are discussed in more details in Chapter 8. A brief discussion of highly-enriched uranium (HEU) is given here.

In February 1993, the United States and the Russian Federation signed The Agreement between the Government of the United States and the Government of the Russian Federation Concerning the Disposition of Highly Enriched Uranium from Nuclear Weapons to blend down 500 metric tons of HEU to low-enriched uranium for peaceful use in commercial reactors over twenty years. The resultant low-enriched uranium represents the equivalent of approximately 153 000 tU and 92 million separative units of enrichment services. In 2002, the United States and the Russian Government approved an amendment to the Agreement's implementing contract, effective January 2003, establishing market-based pricing terms to the planned delivery of low-enriched uranium derived from 30 tonnes HEU per year.

Additionally, the United States has committed to the disposition of about 174 tonnes of surplus HEU with about 153 tonnes planned to be eventually blended down for use as low-enriched uranium (LEU) fuel in research and commercial reactors and 23 tonnes slated for disposal as waste.

Nuclear fuel produced by reprocessing of spent reactor fuels and from surplus military plutonium

The constituents of spent fuel from power plants are a potentially substantial source of fissile material that could displace primary production of uranium. When spent fuel is discharged from a commercial reactor it is potentially recyclable, since about 96% of the original fissionable material remains along with the plutonium created during the fission process. The recycled plutonium can be reused in reactors licensed to use mixed-oxide fuel (MOX).

The use of MOX has not yet significantly altered world uranium demand because only a relatively small number of reactors are using this type of fuel. Additionally, the number of recycles possible using current reprocessing and reactor technology is limited by the build-up of plutonium isotopes that are not fissionable by the thermal neutron spectrum found in light-water reactors and by the build-up of undesirable elements, especially curium.

In January 2005, there were over 35 reactors, about 8% of the world's operating fleet,[4] licensed to use MOX fuel, in Belgium, France, Germany, India, Sweden and Switzerland. Additional reactors could be licensed to use MOX in China and the Russian Federation. The United States has licensed a reactor to use MOX as part of its weapons material disposition programme with initial tests of MOX fuel loading in 2005. Japan is planning to begin using MOX fuel commercially in 2010. MOX fuel fabrication facilities exist or are under construction in Belgium, China, France, India, Japan, the Russian Federation and the United Kingdom.

The Euratom Supply Agency (ESA) reported that the use of MOX fuel in the EU-15[5] displaced 9 280 tU through the use of 77.2 tonnes of plutonium in MOX fuel from 1996 to 2004 [Reference 7.1]. Since the great majority of world MOX use occurs in Western Europe this provides a reasonable estimate of the impact of MOX use worldwide during that period.

Reprocessed uranium

Uranium recovery through reprocessing of spent fuel has been conducted in the past in several countries, including Belgium and Japan, but is now routinely done only in France and the Russian Federation. This is because recycling of reprocessed uranium (RepU) is relatively costly, in part due to the requirement for dedicated conversion, enrichment and fabrication facilities. Changing market conditions are, however, leading to renewed consideration of this recycling option. Very limited information was available concerning how much reprocessed uranium was used in the 2003 Red Book. The available data indicated that it represented less than 1% of projected world requirements annually.

Mixed-oxide fuel produced from surplus weapons-related plutonium

In September 2000, the United States and Russia signed an agreement for the disposition of surplus plutonium. Under the Agreement, both the United States and Russia will each dispose of 34 tonnes of surplus weapon-grade plutonium at a rate of at least two tonnes per year in each country

---

4. In December 2002, Sweden authorised the limited use of MOX fuel at the Oskarshamn nuclear power plant. This decision allowed the use of 900 kg of plutonium separated from spent fuel removed from Swedish reactors prior to 1982. Since 1982, Swedish used nuclear fuel has been placed in storage pending final disposal.
5. Data are for 15 EU countries prior to enlargement in May 2004.

once facilities are in place. Both countries agreed to dispose of surplus plutonium by fabricating it into MOX fuel for irradiation in existing nuclear reactors. This approach will convert the surplus plutonium to a form that cannot be readily used to make a nuclear weapon.

Uranium produced by re-enrichment of depleted uranium tails

Depleted uranium stocks or enrichment tails (material derived from the uranium enrichment process) are another major potential secondary supply source. As of the beginning of 2000 the inventory of depleted uranium was estimated at about 1 200 000 tU, which could provide about 452 000 tU of equivalent natural uranium if it were re-enriched [Reference 7-2].[6] The depleted uranium inventory would increase by up to 57 000 tU annually based on uranium requirements of 65 000 tU per annum. Though tails represent a large potential supply of uranium, that potential will only be realised if there is surplus enrichment capacity with relatively low operating costs. Currently, Russia is the only country that is re-enriching tails.

The 2003 Red Book was the first edition to address availability of these and other secondary supply sources (e.g. reprocessed uranium, military plutonium). How completely this material will be accounted for in future Red Books is uncertain. Similarly, the complex issue of the eventual availability of HEU contained in nuclear weapons is beyond the scope of the Red Book. Though this material represents a very large potential source of future secondary supply, its availability will be determined through bilateral agreements that may take decades to complete.

Uranium production in selected countries

As is almost always the case, while worldwide totals show broad trends, the more interesting statistics and a better perspective of uranium production are associated with the production histories in individual countries. Therefore, as a complement to discussions of global trends, the following sections will describe historical production trends for important countries or political entities (i.e. the former Soviet Union). These sections provide details of country specific production information including comparisons between production and capacity in individual countries that help to better understand global production statistics and also provide insights into the types of information included in the Red Book country reports.

Australia

Figure 7.8 compares Australia's uranium production with its reported production capacity. Australia's production was basically divided into two eras. During the first era, which extended from 1954 to 1971, production was largely for military purposes, though beginning in the late 1960s limited production was available for civilian use. Production ceased between 1971 and 1975. All production after 1975 was for civilian reactor use. The first production during the post-1975 or the "modern era" came from the Mary Kathleen mine in Queensland. Production gradually shifted to the unconformity-related deposits in the Northern Territory; Nabarlek and Ranger began production in 1980 and 1981, respectively. Olympic Dam, currently the world's largest uranium deposit began production in 1988

---

6. OECD Nuclear Energy Agency, (2001) *Management of Depleted Uranium*, Paris, France. This total assumes 1.2 million tU at 0.3% assay re-enriched to produce 336 000 tU of equivalent natural uranium, leaving 864 000 tU of secondary tails with an assay of 0.14%. These secondary tails could then also be re-enriched providing a further 106 000 tU equivalent leaving 758 000 tU of tertiary tails with an assay of 0.06%.

with uranium produced as a co-product of copper and gold mining. Production varied between 2 200 and 4 400 tU annually from the early 1980s through the mid-1990s. Mill capacity was increased at Ranger in 1997 and the capacity of Olympic Dam was increased in 1999. These modifications accounted for the increases in production beginning in 1995. The start of ISL production at Beverley in 2001 also contributed to the rise in post-1995 production.

The ratio between production and capacity in Australia has averaged 0.81. The anomalous spike in the ratio in 1981 occurred during the ramp up in production at Nabarlek and Ranger and may have resulted from processing of high-grade stockpiled ore.

Figure 7.8. Historical uranium production in Australia

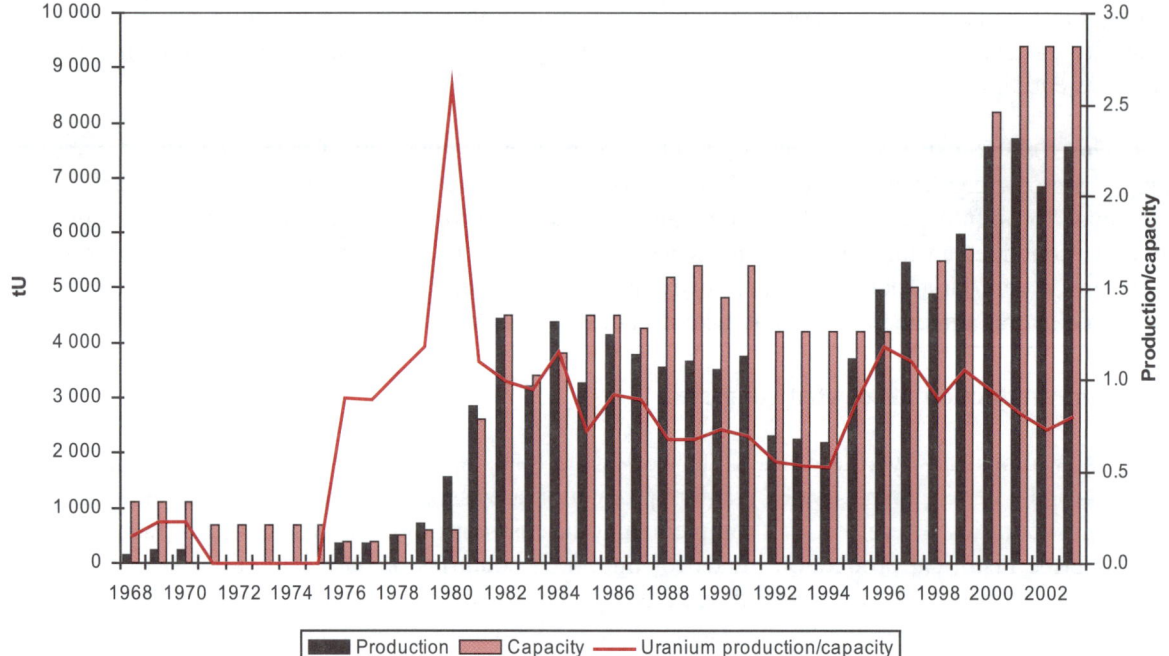

Canada

Canada's uranium industry began in the Northern Territories with the 1930 discovery of the Port radium pitchblende deposit. Exploited from 1933 to 1940, the deposit was re-opened in 1942 in response to demand for uranium for British and United States defence programmes. A ban on private exploration and development was lifted in 1947, and by the late 1950s some twenty uranium production centres had started up in five producing districts. Production peaked in 1959 at 12 200 tU. No further defence contracts were signed after 1959 and production began to decline. Despite government stockpiling programmes, output fell rapidly to less than 3 000 tU in 1966, by which time only four producers remained. While the first commercial sales to electric utilities were signed in 1966, it was not until the mid-1970s that prices and demand had increased sufficiently to promote expansions in exploration and development activity. By the late 1970s, with the industry firmly re-established, several new facilities were under development. Annual output grew steadily throughout the 1980s, as Canada's focus of uranium production shifted increasingly from east to west, with the Key Lake mine opening in 1988. In the early 1990s, poor markets, low prices and high production costs led to the closure of three of four Ontario production centres. The last remaining Ontario uranium centre closed in mid-1996. Since the last Elliot Lake production facility closed in 1996, all active uranium production centres are located in northern Saskatchewan. McArthur River, the world's highest grade uranium deposit and world's largest uranium production centre began operation in 1999 with its ore milled at the Key Lake facility.

Throughout most of its production history, Canada shows a relatively close relationship between capacity and annual output. The average ratio between capacity and production was 0.84; there were only brief periods when the ratio was significantly less than 1.0 and several periods when production exceeded capacity (Figure 7.9). The drop off in production between 1988 and 1991 and in capacity between 1991 and 1992 likely was the result of decreased production and eventual closing of operations in the Elliot Lake district. The sharp decrease in production and accompanying decrease in the production to capacity ratio in 1998-1999 resulted from curtailment of production due to low uranium prices, including temporary closure of the Rabbit Lake mine/mill complex. The decrease in the ratio in 2003 was largely attributable to temporary suspension of production at McArthur River because of mine flooding.

Figure 7.9. Historical uranium production in Canada

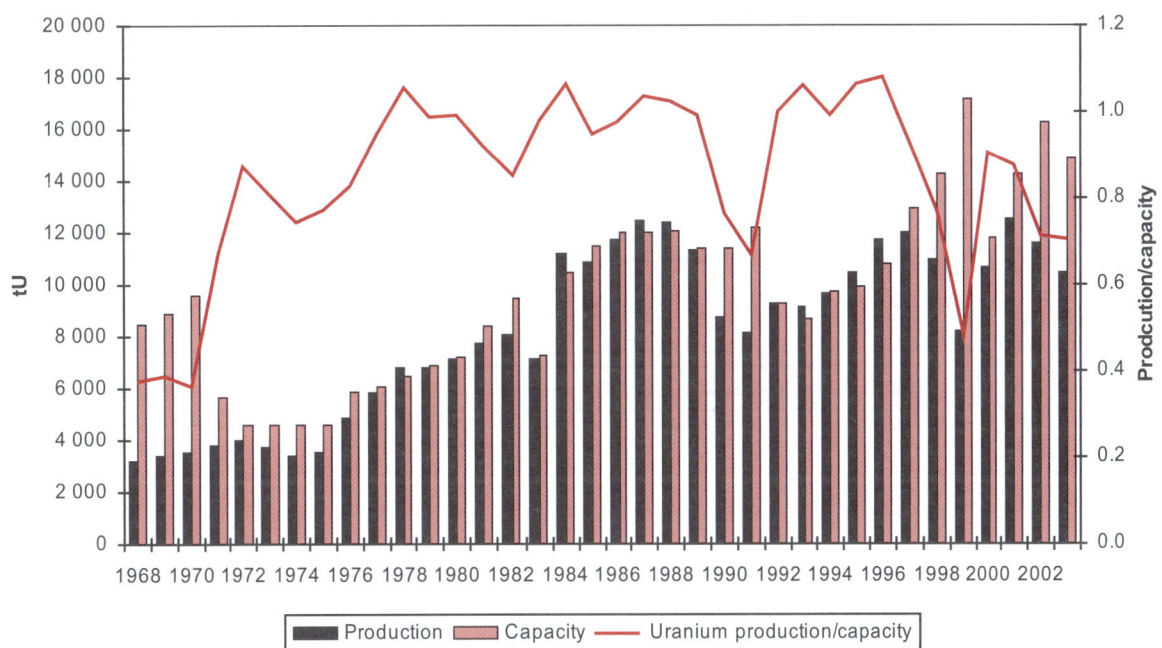

Czech Republic including the former Czechoslovakia

The industrial production of uranium in Czechoslovakia began in 1946. Between 1946 and the dissolution of the Soviet Union, all uranium produced in Czechoslovakia was exported to the Soviet Union. The first production came from Jachymov and Horni Slavkov mines, which completed operations in the mid-1960s. Pribram, the main vein deposit, operated in the period 1950-1991. The Hamr and Straz production centres, supported by sandstone deposits, started operation in 1967. The peak production of about 3 000 tU was reached around 1960 and production remained between 2 500 and 3 000 tU/year from 1960 through 1990, when it began to decline. During the period 1946-2003 a cumulative total of 108 649 tU were produced in the Czech Republic. About 86% of that total were produced by underground and open-pit mining methods while the remainder was recovered using in situ leaching (ISL).

Detailed information is not available on historical annual capacity in Czechoslovakia/ Czech Republic. Nevertheless, its production history is typical of a centrally planned economy where it is assumed that production was close to capacity between 1960 and about 1989, after which mine closures in response to the pressures of moving to a market economy began to limit annual production (Figure 7.10).

Figure 7.10. Historical uranium production in Czechoslovakia/Czech Republic

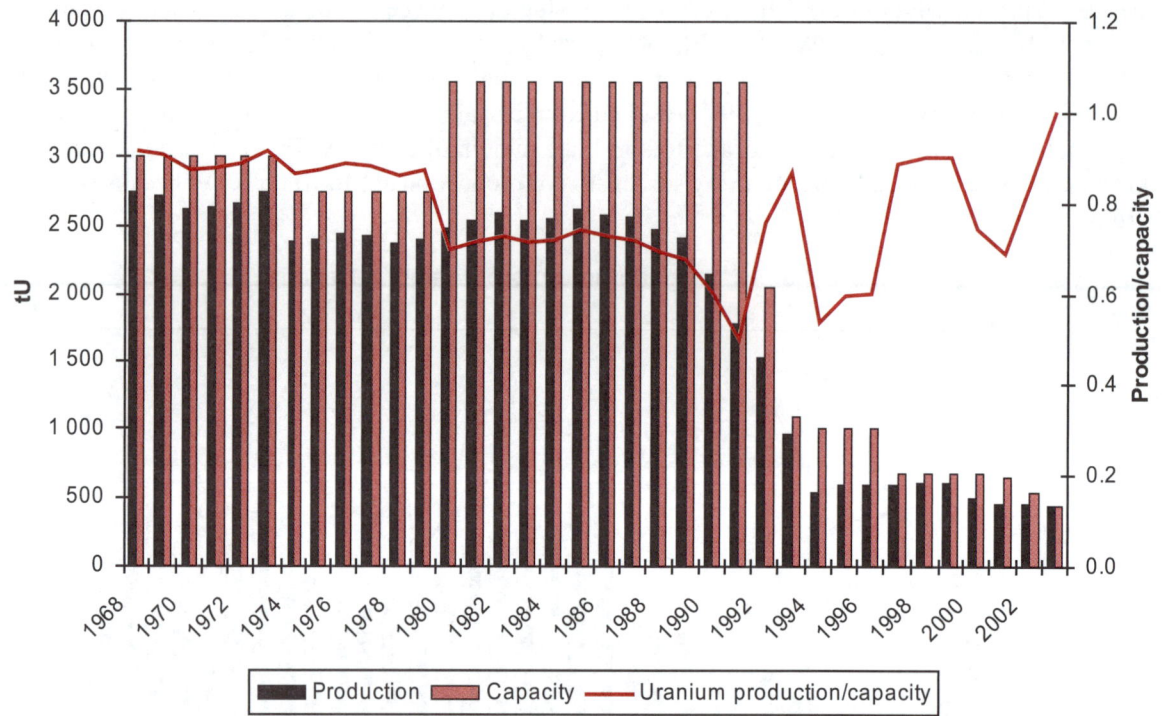

France

Uranium production started in France in 1954 (50 tU) in the Bessines district (vein-type deposits in granite), and in 1955 in the Vendée district (vein-type deposits in granite and metamorphical formations) and the Forez district (vein-type deposits in granite).

By 1969 the total production capacity from national ore was 1 500 tU/year, in three plants owned by the *Commissariat à l'énergie atomique*: Bessines (Limousin), Écarpière (Vendée) and St-Priest (Forez). Chemical concentrates from Gabon were also processed at the Geugnon plant (400 tU). In 1973, in addition to these three plants, a few tens of tonnes of uranium were produced by the *Compagnie française des minerais d'uranium* from the Langogne deposit in the Massif Central.

In 1978, a total of five plants or production units, with a production capacity of 2 410 tU/year, processed all the ore mined in France. These plants were located near the mining centres of Bessines, Écarpière, Saint-Priest, Langogne (vein-type deposits) owned by the *Compagnie française des minerais d'uranium and Saint-Pierre du Cantal* (sandstone and vein-type deposits) owned by the *Société centrale de l'uranium et des minerais et métaux radioactifs*. Ore produced by small mining companies in the Massif Central and in Brittany was processed at the Bessines and Écarpière plants. In addition, the *Compagnie minière Dong-Trieu* constructed a plant, where production started in 1979, at

Maihac-sur-Benaize (north of Massif Central) to process ore from the Bernardan district (vein-type deposits in granite).

Production in the Lodève district (Permian sedimentary formations) started in 1981, following the closure of the Forez district. Production at the Balaures production centre, owned by *Compagnie française de Mokta*, (vein-type deposit in metamorphic rock) started in 1982.

Production capacity reached its peak in France in 1989 at 3 720 tU/year and then started to decrease in 1990 with the closure of the Langogne plant followed in 1991 by the closure of the Vendée production centre. When the Crouzille uranium mining district ceased production in 1994 after producing a total of almost 25 000 tonnes of uranium, production capacity had declined to only 1 570 tU/year. The two last mining districts, Lodève and Bernardan, were closed in 1997 and 2001 respectively. From 1968 to 2003, the average ratio between uranium production and production capacity was 0.77, varying between 0.41 in 1994 to 0.96 in 1975 (Figure 7.11).

Figure 7.11. Historical uranium production in France

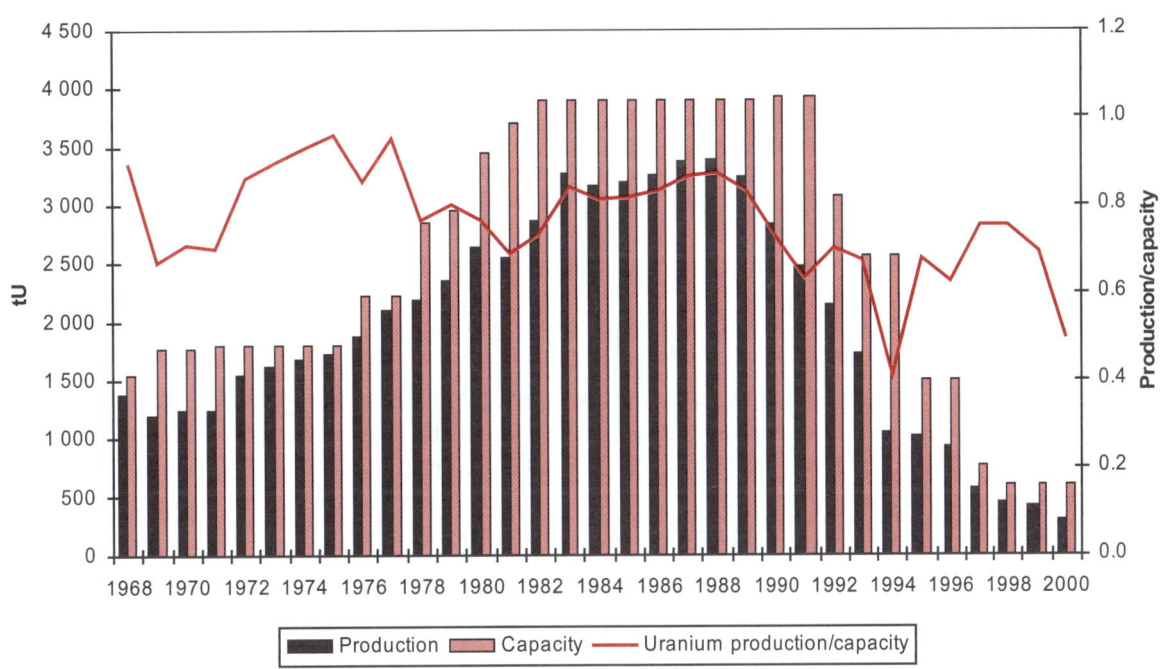

Germany including the former German Democratic Republic (GDR)

Former German Democratic Republic before 1990

Uranium mining was undertaken from 1946 to 1953 by the Soviet stock company, SAG Wismut. These activities were centred on old mining locations of silver, cobalt, nickel and other metals in the Erzgebirge (Ore Mountains) and in Vogtland, Saxony. The mining of uranium first began at the cobalt and bismuth mines near Schneeberg and Oberschlema (a former famous radium spa). During this early period more than 100 000 people were engaged in exploration and mining activities. The richer pitchblende ore from the vein deposits was hand-picked and shipped to Soviet Union for further processing. Lower grade ore was treated locally in small processing plants. In 1950, the central mill at Crossen near Zwickau, Saxony was brought into operation.

In 1954, a new joint Soviet-German stock company was created, *Sowjetisch-Deutsche Aktiengesellschaft Wismut (SDAG Wismut)*. The joint company was held equally by both governments. The entire uranium production either hand-picked concentrate, gravity concentrate, or chemical concentrate was shipped to Soviet Union for further treatment. The price for the final product was simply agreed to between the two partners. At the end of the 1950s, uranium mining was concentrated in the region of Eastern Thuringia. Uranium exploration had started in 1950 in the vicinity of the radium spa at Ronneburg. From the beginning of the 1970s, the mines in Eastern Thuringia provided about two-thirds of SDAG Wismut annual production.

Two processing plants were operated by SDAG Wismut in the territories of the former GDR. A plant at Crossen, near Zwickau in Saxony, started processing ore in 1950. The ore was transported by road and rail from numerous mines in the Erzgebirge. The composition of the ore from the hydrothermal deposits required carbonate pressure leaching. The plant had a maximum capacity of 2.5 million tonnes of ore per year. Crossen was permanently closed on 31 December 1989.

The second plant at Seelingstadt, near Gera, Thuringia, started ore processing operations in 1960 using the nearby black shale deposits. The maximum capacity of this plant was 4.6 million tonnes of ore per year. Silicate ore was treated by acid leaching until the end of 1989. Carbonate-rich ores were treated using the carbonate pressure leaching technique. After 1989, Seelingstadt operations were limited to the treatment of slurry produced at the Königstein Mine using the carbonate method.

Federal Republic of Germany before 1990

A uranium processing centre at Ellweiler, in the state of Baden-Württemberg was operated by Gewerkschaft Brunhilde beginning in 1960. Serving as a test mill for several types of ore its capacity was only 125 tU per year. It was closed on 31 May 1989 after producing around 700 tU.

Prior to reunification, the GDR was the world's third largest uranium producer, after the United States and Canada. This was not generally known before the unification of Germany in 1990. Uranium production in Germany was dominated by output in the GDR. Production was likely set by quotas and is assumed to have been close to capacity (Figure 7.12). The steep drop off in production in 1990 occurred with the closure of all uranium mines in Germany following reunification.

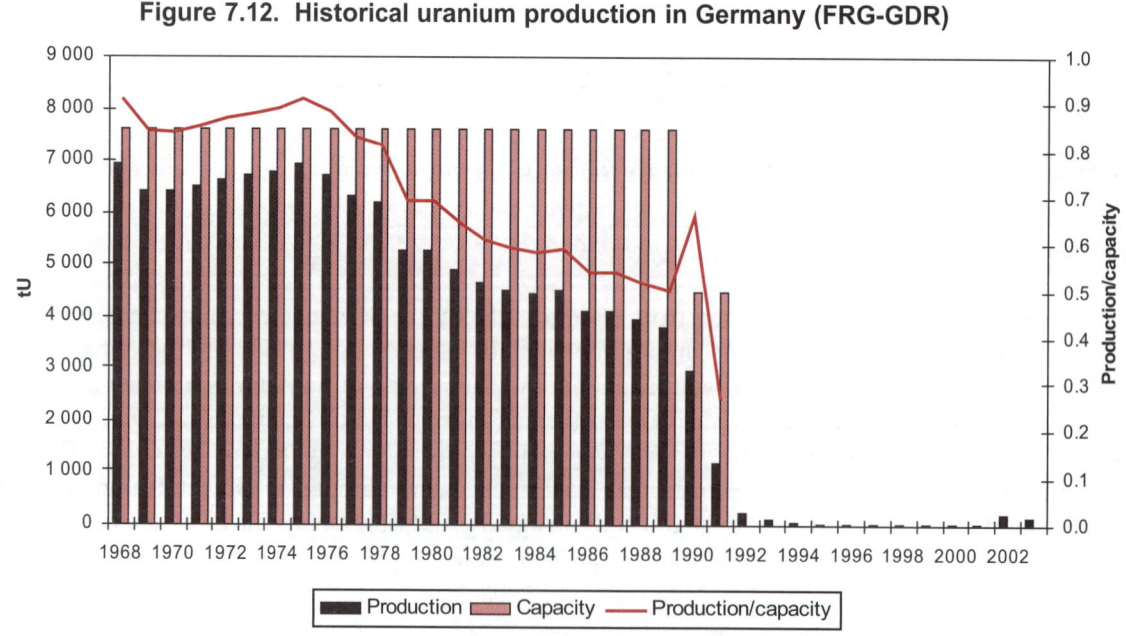

Figure 7.12. Historical uranium production in Germany (FRG-GDR)

Namibia

In 1928, Captain G. Peter Louw discovered uranium mineralisation in the vicinity of the Rössing Mountains in the Namib Desert. Over many years he tried to promote the prospect, but it was not until the late 1950s that Anglo-American Corporation of South Africa prospected the area by drilling and by some underground exploration. Due to erratic uranium values and poor economic prospects for uranium, the Anglo-American Corporation abandoned the search. As a result of an upswing in the uranium market demand and prices, extensive uranium exploration started in Namibia in the late 1960s. Several airborne radiometric surveys were conducted by the geological survey during this period and numerous uranium anomalies were identified. In August 1966, Rio Tinto Zinc (RTZ) acquired the exploration rights for the Rössing deposit and conducted an extensive exploration programme that lasted until March 1973. Surveying mapping, drilling, bulk sampling and metallurgical testing in a 100 t/day pilot plant indicated the feasibility of establishing a production centre. Mine development commenced in 1974, and commissioning of the processing plant and initial production occurred in July 1976 with the objective of reaching full design capacity of 5 000 short tons of U_3O_8/year (3 845 tU/year) during 1977. Due to the highly abrasive nature of the ore, which was not identified during the pilot plant testing stage, the production target was not reached until 1979 after major plant design changes.

Uranium production and capacity in Namibia were directly related to the Rössing mine, the nominal annual capacity of which ranged between 3 700 and 4 000 tU. Production trends at Rössing after 1980 were closely related to restructuring and downsizing to match output with sales commitments (Figure 7.13). In the early 1980s the ratio of production to capacity was equal to or slightly below 1.0 at Rössing. Over time, however the ratio dropped to as low as 0.54 in 1992, recovered to 0.97 in 1997, but began to again decline, reaching 0.51 in 2003.

Figure 7.13. Historical uranium production in Namibia

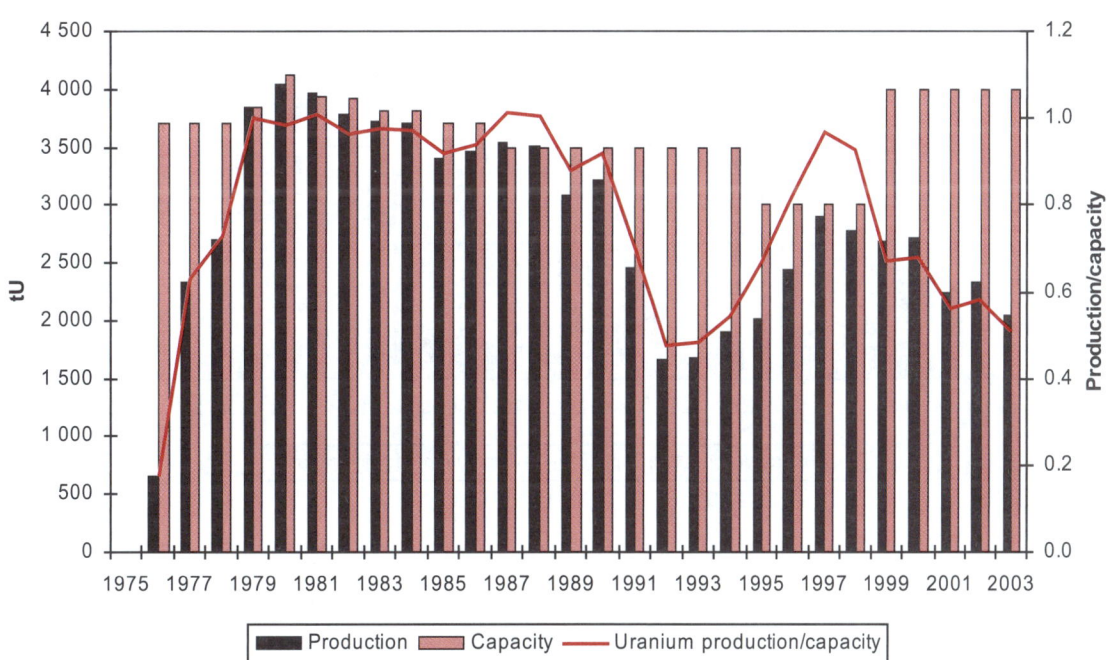

Niger

Uranium exploration in the Arlit area of Niger began in 1956 and was conducted by the French *Commissariat à l'énergie atomique* (CEA), later by COGEMA. Discovery of mineralised sandstone areas eventually led to the mining by the open-pit method of the Arlette, Artois and Ariege deposits by the *Société des Mines de l'Aïr* (Somaïr) beginning in 1970 and mining of the Akouta and Akola deposits using underground methods by the *Compagnie des Mines d'Akouta* (Cominak) beginning in 1978.

The ratio of production to capacity in Niger averaged 0.84 between 1971 when production began and 2003. The period between 1971 and 1980 saw a rapid increase in both capacity and production with development of the Arlit (1971) and Akouta (1978) production centres (Figure 7.14). During this period production and capacity were approximately in balance, with the lowest ratio, 0.57, occurring in the first year of Arlit production. Production peaked in 1981, and then began a steady decline to a low point in 1990, largely in response to low uranium prices. During most of this time capacity remained relatively constant so the production to capacity ratio declined in parallel with production to a low of 0.65 in 1987, before again starting to increase. Production capacity began to decrease in 1988, which marked the end of heap leaching and decreases in plant capacity in response to low uranium prices. Since 1997 production has been approximately in balance with capacity.

Figure 7.14. Historical uranium production in Niger

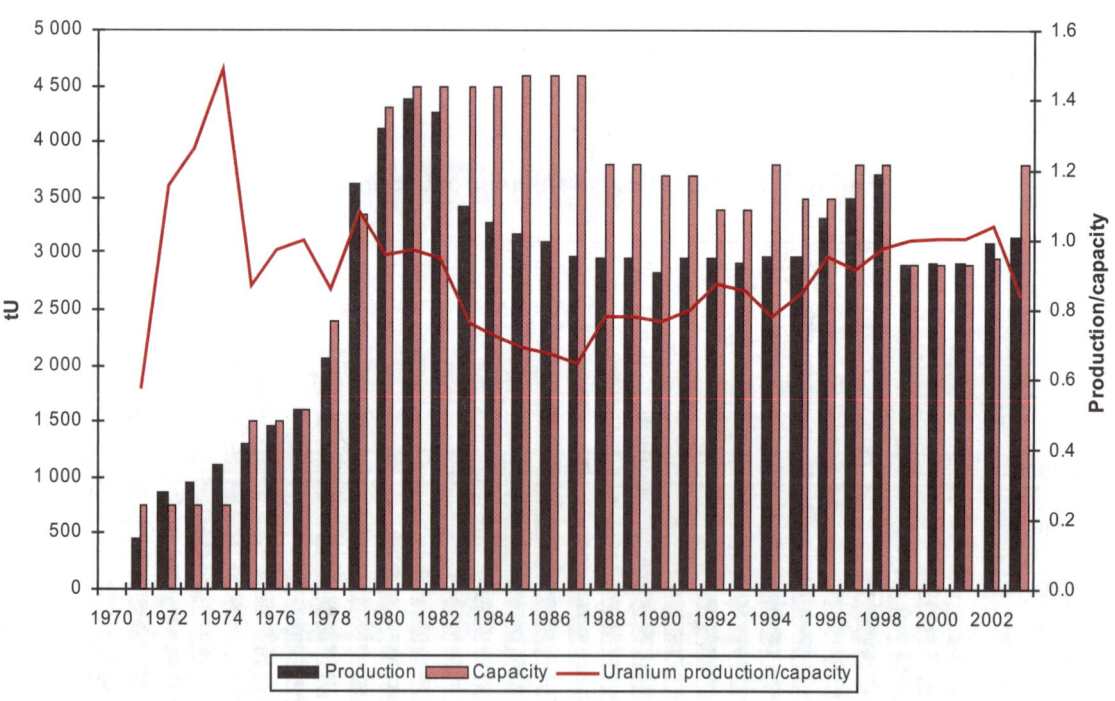

South Africa

Uranium production in South Africa commenced in 1952 with the commissioning of a plant at West Rand Consolidated Mine extracting uranium from quartz-pebble conglomerates of the Witwatersrand Supergroup. During 1953 a further four plants came into production at various centres. Total uranium production peaked in 1959 when 4 957 tU were produced from 17 plants being fed from 26 mines within the Witwatersrand Basin. Production thereafter declined to 2 263 tU in 1965.

The world oil crisis which emerged in 1973 stimulated the demand for uranium as a source of energy. The large tailings stockpiles containing uranium which had accumulated over many decades at the time became a readily available source of uranium. These stockpiles were processed at Welkom (Joint Metallurgical Scheme – 1977), on the East Rand (ERGO – 1978) and at Klerksdorp (Chemwes – 1979) which culminated in a record uranium production of 6 028 tU in 1980.

In 1967 there were seven producers (2 585 tU); this number increased to 14 in 1983 (5 880 tU). From 1983 there was a steady decline in the number of producers with only three remaining in 1994 (1 550 tU). Palabora Mining Company, which commenced uranium production in 1994 outside of the Witwatersrand Basin as a by-product of copper mining, ceased production in 2002 leaving Nuclear Fuels Corporation of South Africa (Pty) Limited (Nufcor) as the sole producer of uranium in South Africa at present.

Uranium has always been produced in South Africa as a by-product, mainly of gold mining operations, with minor by-product contribution from the Palabora copper mine. The ratio between production and capacity averaged 0.79 between 1968 and 2003, varying between 0.45 in 1978 and 1.0 in 1974 (Figure 7.15). There has been a very close correlation between production and uranium prices. The sharp increase in production capacity in 1978 coincided with an increase in the uranium market price. In the period between 1975 and 1978 when capacity reached its highest level, the number of uranium producers increased from seven to 25. As the price of uranium began to fall so too did the number of producers, from 25 in 1978 to seven in 1982. Though changing gold prices may have accentuated the trends they resulted largely from changing uranium prices. For the most part, gold production continued, but as uranium prices fell and uranium recovery became uneconomic the uranium circuits were shut down. In 2003 there was only one uranium producer in South Africa.

Figure 7.15. Historical uranium production in South Africa

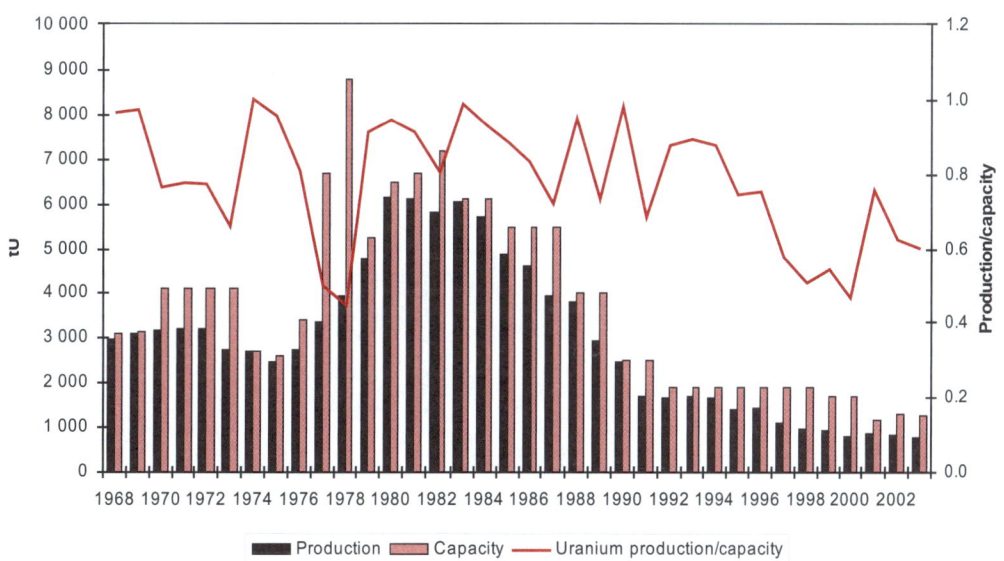

United States

Under the Atomic Energy Act of 1946, designed to meet the US government's uranium needs, the Atomic Energy Commission (AEC) from 1947 to 1970 fostered a domestic uranium industry, chiefly in the western states, through incentive programmes for exploration, development, and production. To assure that the supply of uranium ore would be sufficient to meet future needs, the AEC, in April 1948, implemented a domestic uranium ore procurement programme designed to stimulate a civilian-based domestic mining industry. The AEC also negotiated uranium concentrate procurements contracts, pursuant to the Atomic Energy Act of 1946 and 1954, with guaranteed prices for source materials delivered within specified times. Contracts were structured to allow companies that built and operated mills the opportunity to amortise plant costs during their procurement-contract periods. By 1961, a total of 27 privately owned mills were in operation. Eventually, 32 conventional mills and several pilot plants, concentrators, up graders, heap-leach, and solution-mining facilities were operated at various times. The AEC, as the sole government purchasing agent, provided the only US market for uranium. Many of the mills were closed soon after completing deliveries scheduled under their uranium contracts, although several mills continued to produce concentrate for the commercial market after fulfilling their AEC commitments. The Atomic Energy Act of 1954 made lawful private ownership of nuclear reactors for commercial electricity generation. By late 1957, domestic ore reserves and milling capacity were sufficient to meet the government's projected requirements. In 1958, the AEC procurement programmes were reduced in scope, and, in order to foster utilisation of atomic energy for peaceful purposes, domestic producers of ore and concentrate were allowed to sell uranium to private domestic foreign buyers. The first commercial sale of uranium concentrate occurred in 1966. The AEC announced in 1962 a "stretch-out" programme for post-1966 uranium concentrate delivery. Under this option, a milling company could defer a part of its 1963-1966 contract deliveries until 1967-1968. In return, the AEC would agree to purchase from the company in 1969-1970 an additional amount of concentrate equal to the amount of its deferred deliveries under the stretch-out plan. The government's uranium procurement programme was ended in 1970 and the industry became a private sector, commercial enterprise with no additional government purchases. Uranium concentrate production in the United States has supported the commercial market since 1970. The peak year for US production was 1980 (16 811 tU), subsequently the US industry has experienced generally declining annual production in the period 1981-2003. Since 1991, production from *in situ* leach mining and other non-conventional production methods has dominated US annual production.

Uranium production and capacity in the United States both peaked in 1980 (Figure 7.16). Since then, with the exception of brief periods of stability or modest increases, production has declined steadily from its peak of 16 811 tU in 1980 to a low of 769 tU in 2003. During the period from 1968 to 2003, the ratio of production to capacity averaged 0.54 and ranged from 1.0 in 1983 to 0.11 during 2001-2002. However, the historic average of only 54% utilisation of capacity and the 11% utilisation indicated for 2001 and 2002 is artificially low because it takes into account mills that were placed in a standby status and for which there were no active mines. For example, in 2001 the capacity of active production centres was 1 920 tU compared to a total reported capacity of 9 510 tU. At least 50% of the disparity between the two totals was related to production centres that, though they are officially listed as being on standby status, had very little chance of ever operating again. The decrease in capacity between 2001 and 2002 is an artefact of reporting that resulted from eliminating some of these "suspect" production centres.

At the end of 1980, the year of peak production in the United States (16 811 tU) there were 22 conventional mills in operation, 11 ISL operations, eight plants to recover uranium from phosphoric acid and two plants to recover uranium from oxide copper leach liquors. By 1985 production had declined to 4 352 tU. Most of the decline was due to a reduction in conventional mills from 22 to only eight, most of which were operating at well below capacity. In addition most of the unconventional uranium recovery plants were inactive. In 2003, the historical low point in US production, there were two active ISL operations. One new ISL project was being developed and a conventional mill that had been on standby was preparing to restart to process ore from several small mines in western Colorado.

Figure 7.16. Historical uranium production in the United States

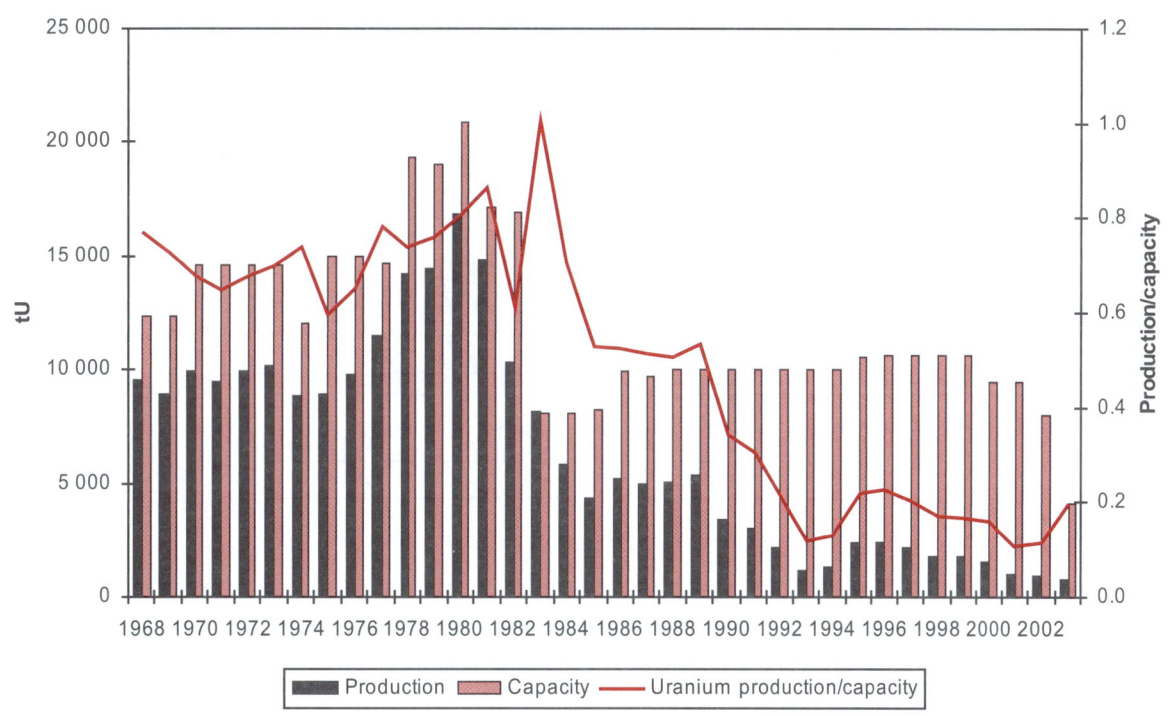

The Soviet Union (USSR), former Soviet Socialist Republics and Commonwealth of Independent States (CIS) Countries

For purposes of this discussion, all uranium production between 1968 and 1991 is allocated to the USSR. Following dissolution of the USSR in 1991, production is allocated to the individual countries that were once part of the USSR. Uranium production in the USSR grew in a stepwise manner between 1972 and 1973 and again between 1976 and 1977 (Figure 7.17). Once capacity of 16 500 tU was reached in 1977 it remained stable until 1991 when capacity declined to 10 500 tU as the individual CIS countries began to close high-cost operations and in the case of Kazakhstan and Uzbekistan to emphasise ISL extraction.

Between 1968 and 2003, the ratio between production and capacity averaged 0.87 and varied from 0.64 in 1996 to 0.97 in 1988. Once production capacity stabilised in 1977, there was a close correlation between production and capacity, with the ratio between the two averaging just under 0.90 or 90% capacity utilisation. From a low of 0.64 in 1996, the production to capacity ratio has steadily improved, reaching 0.88 to 2003. This improvement is widespread as the CIS countries are beginning to benefit from higher uranium prices.

Figure 7.17. Historical uranium production in the USSR and the CIS Countries

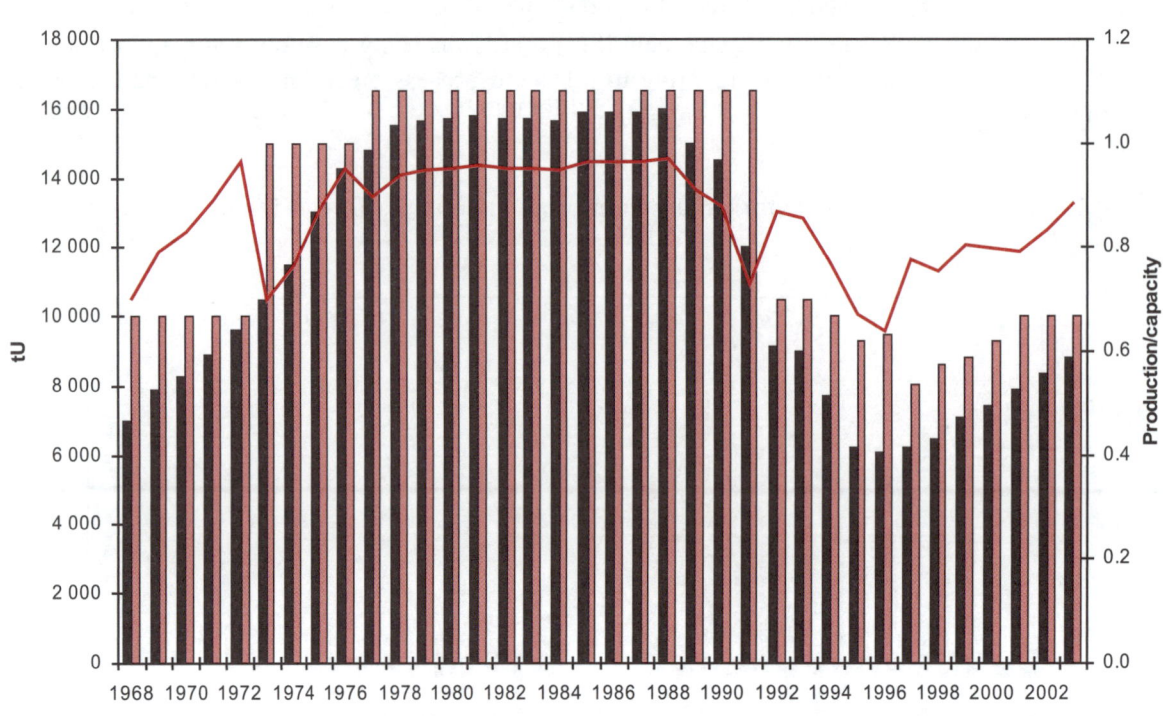

Since the break up of the Soviet Union, information concerning the production of uranium has increasingly become available from the former Soviet Socialist Republics, now independent or independent states within the CIS. Brief production histories of these states are given below.

Estonia

Uranium production in Estonia is associated to the Sillamae Metallurgy Plant, located in north-eastern Estonia at the town of Sillamae, 185 km east of Tallin near the shore of the Gulf of Finland.

The Sillamae plant was constructed in 1948 to process uranium bearing ores. It was first used to recover uranium from alum-shale mined in Estonia. Alum-shale mining continued until 1963, when the mines were closed due to difficulties in recovering the low and variable uranium content of the ore. The ore was extracted from underground mines located to the west of the plant. From 1948 to 1963 about 240 000 tonnes of locally mined alum-shale were processed. Uranium production from the alum-shale is estimated to have been about 65 tU.

After 1963, higher grade uranium ores containing up to 1% U were imported from Eastern Europe for processing. Uranium production continued until about 1977. Most of the ore was brought from Czechoslovakia (2.2 million t) and from Hungary (1.2 million t). Small amounts were also brought from Poland, Romania, Bulgaria and the former German Democratic Republic. An estimated 4 013 000 t of uranium ores were processed at the plant. Assuming an average recovery of about 92% of the contained U, total uranium production of the plant is estimated to be about 23 000 tU.

Kazakhstan

Uranium exploration in Kazakhstan started in 1948 on the Kurdai deposit situated south of Kazakhstan. Uranium mining in Kazakhstan started in 1957 using the open-pit method in the southern part of the country, on the Kurdai deposit. Until 1978 four companies, belonging to the USSR Ministry of Middle Machine Construction, mined uranium by underground and open-pit methods: Kyrgyzski Mining Combine, Leninabadski Mining and Chemical Combine in the south, Tselinny Mining and Chemical in the north and Prikaspiiski Mining and Chemical Combine in the west. About 15 deposits, with an approximate cumulative output of 5 000 tU, were mined. Deposits mined out during these years, were mainly vein-stockwork mineralisation type.

ISL uranium mining of sandstone deposits started in 1978. Mineralisation is represented by roll-type ore bodies tens of kilometres in length. All deposits of the Shu-Saryssuiskaia and Syr-Daryinskaia uranium provinces are sandstone-type deposits.

Russian Federation

The first organisation responsible for uranium production was the Lermontov Complex, presently Lermontov State Enterprise "Almaz". Almaz is located 1.5 km from the town of Lermontov, in the Stavropol region or district. This district included the Bestau and Byk vein deposits, which have been mined out. Their original resources totalled 5 300 tU, at an average grade of 0.1% U. These resources were extracted by two underground mines starting in 1950. Mine 1 (Beshtau) was closed in 1975 and Mine 2 (Byk) in 1990. The ore was processed at the local processing plant using sulphuric acid leaching starting in 1954. From 1965 to 1989 stope or block leaching were also used. From the 1980s until 1991 uranium ore transported from Ukraine and Kazakhstan was also processed at Almaz. Production from local deposits totalled 5 685 tU, with 3 930 tU extracted by underground mining and 1 755 tU by a combination of different leaching technologies.

Between 1968 and 1980, 440 tU were produced by ISL from the Sanarskoye deposit in the Transural district. The Malyshevsk Mining Enterprise operated the project.

The joint Stock Company "Priargunsky Mining-Chemical Production Association" has been the only active uranium production centre in Russia in the last decade. The Priargunsky production centre is located in the Chita region 10-20 km from the town of Krasnokamensk, which has a population of about 60 000 people. The production is based on 19 volcanic deposits of the Streltsovsk uranium district, which has an overall average uranium grade of about 0.2% U. This district has an area of 150 km^2. Mining has been conducted since 1968 by two open-pits (both are depleted) and three underground mines (mines 1 and 2 are active and mine 4 is closed). Milling and processing has been carried out since 1974 at the local hydrometallurgical plant using sulphuric acid leaching with subsequent recovery by a combination of ion exchange and solvent-extraction. Since the 1990s low-grade ore has been processed by heap and stope/block leaching.

More than 100 000 tU have been produced from the Stresovsk deposits at Priargunsky, making it one of the most productive uranium districts in the world. Cumulative production through 2003 in the Russia Federation totalled 116 683 tU, which made it the fifth largest uranium producer in the world based on historical production.

Turkmenistan

Uranium exploration in Turkmenistan started in the 1940s, and was mainly carried out by Uzbek geologists who discovered the Cernoye deposit, which from 1952 to 1967 was mined using open-pit and underground methods. After enrichment by radiometric ore sorting on site, ore concentrates were shipped at the Mungishlak plant in Kazakhstan for yellow cake production. Total production from the Cernoye deposit would have been between 5 000 and 7 000 tU, at an average grade of 0.4-0.5% U, with grades between 0.1 and 20% U.

Ukraine

A decision was made by the government in 1951 to create Vostochnyi mining-processing combine (VostGOK) in the city of Zheltye Vody in the North Krivoy Rog area for mining uranium ores from the Pervomayskoye and Zheltorechenskoye deposits, which had been explored by that period. The Pervomayskoye deposit was completely mined out in 1967 and Zheltorechenskoye ended production in 1989. The Michurinskoye deposit was discovered in 1964 and in 1967 the construction of a mine, called Ingul'skaya, began and it remains in production. The Vatutinskoye deposit was discovered in 1965 and construction of the Smolinsky mine was started in 1973. *In situ* leaching of uranium has been practiced in Ukraine from 1961 to 1983 at two deposits, Devladovskoye and Bratskoye.

Uzbekistan

Uranium production in Uzbekistan began in 1946 at several small volcanic vein deposits in the Fergana valley and Kazamazar uranium district. The mines are no longer in operation and the deposits are depleted. The ore was processed in the Leninabad uranium production centre in Tajiskistan. Commercial uranium mining began at Uchkuduk in 1958 with the development of both open-pit and underground mines. The ore was stockpiled until the completion in 1964 of the hydrometallurgical uranium processing plant in Navoi, located some 300 km southeast of Uchkuduk. ISL experiments conducted at the Uchkuduk deposits started as early as 1963, leading to the commercial application of ISL in 1965. Conventional underground mining operations started at the Sabysai and Sugraly deposits in 1966 and 1977 respectively. In 1975, ISL extraction began to replace the underground mining of the Sabyrsaj mine, and conventional underground mining at Sabyrsaj was stopped in 1983. The Ketmenchin ISL plant began operation in 1978. In 1994, reduction of uranium demand led to the closure of the open-pit at the Uchkuduk mine as well as both underground and ISL Sugraly mines. As of 1 January 2005, three mining divisions are producing uranium by *in situ* leaching: the Northern Mining Division (Uchkuduk), Mining Division No. 5 (Zafarabad) and the Southern Mining Division (Nurabad). The Eastern Mining Division was closed for economical reasons. Uranium concentrates are processed in the hydrometallurgical plant in Navoi.

Uranium production by mining method

The first effort to acquire information on the distribution of uranium production by mining method began with the 1993 Red Book. That effort was progressively expanded, with the most comprehensive data set having been developed during preparation of the 2003 Red Book. Table 7.3 is a compilation of data on production according to mining/extraction method between 1990 and 2003 [note, data for 1991 were not available]. Figure 7.18, shows the percentage distribution of total 1990-2003 production by mining method and Figure 7.19 compares production from 1990 through 2003.

Table 7.3. Uranium production by mining/extraction method

Production method	1990	1991	1992	1993	1994	1995	1996	1997	1998	1999	2000	2001	2002	2003
Open-pit	38	NA	41	42	38	37	39	49	39	35	28	26.1	26.8	27.8
Underground	55	NA	33	32	41	43	40	32	40	36	43	44.2	43.1	41.6
ISL	6	NA	15	15	14	14	13	13	13	17	15	15.5	18.3	18.4
Other*	1	NA	11	11	7	6	8	6	8	12	14	0.5	0.8	0.5
Heap leaching												1.2	1.7	1.9
In place leaching												0.1	0.1	<0.1
Co-/by-product												12.4	9.1	9.7

NA Not available.

* From 1990 to 2000, "Other" include recovery from phosphate, co-/by-product of copper and gold, stope/block leaching heap leaching and mine water recovery.

Open-pit and underground mining accounted for 75% of worldwide uranium production between 1990 and 2003; addition of ISL extraction brings the total to 90% of total output between 1990 and 2003 (Figure 7.18). The co-/by-product category is artificially low in Figure 7.18 because both were included in the "other" category until 2000. Similarly the "other" category is correspondingly high.

Figure 7.19 compares percentage distribution by mining/extraction method in 1990 and 2003 to determine how or if the uranium production industry has changed. In fact, the industry changed quite a lot in response to a long period of depressed uranium prices and political changes. Open-pit and underground mining declined by 10% and 15%, respectively, as mines were closed throughout the world in response to high production costs and low uranium prices. The loss of output that resulted from open-pit mine closures in the CIS, Gabon and the United States, and reduced output from open-pit mines in other countries was partially offset by opening of the McClean Lake mine in Saskatchewan in 1999, the Lagoa Real mine in Brazil in 2000 and expansion of capacity at the Ranger mine in Australia, which was completed in 1997 (Figure 7.19).

Similarly production losses resulting from closure of underground mines in the CIS, France, the Czech Republic, Hungary and the United States, were partially offset by opening of the McArthur River mine in Saskatchewan in 1999. The 15% increase in ISL output between 1990 and 2003 is attributable to expansion of production in Kazakhstan and Uzbekistan and opening of the Beverley mine in Australia in 2000, the Yining mine in China in 1997, the Dalmatavoskoe mine (pilot plant) in 2002 in Russia and the Smith Ranch mine in the United States in 1997. These expansions and new mine additions more than offset closure of several ISL operations in the United States between 1990 and 2000.

Co-/by-product recovery of uranium accounted for 9.4% of 2003 output. To make a direct comparison between 1990 and 2003, however, in-place leaching and heap leaching need to be added to the co-/by-product category as they were in 1990, which results in a combined category total of 12% for 2003. The increase in "other" production from 1% in 1990 to 12% in 2003 is almost entirely attributable to increased output from Olympic Dam in Australia where uranium is recovered as a co-product along with copper. Gold and silver are also recovered as by-products. The expanded output from Olympic Dam more than offset declining production of by-product recovery from gold operations in South Africa and the closure of by-product recovery of uranium from phosphoric acid production in the United States and Belgium in 1999.

Figure 7.18. Percentage distribution of total production between 1990 and 2003 by mining/extraction method

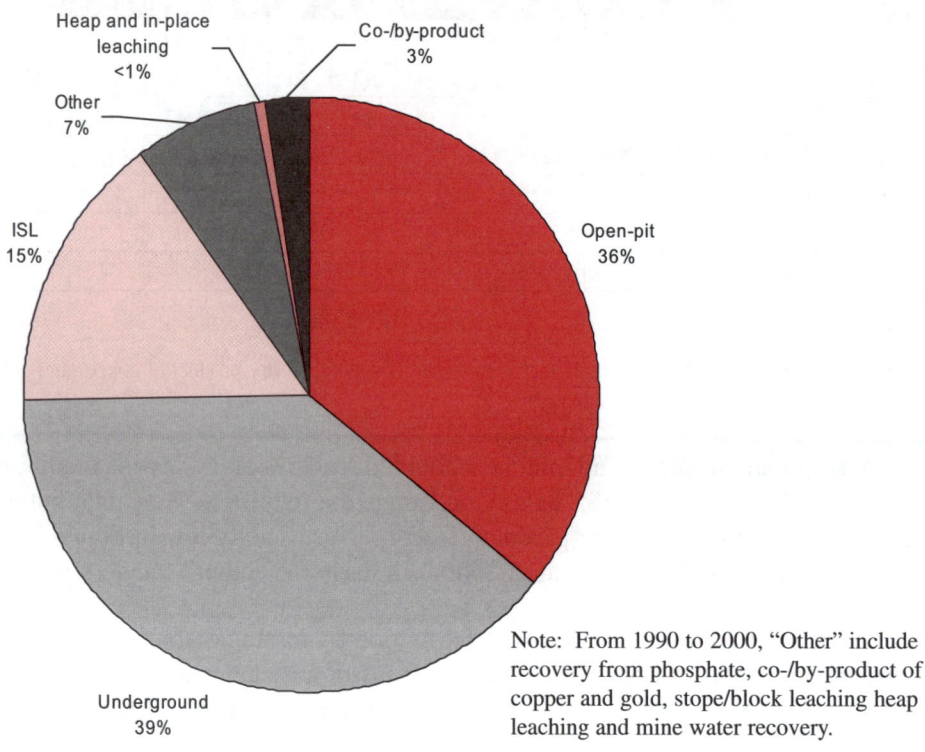

Note: From 1990 to 2000, "Other" include recovery from phosphate, co-/by-product of copper and gold, stope/block leaching heap leaching and mine water recovery.

Figure 7.19. Comparison of 1990 and 2003 production by mining/extraction method (%)

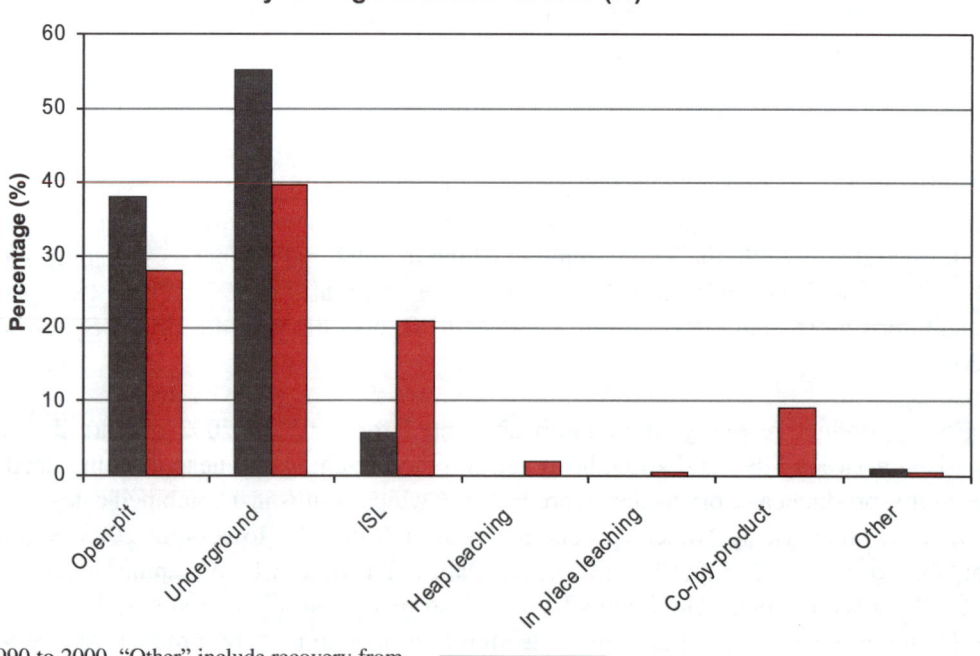

Note: From 1990 to 2000, "Other" include recovery from phosphate, co-/by-product of copper and gold, stope/block leaching heap leaching and mine water recovery.

114

Employment at uranium production centres

Historical data on employment at uranium production centres is incomplete, but a sufficiently complete data set is available for select countries to portray employment trends between 1980 and 2003. Figure 7.20 summarises employment trends in six countries during this period and Table 7.4 shows employment figures in five-year increments for the six countries included in Figure 7.19 as well as for other selected countries; together these 12 countries accounted for approximately 97% of worldwide production in 2003. Production centre employment data by country are provided in Appendix 7.3.

Table 7.4. Employment at uranium production centres (1980-2003)

	1980	1985	1990	1995	2000	2003
Australia	500	460	1 183	413	527	655
Canada	6 100	5 333	2 495	1 350	1 026	965
France	NA	3 508	2 276	468	NA	NA
Germany*	85	27 893	15 710	4 400	3 115	2 444
Niger	NA	3 552	3 173	2 109	1 680	1 606
United States	19 920	18 389	2 446	535	401	204
Russia	NA	NA	NA	14 000	12 500	12 785
Kazakhstan	NA	NA	NA	6 850	4 000	3 870
Namibia	NA	NA	NA	1 246	902	780
Romania	NA	NA	NA	6 000	2 150	2 000
Uzbekistan	NA	NA	NA	7 378	7 331	8 460
China	NA	NA	10 000	8 000	8 500	7 700

* 1985 and 1990 includes GDR employment totals; since 1995 all employment is attributable to rehabilitation activities.

Though the data in Figure 7.20 and Table 7.4 are by no means complete, they do reflect the recent history of the uranium production industry. The precipitous drop in employment in the United States between 1980 and 1985 accompanied an equally precipitous decrease in production – from 16 811 to 4 352 tU in 1980 and 1985, respectively. Germany's dramatic decrease in employment between 1989 and 1993 resulted from closing of its uranium production facilities. Employment after 1993 was largely related to remediation and decommissioning of GDR production sites.

Though they were much less dramatic than the German and United States examples, most other countries depicted in Figure 7.20 and Table 7.4 showed generally declining employment in uranium production. Some of these decreases in employment were accompanied by decreases in production; others were not. For example, despite Canada's decrease in employment between 1983 and 1994 production increased by about 40%, as new high-grade mining operations were opened in the Athabasca Basin. Despite maintaining relatively low employment totals between 1990 and 2003, Canada has become the world's leading uranium producer, accounting for about 30% of worldwide production in 2003 but only about 2% of employment related to uranium production. This seeming contradiction results from the efficiencies of its high ore-grade operations – fewer tonnes of ore have to be mined and processed to recover the same amount of uranium compared to lower grade operations. Lower tonnage requirements to reach a given production level as measured in tU/year result in lower employment requirements.

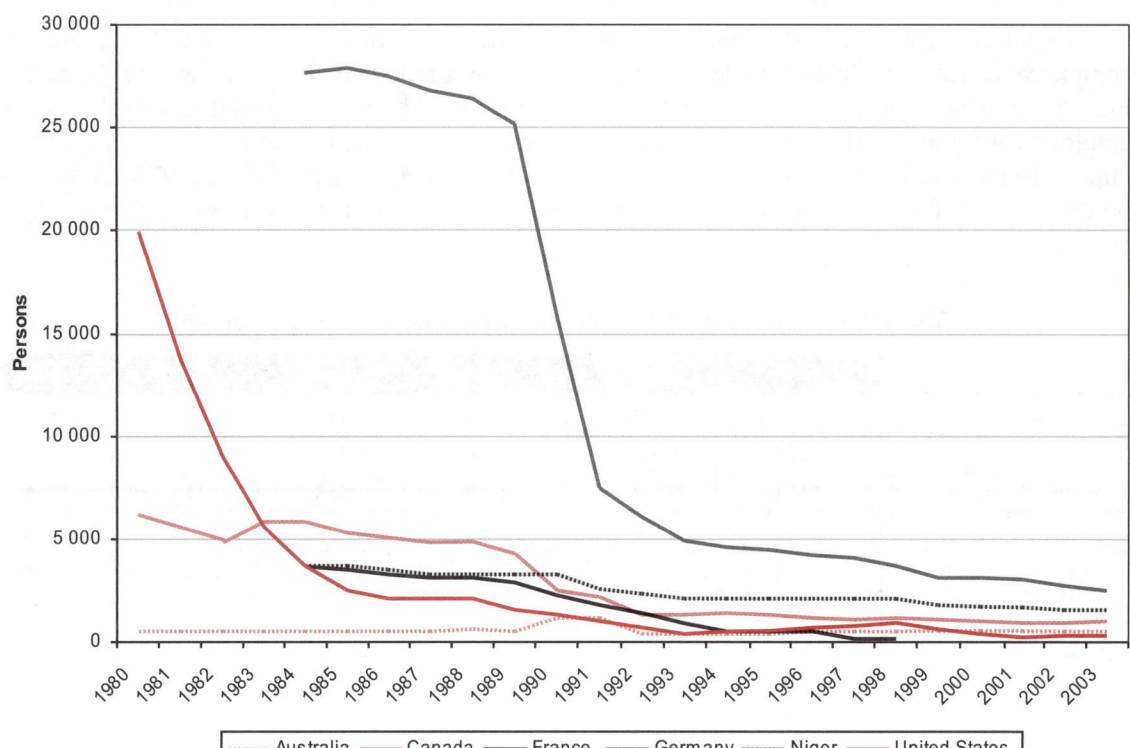

Figure 7.20. Employment at uranium production centres in selected countries (1980-2003)

A key point from the Canadian example is that employment numbers do not always have a direct correlation to production trends. For example, in the case of some developing countries, maintaining stable employment is a key consideration. Therefore, production may decline at higher rates than employment. On the other hand, efficiencies (or higher ore grades) may allow employment to stabilise or decrease even as production increases.

Ownership of uranium production

Ownership of uranium production is subdivided in the Red Book into four categories based on whether the companies operate domestically or in another country and whether they are government- or privately-owned.

Red Book ownership information on which comprehensive year-to-year comparisons can be made has only been available since 1992, when data for China, the CIS and Eastern European Countries were first published. In 1992, ownership for only 85% of worldwide production was available. By 1994 and for all succeeding Red Books, ownership of total worldwide production was accounted for (Table 7.5).

The structure of the uranium production industry has not changed significantly during the past eight years (Figure 7.21). Government and private ownerships were nearly equal in both 1994 and 2002. The most significant change in the industry was in Canada where production controlled by a then partly government-owned domestic company (Cameco Corporation) declined from 40% in 1990 to 0% in 2002. In the interim period, both the federal and provincial governments sold their respective shares in Cameco. However, government-owned (domestic and foreign companies) control of Canadian production still totalled nearly 47% in 2002 because of COGEMA acquisition of an interest in the

116

McArthur River/Key Lake operation and its ownership interests in other production centres including Cluff Lake and McClean Lake, although the latter facility was closed at the end of 2002 and is now being decommissioned.

In 1994, 12 countries reported 100% government control of their production – Brazil, Bulgaria, China, Czech Republic, China, Germany, India, Kazakhstan, Pakistan, Romania, Russia, Ukraine and Uzbekistan. In 2002, 11 countries reported 100% government control of their production. Bulgaria and Kazakhstan (97% government control) were dropped from the list and Hungary was added. Interestingly, only South Africa reported 100% private ownership of its 2002 production. Namibia and the United States reported 97% and 99% private ownership, respectively.

Table 7.5. Ownership of uranium production (1992-2002)

| | Domestic mining companies | | | | Foreign mining companies | | | | Total production |
| | Government | | Private | | Government | | Private | | |
	tU	%	tU	%	tU	%	tU	%	tU
1992	17 309	48.2	8 373	23.3	1 474	4.1	3 521	9.8	35 919
1994	13 415	42.5	10 815	34.3	2 079	6.6	5 246	16.6	31 555
1996	10 666	29.5	13 365	36.9	5 060	14.0	7 104	19.6	36 195
1998	11 187	32.0	15 050	43.0	3 587	10.2	5 212	14.9	34 986
2000	11 105	30.8	11 191	31	6 167	17.1	7 548	20.9	36 111
2002	11 404	31.6	11 800	32.7	6 827	18.9	6 011	16.7	36 033

Figure 7.21. Ownership trends (1994 -2002)

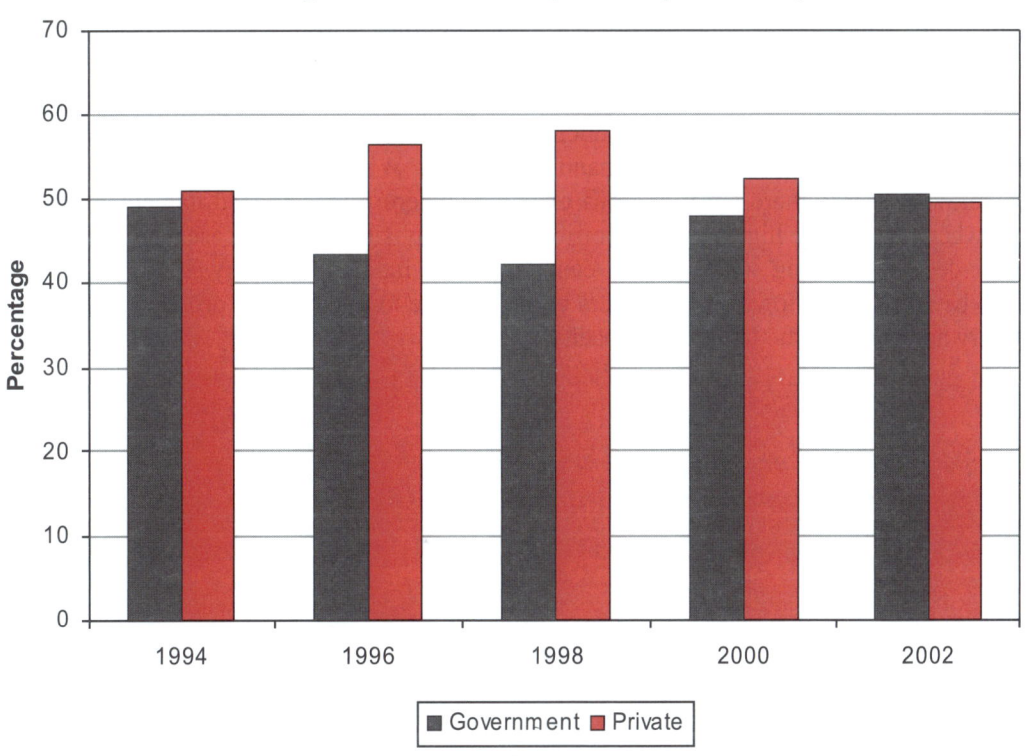

117

Projections of future uranium production capacity

The discussion of uranium production has up to this point largely focused on historical information that is contained in the Red Books. Provision is, however, also made in the Red Book for projections of uranium production capability based on four scenarios:

- "A-I Scenario": Production capability that is supported only by RAR and EAR-I recoverable at costs <USD 40/kgU, which are tributary to existing and committed production centres.

- "A-II Scenario": Production capability that is supported only by RAR and EAR-I recoverable at costs <USD 80/kgU, which are tributary to existing and committed production centres.

- "B-I Scenario": Production capability that is supported only by RAR and EAR-I recoverable at costs <USD 40/kgU, which are tributary to existing, committed, planned and prospective production centres.

- "B-II Scenario": Production capability that is supported only by RAR and EAR-I recoverable at costs <USD 80/kgU, which are tributary to existing, committed, planned and prospective production centres.

The forecasts provided in the Red Book relate to future production capacity based on current production facilities and those that could be built in the future. As we review these projections it is important to make the distinction between actual production and total capacity. Figure 7.22 compares worldwide annual production with capacity between 1968 and 2003. During this 35-year time period, which spanned most of the modern uranium production era and included a wide range of economic conditions, the ratio between production capacity and annual production averaged 0.76, which means that capacity exceeded production by an average of 24%. From 1968 to 2003, the ratio varied from a high of 0.88 in 1981 to a low of 0.57 in 1999.

There are many reasons why, historically, production has never equalled capacity. Technical problems, including shortages of equipment and supplies are one reason. Labour disputes can disrupt production for varying lengths of time. Reductions in ore grade in portions of a mine can result in lower output of yellowcake, even though the tonnage processed remains the same as previous years. Finally, many producers attempt to closely match production with sales commitments, cutting back on output when demand is soft and increasing it as demand strengthens. While it is most common for mines to produce at below capacity, there are numerous examples where mines exceed nameplate capacities, particularly when producing higher grade portions of a deposit. In other cases, improvements in processing or de-bottlenecking within a mill can effectively increase capacity. If these changes are considered to be permanent, however, producers typically have to update their production licenses and gain regulatory approval for the increased capacity.

Figure 7.22. Comparison of worldwide production and production capacity

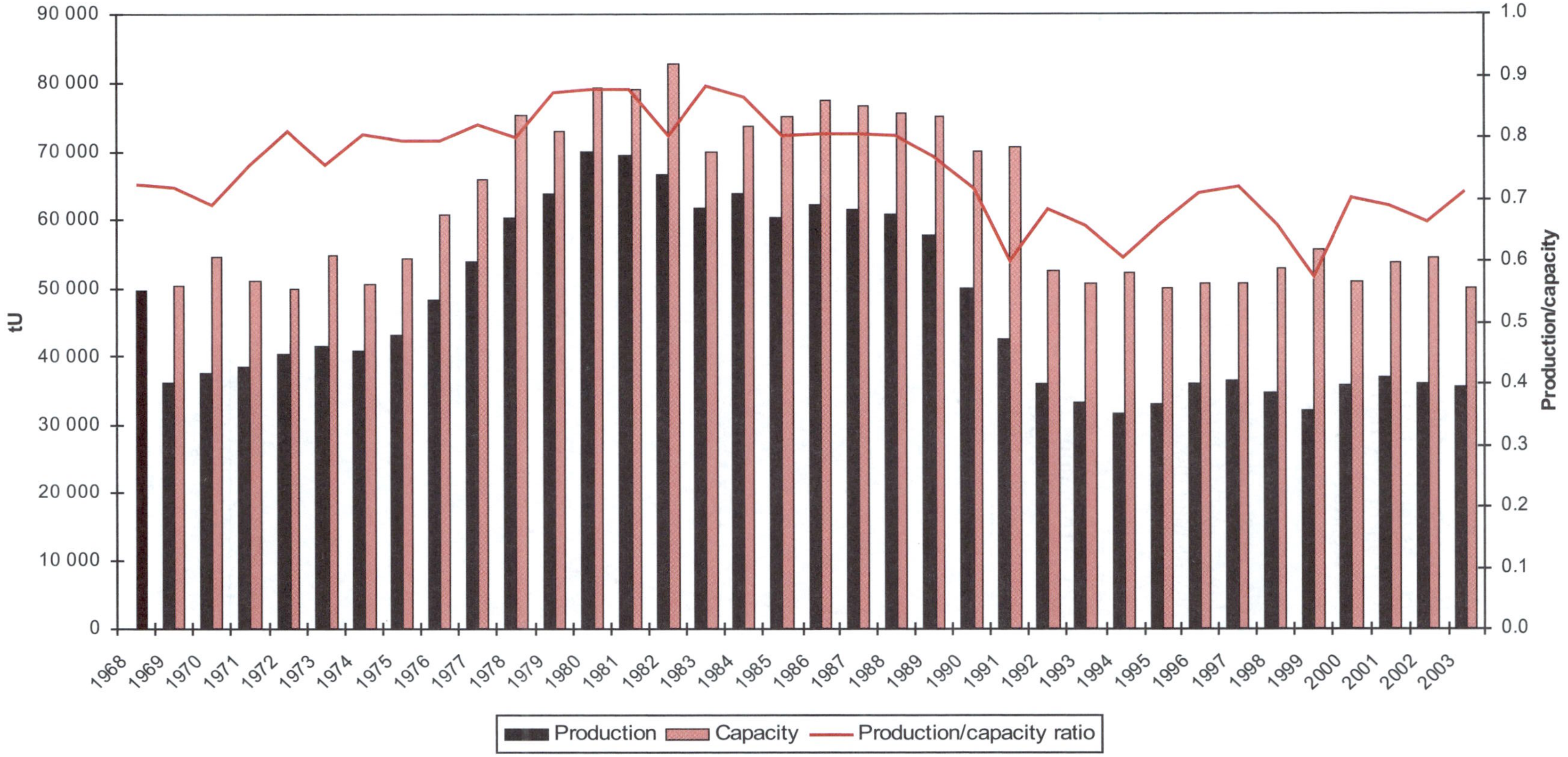

Figure 7.23 summarises the forecasts of production capability for the A-II Scenario for Red Books beginning in 1975 and extending through 2003. The market price curve is included to show the market conditions extant at the time the projections were made. The market price when the 1975 forecast was made was nearly USD 62/kgU. The 1975 forecast projected that capacity would grow from 25 600 tU in 1975 to 60 000 tU in 1980 and to 87 000 tU in 1985. The 1977 and 1979 projections, when the market price was USD 111/kgU, showed similarly optimistic projections for growth in production capability. The 1979 forecast, which was made when the uranium market price was at an all time high, projected production capability in 1995 of 123 000 tU for the A-II scenario, the highest capacity projection recorded in any Red Book regardless of Scenario.

This optimism preceded the Three Mile Island accident and its impact on the market price. Table 7.6 compares projections made in 1979 with same year forecasts made in 1982 through 1993. With few exceptions lower market prices at the time of a forecast inevitably resulted in lower production capacity forecasts than those made for the same year when prices were high.

Figure 7.23. Projections of production capability for the A-II scenario (1975 through 2003 Red Books)

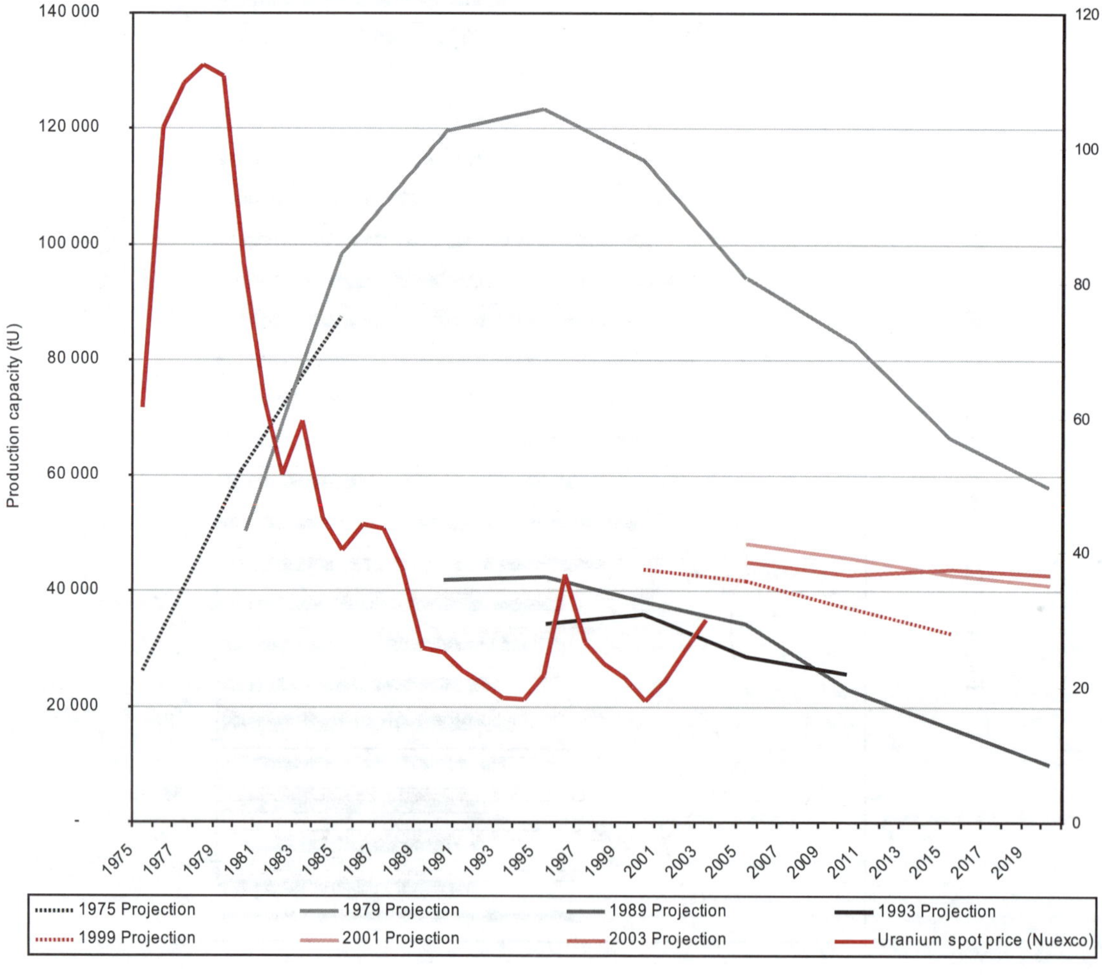

Table 7.6. Comparisons of production capacity projections (1979-1993)

Year of projection	Spot market price - year of projection (USD/kgU)	Projection for 1985 (tU)	Projection for 1990 (tU)	Projection for 1995 (tU)
1979	111.00	87 000	119 300	123 000
1982	51.74	92 000	70 230	
1986	44.20		48 530	45 730
1988	37.83		42 150	41 760
1989	28.00		42 400	42 760
1991	22.60		32 930	37 230
1993	18.24			34 740

Table 7.7 compares projected and actual production capability for Red Book editions from 1975 to 1993. Inevitably the closer the forecast year is to the year being forecasted, the closer the projections are to actual production capability. Setting that inevitability aside, there is a wide range between projected and actual capability. Projections made when prices were at historical highs turned out to have overstated actual capability. For example, the 1979 projection for 1995 turned out to be four times actual capability in 1995. Projections for production capability in 1995 made between 1979 and 1993 exceeded 1995 actual capability, though as expected the projections were much closer to reality in the later forecasts.

By comparison, projections made in the 1988 through 1995 Red Books for the year 2000 understated actual 2000 capability under Scenario A-II, which is limited to resources tributary to existing and committed production centres. However, broadening the scope of the analysis to include Scenario B-II, which also includes planned and prospective production centres, narrows the gap between projected and actual capability (Table 7.8). Table 7.9 compares the ratio of projections versus actual capacities for both Scenarios A-II and B-II between 1988 and 2003. A ratio lower than 1 indicates that actual capacity exceeded projected capacity; for ratios greater than 1 this relationship is reversed.

Among the various comparisons between projected and actual production capacity there is no simple explanation as to the range of ratios (Table 7.9). Ratios are closer to 1 the shorter the time period for the projection. High prices when the projections are made tend to result in optimistic projections; low prices have a dampening affect on projections. While there is no obvious pattern among the ratios, for both the 2000 and 2005 projections, the average ratios for Scenario B-II are closest to 1.0. The B-II Scenario, however, also produces the broadest range of ratios relative to ranges within Scenario A-II. Perhaps the strongest message that comes from these comparisons is that given the fact that the projections have a direct bearing on the projected future balance between supply and demand, they should be used with caution.

Table 7.7. Comparison of projected and actual production capacity Scenario A-II (tU)

Year of Projection	1980 Projection	1980 Actual	1985 Projection	1985 Actual	1990 Projection	1990 Actual	1995 Projection	1995 Actual	2000 Projection	2000 Actual	2005 Projection	2005 Actual
1975	60 000	51 000	87 000	44 400	110 000	32 630						
1977	53 000	51 000	92 000	44 400								
1979	50 100	51 000	98 000	44 400	119 300	32 630	123 000	31 024	94 000	43 883		
1982			72 570	44 400	70 230	32 630						
1986					48 530	32 630	45 730	31 024				
1988					42 150	32 630	41 760	31 024	35 660	43 883		
1989					42 400	32 630	42 760	31 024	38 300	43 883		
1991					32 930	32 630	37 230	31 024	32 350	43 883	34 200	48 645
1993							34 740	31 024	36 160	43 883	30 800	48 645
1995									30 584	43 883	20 606	48 645
1997									45 201	43 883	33 361	48 645
1999									43 833	43 883	43 750	48 645
2001											48 319	48 645
2003											45 295	48 645

Table 7.8. Comparison of projected and actual production capacity A-II and B-II Scenarios (tU)

Year of projection	2000 Projection A-II	2000 Projection B-II	2000 Actual	2005 Projection A-II	2005 Projection B-II	2005 Actual
1988	35 660	40 340	43 883			
1989	38 300	45 730	43 883	34 195	42 015	48 645
1991	32 350	37 200	43 883	30 800	34 300	48 645
1993	36 160	48 184	43 883	28 622	37 636	48 645
1995	30 584	41 465	43 883	20 606	38 971	48 645
1997	45 201	59 835	43 883	33 361	61646	48645
1999	43 833	45 752	43 883	43 750	61 192	48 645
2001				48 319	56 074	48 645
2003				45 295	51 155	48 645

**Table 7.9. Ratio between projected and actual production capacity
A-II and B-II Scenarios**

	2000			2005		
	Projection A-II	Projection B-II	Average Ratio	Projection A-II	Projection B-II	Average Ratio
1988	0.81	0.92	0.87			
1989	0.87	1.04	0.96	0.70	0.86	0.78
1991	0.74	0.85	0.79	0.63	0.71	0.67
1993	0.82	1.10	0.96	0.59	0.77	0.68
1995	0.70	0.94	0.82	0.42	0.80	0.61
1997	1.03	1.36	1.20	0.69	1.27	0.98
1999	1.00	1.04		0.90	1.26	1.08
2001				0.99	1.15	1.07
2003				0.93	1.05	0.99
Average	0.85	1.04	0.93	0.73	0.98	0.86

Conclusions regarding uranium production

- Uranium production has been reported by 35 different countries since 1945. Uranium production was 507 tU in 1945, peaked at 69 692 tU in 1980 before beginning a steady decline to 31 503 tU in 1994. Annual production from 1994 to 2003 has stayed within a relatively narrow range of 31 503 tU to 37 020 tU.

- Cumulative production from 1945 to 2003 was 2 204 732 tU. Cumulative reactor requirements during that period were about 1 513 300 tU.

- Newly mined and processed uranium (primary supply) exceeded reactor-related uranium requirements until 1991. From 1991 through 2003, the gap between primary supply and uranium requirements has been filled by secondary supply. In 2003, requirements were met almost equally by primary and secondary supplies.

- Between 1990 and 2003, open-pit and underground mining together accounted for 75% of worldwide production. In situ leach (ISL) extraction accounted for 15% of worldwide output. Between 1990 and 2003, both open-pit and underground mining decreased in percentage of total production; both ISL and co-/by-product recovery increased in percentage of worldwide production.

- Employment associated with uranium mining and processing among the major producing countries has steadily decreased during the past 20 years. This decrease has resulted both from mine closures and from increased efficiencies at ongoing operations.

- Projections of future uranium production capacity have historically been influenced by market price. High prices lead to optimistic forecasts that have exceeded actual production capacity; low prices have led to more pessimistic assessments, which have understated capacity.

- Historically, uranium production has ranged between 57 and 88% with an average of 76% of capacity.

- **Australia.** Though a minor producer between 1965 and 1975, Australia has since become one of the world's leading uranium producers, accounting for 21% of worldwide output in 2003. With proposed expansion of capacity at Olympic Dam, Australia is likely to maintain its position as a leading producer.

- **Canada.** Canada has been a leading producer throughout uranium production history. It accounted for 11% of output in 1965, 18% in 1985 and 29% in 2003. Canada is home to the world's largest known high-grade uranium deposit and is unlikely to relinquish is position as the worldwide leader in uranium production for at least the next 10 years.

- **France and Germany.** Both France and Germany (GDR) were leading producers in 1965, and both maintained prominence through about 1990, when their productions began to decline dramatically. France terminated production in 2001 as economic resources were depleted. After its reunification in 1990, Germany terminated uranium production except for that recovered during remediation of production facilities.

- **Czech Republic.** Production in the Czech Republic declined dramatically after the dissolution of the USSR. It currently has only one operating conventional mine though it continues to recover uranium by ISL during site remediation.

- **United States.** The United States went from being the world's leading producer in 1980, with about 24% of the worldwide total, to just 2% of the total in 2003. The decrease in prominence among worldwide producers came about as US producers closed conventional mines and mills and concentrated on ISL production. Production in the United States could increase in the future, but it is unlikely to ever regain anywhere near the prominence it achieved in 1980.

- **CIS.** The countries of the CIS were collectively a dominant force in uranium production between 1965 and the 1991 dissolution of the Soviet Union. Kazakhstan has since emerged as a key producer, accounting for about 9% of worldwide output in 2003. Because of its large, relatively low-cost ISL-amenable resource base, Kazakhstan could challenge Canada as the world's leading uranium producer within the next 10-15 years.

- **South Africa.** Like the United States, South Africa has experience a significant reduction in uranium output since its production peaked in 1980. South Africa could see some increased production in the near-term as increased uranium and gold prices will justify development of new mines and perhaps additions of uranium processing circuits at more existing operations.

8. NATURAL AND ENRICHED URANIUM INVENTORIES

Until 1990, uranium production exceeded reactor-related uranium requirements. Since then, though requirements have exceeded production (Figure 7.5). By 2003, cumulative production still exceeded cumulative requirements by about 691 400 tU (Figure 7.6). Production of uranium for military purposes is, however, likely included in the cumulative production total, while military material is not reflected in cumulative demand. The exact amount of material that was dedicated to military purposes is not known, nor is it known how much of this material will remain dedicated to military purposes and how much will ultimately become available for civilian use. Therefore, though production exceeded requirements this does not represent a true inventory of excess material that will ultimately be available for civilian use.

As the term was used in the Red Book, "stocks" refers to various categories of within inventories and stockpiles. For example, civilian inventories, most of which are held by utilities, typically include strategic stocks to protect against supply disruptions, pipeline inventory and excess or discretionary stocks that are available to the market. Strategic inventories of natural uranium typically range between one and two years of requirements for utilities and approximately one year for producers. Brokers and speculators also hold inventories to take advantage of market price fluctuations. Governments retain stocks of natural uranium for a variety of strategic reasons, but this material is not well represented in reporting of inventory and stocks in the Red Book, partly for national security reasons. Estimates are made of this material to provide a broad overview of total inventory. Appendix 8.1 provides country by country information on uranium stocks/inventory while Appendix 8.2 summarises uranium stock/inventory polices for selected countries.

History of reporting uranium stocks and inventories

Information on uranium inventories was first reported in the 1976 Red Book, when eight countries reported a total of 82 595 tU equivalent, with governments, producers and users controlling 76%, 20% and 4% of the total, respectively. The 1976 inventory total did not, however, necessarily represent total inventory, as many countries listed "not available" for one or more of the categories, implying that there may have been material in a given category that was not reported. Though uranium stocks are an important source of secondary supply many countries have never provided complete reports of their inventories in the Red Book. Countries that held important inventories consistently were unable to report on this material primarily due to national security or commercial confidentiality reasons. Table 8.1 lists non-depleted uranium inventories reported between 1976 and 2003 along with calculated inventory/requirements ratios. Table 8.2 provides depleted uranium inventories [Reference 8-1]. Figure 8.1 displays the ratio between inventory and requirements.

Table 8.1. Non-depleted uranium inventories and reactor requirements (natural uranium equivalent)

	Total Inventory (tU)*	Requirements (tU)	Inventory/ Requirements ratio
1976	83 000	23 137	3.6
1977	98 617	25 251	3.9
1979	140 303	29 301	4.8
1983	148 985	29 870	5.0
1986	154 386	37 421	4.1
1988	125 930	42 007	3.0
1989	125 717	51 446	2.4
1991	107 286	55 084	1.9
1993	88 124	58 437	1.5
1995	109 177	61 377	1.8
1997	89 678	63 757	1.4
1999	108 122	61 589	1.8
2001	69 033	64 329	1.1
2003	65 103	68 435	1.0

* Including estimated inventories according to stock policy.

Table 8.2. Depleted uranium inventories

Country	(tU)
China, P.R.	2 000
France	190 000
Japan	10 000
Korea	200
Russian Federation	460 000
United Kingdom	30 000
United States	480 000
URENCO (Germany, Netherlands, United Kingdom)	16 000
Total	**1 188 200**

Source: OECD, Management of Depleted Uranium, Paris, 2001.

Figure 8.1. Ratio of non-depleted uranium inventories to requirements (1976-2003)

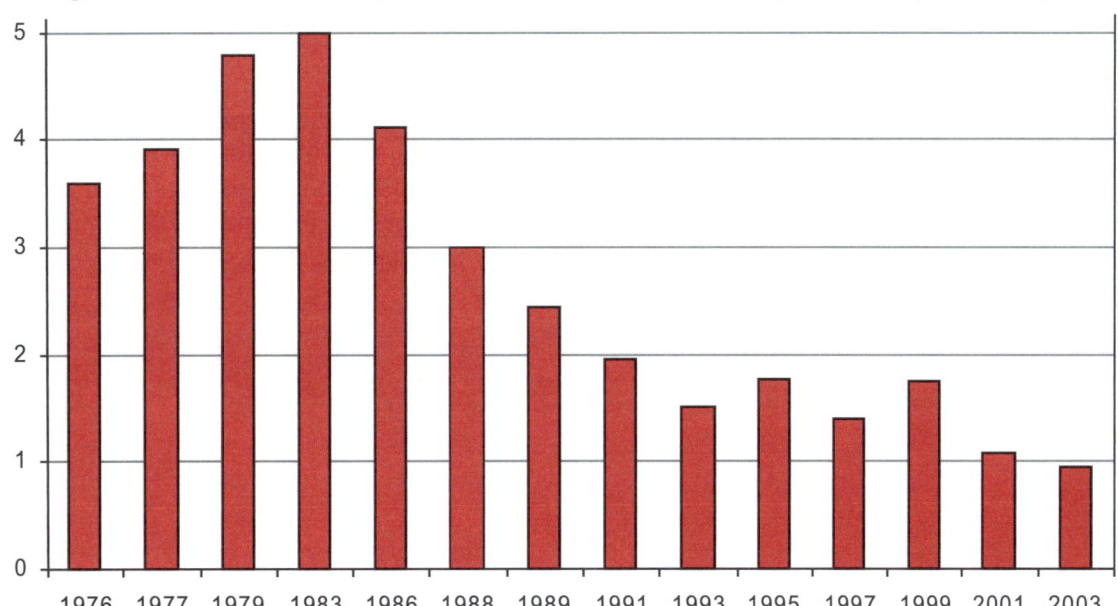

Incomplete reporting of uranium stocks and the need to estimate inventory levels to provide a relatively complete picture precludes any in-depth analysis of stocks and inventory. The number of countries reporting uranium inventories ranged from seven to 17. Trends in the ratio between inventory and requirements were not, however, significantly affected by the number of countries reporting information, probably because estimated total inventories helped offset incomplete reporting. The decline in the ratio between inventory and requirements that began between 1983 and 1986 likely resulted from a combination of inventory drawdown and increasing requirements. The start of this decline approximately coincided with the decline in production that began in 1982. It could also indicate a maturing of the uranium fuel market, moving to more just-in-time deliveries of fuel, such that large inventories were no longer considered necessary. As previously noted inventory drawdown displaced primary supply requirements and contributed to the decline in production in the aftermath of the Three Mile Island and Chernobyl accidents.

Since 1991, the ratio between non-depleted uranium inventory and requirements has averaged about 1.5 varying in a narrow range of between just below 1.0 and 1.9. This stability likely reflects the fact that inventories had reached strategic levels established by utilities and producers to protect against potential supply disruptions.

Conclusions relating to uranium inventories

World cumulative production exceeded requirements by about 691 400 tU between 1945 and 2003. This stock of already-mined uranium, however, does not represent an exact inventory of excess material that will be available for civilian use, but, rather as an upper bound. Nevertheless, natural and enriched uranium stocks and inventories represent a significant resource that has long-term potential to supplement primary supply.

Reporting of stocks and inventories began in 1976. Though the number of countries reporting stocks has increased, reporting is still far from complete. Many countries that hold important inventories remain unable to report on this material primarily due to national security or commercial confidentiality reasons. Therefore, though the Red Book is a source of information on secondary supply in the form of stocks and inventories, because of incomplete reporting caution should be used as to how this information is utilised in making supply-demand projections.

Beginning in the early 1980s, drawdown of inventories of natural and enriched uranium held by utilities has reduced primary supply requirements. As a consequence, the ratio between non-depleted uranium inventory and requirements has decreased from 5.7 in 1983 to just less than 1.0 in 2003. The ratio varied within a narrow range at about 1.5 since 1991, suggesting that utility-held inventories are approaching strategic levels, with much less discretionary material available to the market.

9. UNCONVENTIONAL URANIUM RESOURCES

Distinction between conventional and unconventional resources

Prior to the 2003 Red Book uranium resources were broadly classified as either "conventional" resources or "unconventional" resources. Conventional resources were defined as resources from which uranium is recoverable as a primary product, a co-product or an important by-product (e.g. from the mining of copper and gold). Very low-grade resources or those from which uranium is only recoverable as a minor by-product were termed "unconventional resources".

The distinction between conventional and unconventional resources is not entirely clear cut, but is instead somewhat transitional. For example uranium recovered as a by-product of gold mining operations in South Africa is considered as coming from conventional resources, even though the average ore grade processed in South Africa in 1985 and in 1990 was about 0.015% U. The uranium content in phosphate rock ranges between 0.001 – 0.07% U and averages approximately 0.01% U, not a great deal different from the South African gold-uranium deposits. Even so, uranium recovered from phosphate rocks has been typically classified as coming from "unconventional" resources (a few deposits, where uranium is associated to phosphates, are considered as conventional: Itataia in Brazil, Bakouma in Central Africa).

In early Red Book editions, as the industry was attempting to gain an understanding of the availability and/or adequacy of uranium resources, the distinction between conventional and unconventional resources was not made. In the 1967 Red Book for example, uranium contained in phosphate deposits and deposits in the "other" category[7] were included in the tabulation of Reasonably Assured and Estimated Additional Resources.

By 1970, however, the distinction between conventional and unconventional resources was beginning to take shape, but it was still in the evolutionary stage. Resources in Morocco, which were almost entirely hosted in phosphate deposits, were not included in the resource estimate; however, resources associated with the low-grade Ranstad Shale in Sweden were included. The conventional-unconventional resources distinction has been adhered to since the mid-1970s. Though emphasis clearly was placed on conventional resources, Red Books through 1993 continued to include tables that listed "unconventional and by-product resources"[8] to ensure that information on their availability was not lost. The 1976 Red Book, noting the "enormous quantities" of uranium in unconventional resources, called for more exploration in geologic environments likely to host resources in the 0.01 – 0.10% U_3O_8 grade range.

---

7. Other resources included uranium contained in pegmatite, schist, urano-thorite, copper leach by-product, hyper-alkaline silicates, lignite, monazite and volcanics.

8. Unconventional resources are subdivided into phosphates, non-ferrous ores, carbonatites, and black schist and lignite.

Recovery of uranium from seawater

In the 1965 Red Book, recovery of uranium from seawater was discussed under a major heading comparable to the headings accorded to uranium and thorium. The conclusion in that first Red Book was that uranium resources associated with the world's oceans total about 4 000 million tU, and there was no intrinsic reason why some of that resource should not be extracted from various coast lines at a total rate of thousands of tonnes annually. Thirty-six years later, the estimated resource potential of seawater remained unchanged at 4 000 million tU. Economics, however, have not improved measurably for recovery of uranium from seawater. It is estimated that it would require the processing of about 354 000 tonnes (342 million litres) of seawater with an average U content of 4 parts per billion to produce one kgU. Laboratory scale research conducted in Japan indicated costs of about USD 300/kgU for recovery from seawater, still several times the current market price.

Distribution of unconventional resources by source

Table 9.1 summarises ranges of unconventional resources reported in Red Books dating between 1965 and 1993. Red Books subsequent to 1993 included only summary information because few countries reported on unconventional resources and totals from year to year remained relatively unchanged. According to their current definition, unconventional resources have very low uranium content, and typically uranium is only recoverable as a minor by-product. Therefore, potential recovery of uranium from unconventional resources must take into account the economics and market trends of the primary commodity or alternatively very large operations in which economies of scale partially offset the low-grade of the ore.

Table 9.1 is not all inclusive as a listing of worldwide unconventional resources. For example, the 2003 Red Book stated that resources contained in phosphate deposits potentially total 22 million tU, compared to the range of 7 006 000 to 7 271 000 tU identified in Table 9.1. Additionally, the Chattanooga and Ronneburg black shales in the United States and Germany, respectively, which have combined resources totalling 4.2 million tU, are not listed. Neither are large uranium resources associated with monazite-bearing coastal sands in Brazil, India, Egypt, Malaysia, Sri Lanka and the United States. Though these resource totals represent more of a mineral inventory than rigorous resource estimates, they nevertheless call attention to the fact that extensive potential sources of unconventional uranium resources are not accounted for in Table 9.1.

Phosphate deposits

Uranium contained in phosphate deposits clearly dominates unconventional resources. As previously noted, worldwide totals of phosphate-related resources could total as much as 22 million tU. Exploitation of these resources depends on higher uranium prices than were available between 1981 and 2003 and, as a consequence, production from this source was discontinued in 1999. Annual uranium production capacity from this source primarily depends on the phosphate market. Recovery of uranium from phosphates is, however, based on well known technology that was in use until 1998.

Worldwide production totals for uranium recovery from phosphate deposits are not available. However, it is known that about 690 tU were recovered from processing Moroccan phosphate rock in Belgium, about 17 150 tU were recovered in the United States from Florida phosphate ores during the period from 1975 to 1999. In 1980 there were eight plants recovering uranium from phosphoric acid manufacturing in the United States alone. About 6 000 tU were recovered from processing marine organic deposits (essentially concentrations of ancient fish bones) in Kazakhstan. A small amount of uranium was also recovered in Canada from processing US phosphate rock. When uranium prices reach USD 100-120/kgU[9], by-product recovery of uranium from phosphate deposits becomes economical and could experience a rebirth, perhaps even exceeding production from gold mining operations. It is, therefore, possible that a combination of economics and worldwide production capacity may transform uranium resources associated with phosphate deposits into "conventional" resources simply because they have become an important, economically viable source of uranium.

Table 9.1. Unconventional uranium resources (1 000 tU) reported in Red Books 1965-1993

Country	Phosphates	Non-ferrous ores	Carbonatite	Black schist, lignite
Brazil	28.0-70.0	2	13	
Chile	0.6-2.8	4.5-5.2		
Colombia	20.0-60.0			
Egypt*	35.0-100.0			
Finland			2.5	3.0-9.0
Greece	0.5			4
India	1.7-2.5	6.6-22.9		
Jordan	100-123.4			
Kazakhstan**	58			
Mexico	100-151	1		
Morocco	6 526			
Peru	20	0.14-1.41		
Sweden				300
Syria	60-80			
Thailand	0.5-1.5			
United States	14.0-33.0	1.8		
Venezuela	42			
Vietnam				0.5
Total (rounded)	**7 006 - 7 271**	**16 - 34**	**15**	**308 - 314**

* Includes an unknown quantity of uranium contained in monazite.

** Production of estimated 6 000 tU between 1959 and 1992 has been deducted from reported total.

Black schist and lignite

Though listed as a single category, black schist and lignite are two distinct deposit types. In addition, black "schist" is an incomplete designation, which more properly should be "black shale and black schist". The Ranstad black shale in Sweden dominates the "black schist" category in Table 9.1. Approximately 200 tU were recovered from bituminous shale in the 1960s.

As previously noted, uranium associated with the Chattanooga Shale in the eastern United States and Ronneburg black schist and black shale in Germany are not included in Table 9.1. The Chattanooga Shale is projected to host 4-5 million tU at an average grade of 0.006% U. However, in addition to its very low grade, the environmental footprint of extracting such low-grade

9. As of 1 April 2006, the spot price of uranium was USD 105/kgU (Ux Consulting Compagny, LLC, www.uxc.com).

resources (very large mines and processing plants in populous areas) is a critical deterrent to their development. At the same time, its potential should not be entirely overlooked as new technology could make it viable at higher uranium prices.

The same could be said for the Ronneburg shale/schist units near the Germany-Czech Republic border. Significant resources remain in these deposits (estimated at 169 230 tU), but the Ronneburg area is part of the Wismut multi-billion US dollar environmental remediation programme. Proposals to resume mining in this area would no doubt meet strong opposition on environmental grounds. At a minimum, however, the large mineral inventory associated with the Ronneburg deposits should not be overlooked for the future.

Lignite and sub-bituminous coal with elevated uranium concentrations have been exploited in Williston Basin of North and South Dakota in the United States, the Freital coal deposits near Dresden in Germany, the Ily Basin in eastern Kazakhstan and the Yining area of north-western China. A total of 3 700 tU was produced from the Freital coal during 1947 to 1955 and 1968 to 1989. No information is available on production from the other areas. None of these resources are listed in Table 9.1. At best, uranium resources associated with lignite should be considered only a very low-grade, high-cost mineral inventory. The lignite must first be burned at high temperature and then the uranium, which is in the form of a highly refractory silicate, is leached from the ash.

Uranium from non-ferrous ore

Uranium resources associated with non-ferrous ore are estimated to range between 16 000 and 34 000 tU. Most of these resources are associated with the potential to process tailings from copper mining and processing operations to recover uranium as by-product. Uranium has been recovered from copper concentrator operations in India, South Africa and the United States. Uranium resources recoverable as a co-product of copper at Olympic Dam in Australia are listed as conventional resources.

Carbonatites and igneous rocks

Uranium resources associated with carbonatites total 15 000 tU and are associated with the Araxá deposit in Brazil and the Sokli deposit in Finland (Table 9.1). Large acidic granite batholiths, such as the Conway granite in the eastern United States contain potentially large resources, but at grades in the range of 0.015% U. Recovery of these resources will require very high uranium prices and will likely face opposition because of the large environmental impact such operations will entail.

Monazite

Monazite (Ce, La, Y, Th)PO4 placer deposits are currently being exploited in various countries, e.g. Australia, Brazil, India, Malaysia and Sri Lanka to recover rare earths, tin, thorium and zircon. Some of these placer deposits as well as those in Egypt, the Republic of Korea and the United States contain elevated uranium values (0.01% to 0.1% U) and are a potential future source of uranium. More importantly, these deposits hold large amounts of thorium, which has potential use as a nuclear fuel. Chapter 10 gives additional information on thorium.

Conclusions regarding unconventional resources

Unconventional uranium resources are characterised by having very low uranium content; uranium is typically only recoverable as a minor by-product from unconventional resources.

Historically phosphate deposits are the only unconventional resource from which significant quantities of uranium have been recovered. When uranium prices reach USD 100-120/kgU, by-product recovery of uranium from phosphate deposits becomes economically viable and could again become an important, competitive source of uranium.

Though, because of high extraction costs associated with their very low grades, they have not accounted for significant production totals in the past, information on unconventional resources should be preserved. A combination of new technology and sharply higher uranium prices may make them potentially viable resources in the future.

10. THORIUM[10]

Reporting history

The first Red Book, published in 1965, was titled World Uranium and Thorium Resources. Information on thorium resources was published in Red Books between 1965 and 1982, typically using the same terminology used for uranium resources, e.g. Reasonably Assured Resources (RAR) and Estimated Additional Resources (EAR). With the separation of EAR into EAR – Category I and EAR – Category II in 1983, thorium resources were correspondingly reported as EAR-I and EAR-II in the 1983 and 1986 Red Books. No revised estimates have been published in Red Books since 1986. An overview of thorium deposits and resources was published 1991.

Thorium as nuclear fuel

The average content of thorium in the upper earth's crust reaches 6-10 parts per million, three times the average content of uranium. Thorium is widely distributed in rocks and minerals, usually associated with uranium, elements of the rare-earth group and niobium and tantalum in oxides, silicates and phosphates. In vein deposits, thorium may be present as thorite (thorium-silicate) or thorianite (thorium oxide). The Ce-La-Y-phosphate mineral monazite, with a thorium content of 8-10% is the source of most thorium currently being produced and is the most commercially interesting source of thorium for the future.

Though the isotope ^{232}Th itself is not fissile, it is fertile and through interaction with neutrons the fissile isotope ^{233}U is formed. Because no ^{239}Pu is generated, thorium may be preferable to uranium as a source of nuclear fuel under certain conditions.

Thorium-based fuel cycles have been investigated in the United States, Germany, Russia, India, Japan and the United Kingdom [References 10-5 and 10-6]. Reactors that utilise thorium include high-temperature gas-cooled reactors (HTGR) and "pebble-bed" reactors (THTR), developed and built in the 1960s and 1970s in Germany (Jülich and Schmehausen) and in the United States (Peach Bottom and Fort St. Vrain). In the late 1980s both Germany and the United States decided to shut down their thorium-fuelled reactors. Experimental thorium-fuelled reactors were operated in the United Kingdom (Dragon) and in India. In high-temperature reactors, helium is used as the cooling gas. Temperatures as high as 800-1 000°C are used for generating electricity and for chemical processes (e.g. gasification of coal). However, high costs of fuel fabrication and unsolved technical problems have slowed further developments. Publications in the 1970s indicated a 1 000 MW reactor requires an initial loading of ~40 tTh and ~10 t of highly-enriched U (~90% ^{235}U) and annual re-loadings of ~10 tTh.

---

10. The proceedings of the IAEA Technical Meeting on "Fissile Material Management Strategies for Sustainable Nuclear Energy", September 2005, were used extensively in the preparation of this Chapter [Reference 10-1].

For countries having limited access to uranium resources thorium-fuelled reactors remain a future option, and research for advanced thorium reactor types continues in some countries. Recently, thorium has been tested in a molten-salt concept in Japan. In India, where there are large resources in coastal placer deposits, thorium is used in Kakrapar and is planned to be used with HEU and Pu as a major source of nuclear fuel in the future [Reference 10-7].

Major thorium deposit types

Classification of major types of thorium deposits is based on their genesis and on descriptive characteristics. In general, thorium deposits can be divided into those associated with the magmatic cyle (endogenous deposits) and those of sedimentary origin (exogenous deposits) [References 10-2 and 10-3]. Endogenous deposits include:

- granites and pegmatite (Jos, Nigeria; Bancroft, Canada);

- alkaline rocks, e.g. nepheline syenite (Ilimaussaq, Greenland; Langesundfjord, Norway; Ulug Tansek, Russian Federation);

- hydrothermal veins (Wet Mountains, Powderhorn, United States; Steenkampskaal, South Africa);

- carbonatites (Araxá, Brazil; Mountain Pass, United States; Palabora, South Africa).

Exogenous deposits include:

- alluvial placers in river valleys (North and South Carolina, United States);

- coastal placers (Kerala, India; Bahia, Brazil);

- ancient, metamorphosed placers (Blind River, Canada).

Thorium resources

In the early 1970s, orders for high temperature reactors that can operate on the thorium fuel cycle prompted assessment of worldwide thorium resources. Commercial interest in the thorium fuel cycle did not, however, develop and as a consequence there has never been a large market for thorium. Because thorium has had a limited market, there has been little incentive to explore for new thorium deposits or to develop detailed information on currently known thorium deposits. However, some experts (see references) continued to provide resource estimates after 1986, the highest of which totals 6.1 million tonnes thorium including undiscovered resources.

Worldwide thorium resources, which are listed by major deposit type in Table 10.1, are estimated to total about 6 078 000 tTh though no economic potential is implied for the resources listed. Additionally, the resource totals include a broad range of confidence levels from RAR to Speculative Resources.

Table 10.1. Thorium resources by deposit type

Major deposit type	Resources (1000 t Th)	Percentage
Carbonatite	1 900	31.3
Placer	1 500	24.7
Vein-type	1 300	21.4
Alkaline rocks	1 120	18.4
Other	258	4.2
Total	**6 078**	**100.0**

Table 10.2 lists uranium resources on a country-by-country basis. Listing of thorium resources by country is, however, subjective due to variability in data quality – some data are more than 20 years old. Accordingly, confidence levels for the estimates and their economic attractiveness are not considered in Table 10.2. Instead the resources listed in Tables 10.1 and 10.2 represent a worldwide resource base; a starting point from which to determine thorium resource distribution and recovery potential.

The World Nuclear Association, referring to reserves in the United States Geological Survey, Mineral Commodity Summaries, lists worldwide resources totalling 1 200 000 tTh [Reference 10-5]. The same figures are reported in [References 10-8 and 10-9], and are supplemented with additional resources to account for a reserve base of 1 400 000 tTh. Thorium resources in the Commonwealth of Independent States totalling 1 700 000 tTh were reported in a 1997 publication of the IAEA [Reference 10-4].

Table 10.2. Estimated thorium resources by country

Country	Thorium resources (1 000 t Th)
Brazil	1 306
Turkey	880
United States	432
Australia	340
India	319
Egypt	295
Norway	180
Canada	173
South Africa	115
Commonwealth of Independent States	1 650*
Other	388
Total	**6 078**

* Adjusted for "off-grade". About 53 000 tTh in the CIS are "off-grade" [Reference 10-4]. Deposits with grades between 0.1 and 2% Th, that total 75 000 tTh, are economically interesting and estimated as Reasonably Assured Resources.

Most thorium resources have been discovered and evaluated during exploration for uranium, rare earth elements (e.g. monazite in coastal placer deposits) or tantalum, niobium etc. Estimates for resources and production costs of thorium resources have been included in Red Books dating between 1965 and 1986. Thorium resources were typically limited by a maximum production cost (e.g., in 1965 resources were listed in a cost range of USD 5 to 10/lb. of ThO_2; in 1979 and 1986 resources projected to be recoverable at <USD 75/kg Th and in <USD 80/kg Th), respectively were listed.

Table 10.3. World thorium resources by confidence and cost categories (1 000 tTh)

Country	RAR <USD 80/kg Th (t Th)	EAR-I <USD 80/kg Th (t Th)	EAR-II* (t Th)
Turkey	344	NA	400-500
India	319	NA	NA
Brazil	171	50	329-700
United States	122	278	274***
Russian Federation**	75	NA	NA
Greenland	54	NA	32***
South Africa	18	NA	130****
Australia	13	<1	300***
Venezuela	NA	300	NA
Norway	NA	132	132
Egypt	NA	100	280***
Canada	NA	44	128***
Others	23	10	81
Total	**1 139**	**914**	**2 086-2 557**

*	Costs of recovery not available.
**	Estimate based on Reference 10-4.
***	Earlier estimate.
****	Preliminary estimate in 1983.

Worldwide thorium resources in the RAR, EAR-I and EAR-II categories total 4.1-4.6 million tTh (Table 10.3). This represents about 67% of total worldwide thorium resources, including those unspecified as to confidence category or production cost (Table 10.2). Differences among various resource estimates are the result of the approaches or the resource categories used in the estimates. For example, total unspecified resources are higher than the resources listed in Table 10.3 because they include resources recoverable at costs higher than USD 80/kg Th and resources in categories of lower degrees of confidence than EAR-II (e.g. Speculative Resources). Therefore, resource totals in Australia, Brazil, South Africa and the CIS are higher in Table 10.2 than in Table 10.3, mainly because they include resources in placer deposits and other deposits not considered in Red Book estimates.

Recovery of thorium

Thorium is often associated with other minerals that can be economically mined as the primary commodity, with thorium being recovered as a co-product or by-product. The majority of the costs for mining and milling would generally be borne by the principal commodity. Recovery from ores where thorium is the principal or single beneficial element would generally be restricted to specific circumstances, e.g. high demand or very high-grade deposits.

Placer-type deposits are the principal sources for thorium, mainly concentrations of heavy (rare-earth) minerals in coastal sands, from which monazite and other Th-bearing minerals are recovered.

Rare-earth ores are recovered for their content of cerium, lanthanum, yttrium and other elements used in catalysts, ceramics, televisions, computer industry etc. Placer deposits have varying grades of valuable minerals; generally only those with concentrations measured in several percent are of economic interest.

Production data for monazite, which contains 8-10% Th, can be useful as a theoretical measure of thorium availability. Worldwide monazite production over the last five years has fluctuated between 5 500 and 6 500 t annually [References 10-10 and 10-11] from which theoretically between 300 and 600 tTh could be recovered annually. India, which accounts for about 90% of worldwide monazite output annually, is investing in the thorium fuel cycle and has completed comprehensive assessments of its domestic resources [Reference 10-5]. Other monazite-producers include Brazil, Malaysia and Sri Lanka (past producer). Production has also been reported from China, Indonesia, Nigeria and the CIS.

Carbonatites and alkaline syenites are potential economic sources for niobium and tantalum. They also often contain thorium concentrations of a few tenths to a few hundredths of a percent. A typical example of a carbonatite with commercial monazite content is Araxa in Brazil, which is currently in production. The Ulug Tansek deposit in the Russian Federation is an example of a potentially economic monazite deposit associated with an alkaline syenite [Reference 10-4]. Worldwide a few tens to a few hundred t Th could theoretically be recovered annually from mining and processing of ore from deposits hosted in carbonatites and alkaline syenites.

Vein deposits with anomalously high thorium content also have economic potential if other commodities are recoverable as co- or by-products. In the 1970s vein deposits in the United States were evaluated for their economic significance. The largest low-cost reserves (~ 100 000 tTh) are located in the Lemhi Pass (Idaho/Montana) and Wet Mountains (Colorado) deposits. In 1979 their costs of recovery were estimated to be <USD 40/kgTh [References 10-2 and 10-3]; no commercial operations have ever been developed at either of these deposits.

Stockpiled thorium supplemented with thorium recovered as a by-product of placer and carbonatite mining operations is adequate to satisfy current worldwide demand, which is estimated at a few hundred tonnes annually. Resources are adequate to satisfy future demand even if requirements were to increase substantially. Increased output could be obtained as a by-product of existing or newly installed facilities.

11. MINE START-UP AND CLOSURE HISTORIES

The initial motivation behind the concept of the Red Book was concern about the adequacy of uranium resources to sustain the nuclear fuel cycle. By 1970, the importance of adequate production capacity along with adequate resources was acknowledged when capacity information was first published in the Red Book in individual country reports. The 1973 Red Book expanded the reporting of production capacity by summarising capacity as of 1973 and also listing projected capacities for 1975 and 1978. Subsequent Red Books continued to expand the time frames included in future projections of production capacity; the 1988 Red Book included capacity projections between 1990 and 2030.

As would be expected under market-based conditions, projected reactor-related uranium requirements were the driving force behind projections of uranium production capability. It was tacitly assumed that production capacity would expand in proportion to requirements for newly mined and produced uranium, after adjustments were made to accommodate for the impact of secondary supply such as inventory drawdown and HEU. Delays in project development were well known to individual project owners and operators, but they were not widely discussed as possible deterrents to security of supply. The Uranium Group and the Red Books were not, however, silent on the issue of potential delays in project development. The 1976 Red Book referred to "…politico-economic, social and environmental constraints limiting…freedom to develop, produce and export from identified deposits". Similarly the 1978 Red Book noted that environmental review processes: "tend to be time consuming and, even assuming favourable results, can considerably delay development of projects".

The 1995 Red Book included a description of the environmental review process for new mine development in Canada. Though potential delays in project development relating to more stringent licensing procedures were indirectly mentioned, their implications were not directly stated. The 2001 Red Book was the first to formally acknowledge that stricter environmental and safety standards have the potential to contribute to substantially longer lead times for developing new mines and they need to be considered in projecting future uranium production capability. The 2001 Red Book was also the first to directly link delays in new project development brought about by protracted environmental and safety review procedures with the potential for extended production shortfalls leading to destabilisation of the uranium market and significant upward pressure on uranium prices.

The 2003 Red Book took the potential impact of delays in project development a step further by comparing the timeframes between the start of exploration and discovery of new deposits and between discovery and the start of production for several examples around the world. This comparison showed that uranium production, which starts with exploration and extends through project development, is the culmination of a process that can take decades.

For example, the Jabiluka deposit was discovered in 1971 and is still not in production. National politics, opposition to uranium mining and more that two decades of depressed market prices have combined to delay development of a deposit with resources estimated at 198 000 tU at an average grade of 0.44% U with a projected annual production capacity of 2 290 tU. While Jabiluka is an extreme case, there are deposits that are currently producing for which the time between discovery and the startup of production exceeded 20 years (Table 11.1).

This table spans the modern history of uranium exploration and mining, from the 1940s through the projected startup of Cigar Lake in Canada in 2007 and provides valuable perspective regarding the difficulty of discovering new uranium deposits as well as the time needed to bring them to production. Exploration in the Cigar Lake area began in 1969; discovery of ore grade mineralisation came 12 years later in 1981. The time between the start of exploration and discovery has varied considerably, ranging from one year in the case of Ranger to between 10 and 20 years elsewhere. Discoveries of new deposits will likely become more difficult, more costly and more time intensive, as remaining exploration targets are likely to be deeper or in more complex geologic environments. However, new analytical models and exploration techniques could alleviate some of these difficulties.

A number of factors control the lapsed time between the start of exploration and discoveries – market price, availability of exploration capital, regulatory requirements and political stability – to mention a few. Missing from these tables are the exploration programmes that collectively cost tens of billions of dollars, which were not successful. Revenue from the discoveries must cover the cost of the failures, underscoring the risks inherent in minerals exploration that must be recognised in the market price.

The message to be gained is that exploration for and development of new deposits takes time – time for exploration, time for licensing and time for development and construction. These time frames are likely to expand as exploration targets become more elusive and as environmental and safety requirements become more stringent. To ensure that relatively low-cost resources and production capacities are adequate to maintain a long-term balance between uranium supply and demand, the industry must build adequate timeframes into production planning. In addition, the uranium market must recognise that the inherent risks of exploration and project development require a market price that is sufficiently high to account for the risk of failure and to provide incentives for explorers and producers to undertake those risks.

Table 11.1. Key dates in the development of selected mines

Country	Deposit / mine	Exploration begins	Discovery of deposit	Beginning of production	Mining method	End of mining
Australia	Beverley	1968	1970	2000	ISL	
	Honeymoon	1968	1972	not yet announced	ISL	
	Jabiluka	1968	1971	not yet announced	UG	
	Olympic Dam	early 1970s	1976	1988	UG	
	Ranger	1968	1969	1981	OP	
Brazil	Lagoa Real	1974	1981	2000	OP	
Canada	Cigar Lake	1969	1981	not before 2007	UG	
	Cluff lake	1960s	1975	1980	OP/UG	2002
	Key Lake	1968	Gaertner: 1975	1983	OP	1999
			Deilmann: 1976	1989	OP	
	McArthur River	1981	1988	1999	UG	
	McClean Lake	1974	1979	1999	OP	
	Rabbit Lake	1965	1968	1975	UG	
Czech Republic	Jachymov	1945	19th century	1946	UG	1964
	Pribram	1947	19th century	1950	UG	1991
	Rozna	1954	1954	1957	UG	2005
	Straz	1965	1965	1967	ISL	1996
France	Bellezanne	1946	1966	1975	OP/UG	1992
	Ecarpière	1950	1952	1957	OP/UG	1990
	Fanay	1946	1949	1953	OP/UG	1992
	La Commanderie	1950	1954	1955	OP/UG	1990
	Le Bernardan	1955	1964	1977	OP/UG	2001
	Le Chardon	1950	1957	1957	OP/UG	1991
	Margnac	1946	1949	1954	OP/UG	1995
	Mas D'Alary	1957	1958	1978	OP	1985
	Mas Lavayre	1957	1964	1978	UG	1997
Gabon	Mounana	1955	1956	1961	OP/UG	1975
	Oklo	1955	1968	1970	OP/UG	1999
Germany	Culmitzsch	1948	1948	1951	OP	1965
	Freital	1946	1946	1947/1968	UG	1955/1989
	Johanngeorgenstadt	15th century	1945	1946	UG	1958
	Koenigstein	18th century	1963	1963/1971	UG/ISL	1983/1990
	Lichtenberg	1948	1958	1958	OP	1977
	Niederschlema/	15th century	1946	1949	UG	1990
	Oberschlema	15th century	1946	1949	UG	1958
	Paitzdorf	1948	1950	1950	UG	1990
	Schmirchau-Reust	1948	1950	1950	UG	1990

Table 11.1. Key dates in the development of selected mines (cont'd)

Country	Deposit / Mine	Exploration begins	Discovery of deposit	Beginning of production	Mining method	End of mining
Hungary	Mecsek	1952	1954	1956	UG	1997
Kazakhstan	Inkay	1976	1979	2001	ISL	
	Kanzhugan	1972	1974	1988	ISL	
	Melovoye		1956	1959	OP	1993
	Moynkum		1976	2001	ISL	
	Mynkuduk	1973	1975	1987	ISL	
	Uvanas	1963	1969	1977	ISL	
	Zaozernoye		1955	1961	UG	1990
Namibia	Rössing	1966	1973	1976	OP	
	Langer Heinrich	1970	1973	2006 (projected)	OP	
Niger	Abkorum	1956	1979			
	Akouta	1956	1972	1978	UG	
	Arlit	1956	1965	1971	OP	
	Ebba	1956	1982			
	Imouraren	1956	1977			
	Techili	1956	1988			
Romania	Avram Iancu/Bihor	1950	1954	1962	UG	1999
Russia	Antei		1964	1975	UG	
	Dalmatovskoye		1975	1999	ISL	
	Luchistoye		1966	1972	UG	
	Martovskoye	1964	1964	1974	UG	1995
	Oktyabrskoye	1967	1967	1974	UG	
	Streltsovskoye		1963	1969	UG	
	Tulukuevskoye		1965	1968	OP	
Ukraine	Michurinskoye	1964	1967	1971	UG	
	Severinskoye	1968	1978			
United States	Crow Butte	Late 1970s	1980	1991	ISL	
	Lucky Mc		1953	1954	OP	1988
	Shirley Basin		1957	1959	UG/ISL/OP	1992
	Smith Ranch	1966-1967	1967	1977 1997	UG ISL	1978
	Jackpile-Paguate	Late 1940s	1951	1953	OP	1982
	Highland	1967	1968	1973 1988	OP/UG ISL	1983
	Mt. Taylor	1968	1970	1986	UG	1989
Uzbekistan	Kostcheka	1961				
	Sugraly	Early 1950s		UG :1977 ISL 1958	UG/ISL	UG: 1994
	Uchkuduk	Early 1950s			OP, UG	1994

Figure 11.1 shows graphically the time elapsed between discovery and the start of mining for deposits listed in Table 11.1. This time has been variable during the past 50 years, but it has been constantly increasing, from a few years in the early 1950s, to as much as 20-30 years at the end of the 20th century.

Prior to 1975, the time between discovery of a new deposit and the start of mining was less than 10 years, after which there was a significant increase. At least three factors contributed to this increase: (i) the easiest deposits to develop had already been put into production by 1975; (ii) increasingly stringent environmental constraints and regulations have added significantly to the timeframe between discovery and the start of mining; (iii) the impact of generally depressed uranium market prices during the past three decades has contributed to delays in starting new mines as has competition from secondary supply such as inventory drawdown and HEU from dismantling of nuclear weapons, which reduced demand for primary supply.

Figure 11.1. History of elapsed time between discovery and start of mining for all mining methods

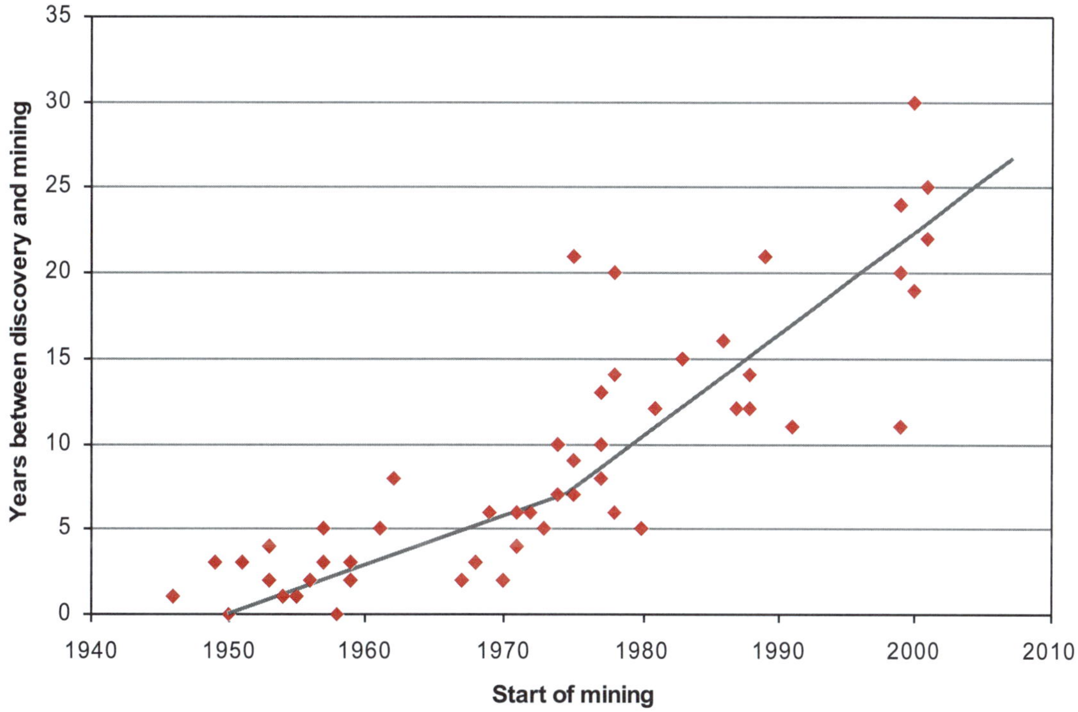

ISL mines

Nine ISL mines were put into operation between 1967 and 2001. The time between deposit discovery and the start of mining varied from two years (Straz, 1967) to 30 years (Beverley, 2000) (Figure 11.2). The time between deposit discovery and start of mining for recently developed ISL operations is about 25 years. Most of the reasons for the lengthening of elapsed time noted above for all mining methods also apply to ISL operations.

Figure 11.2. History of elapsed time between discovery and start of mining for ISL mines

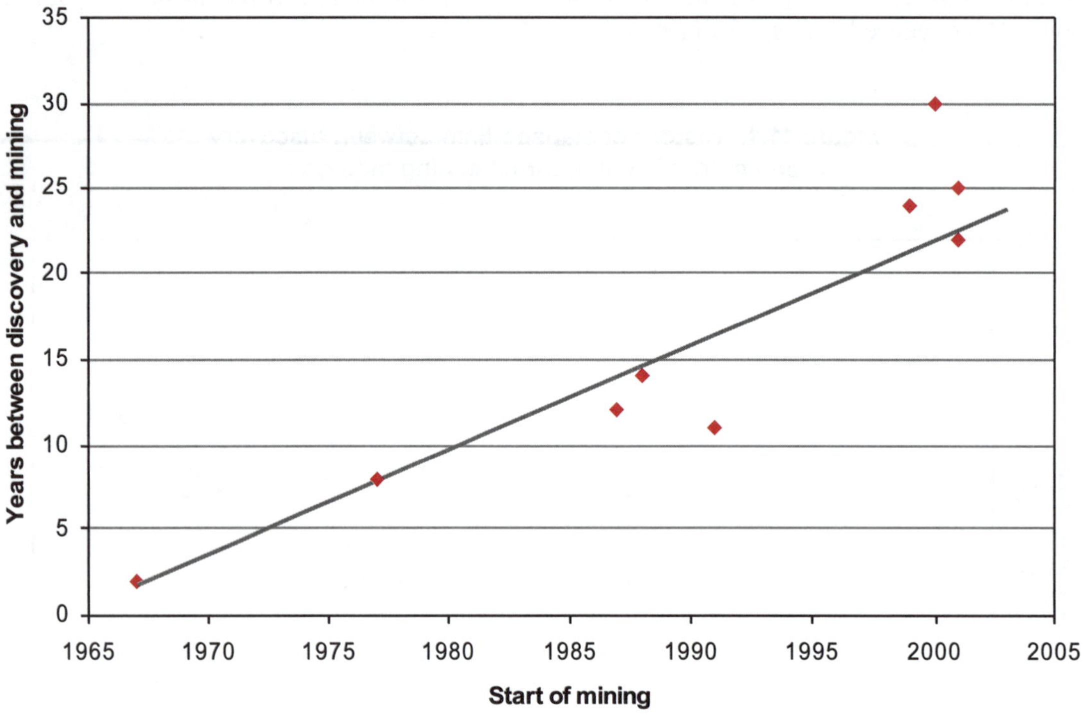

Open-pit mines

Between 1951 and 2000, 16 open-pit mines were put into operation, with the time between discovery and start of mining varying from less than one year (Lichtenberg, 1958) to 21 years (Key Lake Deilmann pit, 1989) (Figure 11.3). Projects where mining started before 1975 were put into operation less than six years after discovery; after 1975, the time between discovery and start of mining exceeds 12 years. For the most recent open-pit mines to be developed, McClean Lake in Canada and Lagoa Real in Brazil, the time between discovery and start of mining was 20 and 19 years, respectively.

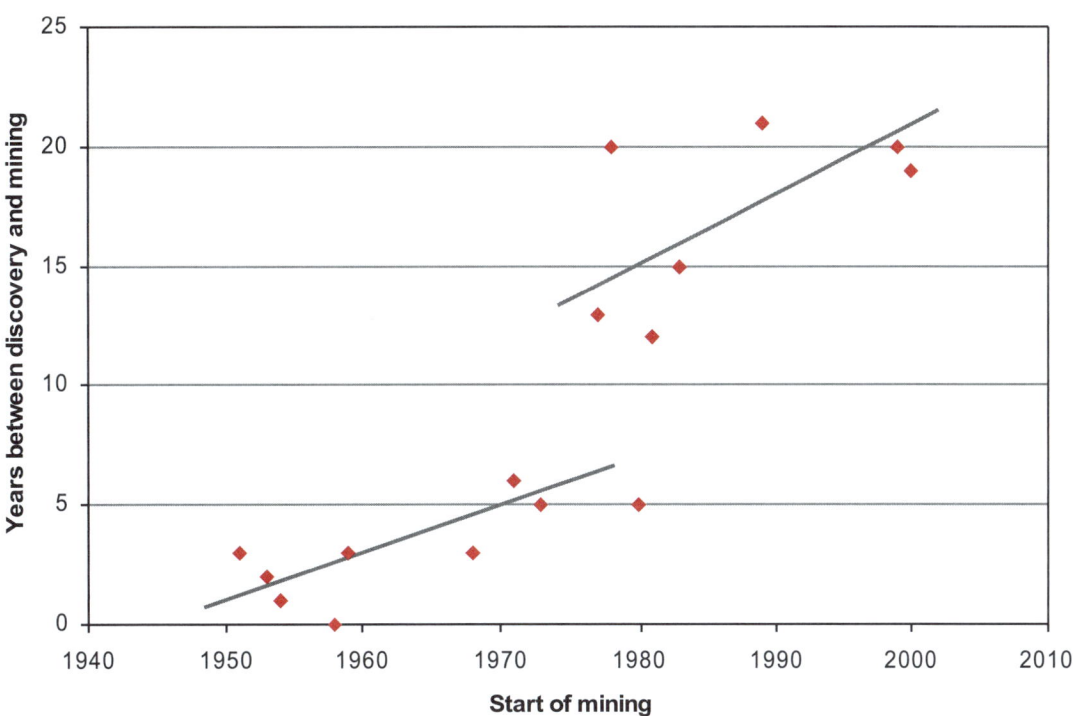

Figure 11.3. History of elapsed time between discovery and start of mining for open-pit mines

Underground mines

A total of 27 underground mines were put into operation between 1946 and 1999, with the elapsed time between discovery and start of mining varying from less than one year (Paitzdorf, 1950) to 21 years (Antei, 1975) (Figure 11.4). Only three major underground mines have started operations during the past 20 years, with an average time of 13 years between discovery of their respective deposits and start of mining. Underground mining at the Cigar Lake deposit in Canada, which was discovered 1981, is expected to began in 2007, 26 years after discovery.

Figure 11.4. History of elapsed time between discovery and start of mining for underground mines

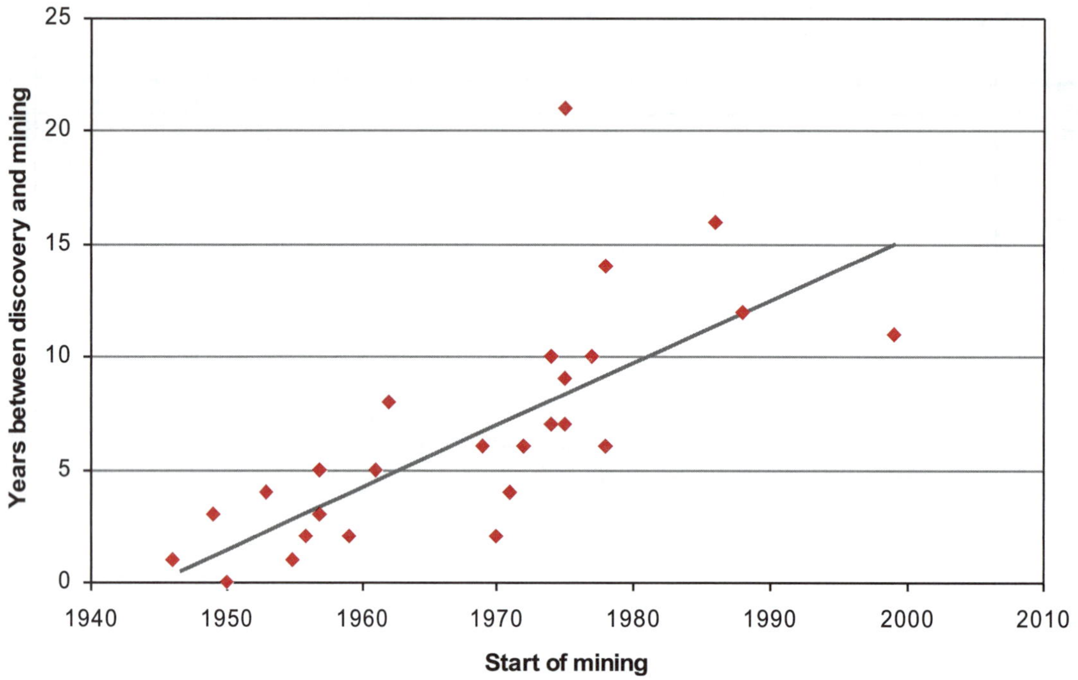

12. ENVIRONMENTAL ASPECTS OF URANIUM MINING AND PROCESSING

There are two different aspects of the environmental impact of the nuclear fuel cycle. As is noted in the 1989 Red Book, nuclear power offers an alternative to increasing emission of greenhouse gases. As part of the nuclear fuel cycle, uranium mining and processing contributes to the benefits of nuclear power. There is, however, the other environmental aspect of uranium production that is more on the public's mind, namely the physical impact of mining and disposal of the waste products from processing uranium ore.

As early as the mid-1970s, the Red Book had already begun to raise awareness about the impact of environmental requirements and review processes on development costs and lead times. The 1983 Red Book correctly noted that, "Most such requirements have resulted from an increased awareness of the impact that mining and ore processing activities can have on the natural environment." The growing consequences of public environmental awareness were accompanied by requirements that the uranium industry assess potential environmental consequences of uranium mining and design processing systems to mitigate those consequences. These requirements in turn led to a growing body of science for environmental management of uranium mining and processing and an increased exchange of information among industry participants. The status of management of waste generated by the uranium mining and milling industry was well documented in a joint IAEA and NEA 1982 symposium [Reference 12-1].

In keeping with the growing importance of environmental awareness in uranium production, a section was added to the Red Book in 1995 on radiation safety and environmental aspects of uranium mining and production. This section documented environmental activities by uranium mining companies in four main areas:

- Rehabilitation of mine and mill sites no longer in operation, in many instances where project operators no longer exist and where legal provisions for proper decommissioning and rehabilitation were insufficient. Many of these sites were abandoned without taking into consideration any safety, restoration or reclamation measures.

- Environmental protection and environmental monitoring at ongoing and planned operations, as well as the decommissioning of recently closed sites.

- Updating or establishing legislation and a regulatory framework that is consistent with recently introduced international standards.

- Overlaid on the latter two activities is the increased use of environmental assessments as a planning tool for evaluating all phases of uranium operations prior to the approval, start-up or closure of an operation. These assessments provide insight into the potential impact of radiation safety and environmental protection considerations on existing uranium production facilities and on the schedule and design of new production facilities.

The section on environmental aspects of uranium production was included in Red Book editions through 2001. Though it was discontinued in 2003 as a summary section under Uranium Supply, information was still collected from individual countries and published in the country reports as "Environmental Activities and Socio-Cultural Issues." In 1996, the Joint OECD/NEA-IAEA Uranium Group broadened its mandate to foster the exchange of information on environmental effects and environmental technologies associated with uranium mining and ore processing. A Working Group on "Environmental Issues in Uranium Mining and Milling" was established to implement this expanded mandate and in 1999 published *Environmental Activities in Uranium Mining and Milling*, as a joint NEA/IAEA Report, supplementing the 1998 Red Book [Reference 12-2]. In 1999, the Uranium Group formed a Working Group on Environmental Restoration of World Uranium Production Facilities, which coordinated publication of *Environmental Remediation of Uranium Production Facilities* [Reference 12-3].

Below are brief overviews of major environmental issues related to past and ongoing uranium mining and milling operations by geographical region. These overviews indicate the status of past and current site restoration and decommissioning projects, and highlight progress that the uranium industry has made in following through on environmental stewardship relating to its current and past operations.

North America

Canada

Since 1993, environmental activities within the uranium mining industry mainly focused on environmental assessments related to new projects. In 1993-1994, six new uranium mining projects in Saskatchewan were reviewed by an independent panel according to the Federal Environmental Assessment and Review Process Guidelines Order (McArthur River project, Extension of Dominique Janine at Cluff Lake, Midwest project, McClean Lake project, Eagle Point underground mine, Collins Bay A and D open-pits). Public hearings for Cigar Lake and McArthur River began in September 1996. The McArthur River project was permitted to proceed subject to a number of conditions in May 1997. The decommissioning plan for Elliot Lake uranium tailings was approved by the Minister of Environment in April 1997. In April 1998, the Midwest and Cigar Lake mines were allowed to proceed to the licensing stage, subject to certain site specific conditions. In 1999-2000, environmental assessments required by the Canadian Environmental Assessment Act were in various stages of preparation or review for: Licensing of historic mines in the Elliot Lake region; suspension of operations at the Cluff Lake production centre; renewal of the McClean Lake operating license, with a request to increase annual production capacity; request to dispose of potentially acid generating waste from the Cigar Lake and McClean Lake mines; and proposal to process approximately 57% of Cigar Lake ore at the Rabbit Lake mill.

Since 1996, uranium mining companies have committed over CAD 75 million (as of January 2003) to decommission all mines, mills and waste management areas at Elliot Lake alone. A comprehensive environmental monitoring programme has demonstrated the success of these efforts.

United States

As of 1 January 1995, out of 26 mills licensed for commercial operation, 20 were under different stages of decommissioning and six were on standby. The report *Decommissioning of US Uranium Production Facilities* examined 25 conventional and 17 non-conventional sites in the United States. Average costs for decommissioning of mills were USD 14.1 million, including USD 7.7 million for tailings reclamation, USD 2.3 million for groundwater restoration, USD 0.9 million for mill dismantling, and USD 3.2 million for indirect costs. The average for non-conventional facilities (ISL operations) was USD 7.0 million, including USD 2.8 million for groundwater restoration, USD 0.9 million for well field reclamation, USD 0.6 million for dismantling of buildings and plant structures, USD 1.2 million for reclaiming evaporation ponds, disposal wells and radiometric surveys, and USD 1.4 million for indirect costs.

Central and South America

Argentina

Decommissioning of the Malargue mill has been completed at a total cost of about USD 12 million for clean-up and restoration. Reclamation activities are now focused on rehabilitation of the Los Gigantes mine and mill complex. Hydrogeochemical studies were performed to define baseline conditions at the Cerro Solo U-Mo deposit. In addition, as part of the Sierra Pintada feasibility study, studies were carried out to improve surface and groundwater monitoring and waste and tailings management.

Brazil

Site monitoring and development of a decommissioning plan for the Poços de Caldas mine and mill complex are ongoing. An environmental impact assessment of the Lagoa Real production centre was completed, which became part of the operating plan for the mine-mill complex.

Western Europe and Scandinavia

Finland

The former small uranium mine of Paukkajanvaara, which was closed in 1962, has been monitored since 1974 and finally covered with soil in 1993.

France

Following closure of the last mines in the Hérault (1997) and Bernardan (2001) districts, reclamation and decommissioning efforts in France focused on these sites. Regulations require decommissioning and rehabilitation of closed mining and milling facilities including waste dumps and tailings facilities so that release of potentially harmful substances is reduced to a minimum and a broad range of specific standards are met. Total expenditures for decommissioning the Forez, Hérault, La Crouzille, Vendée and other sites totalled nearly FRF 793 million (USD 134 million) through 2000. Monitoring of air and water quality in the vicinity of these facilities is ongoing.

Germany

Since commercial uranium production was terminated in Germany in 1990, substantial programmes have been carried out for decommissioning mines, mills and adjacent facilities. By year end 1998, about 90% of underground rehabilitation work has been completed. Remediation of waste rock piles, stabilisation of mine spoil, chemical processing of remaining uranium ores at the mining facilities, rehabilitation of tailings and disposal facilities, demolition of production plants and buildings, water treatment, and monitoring of air and water quality in the vicinity of these facilities, which together comprise the largest uranium restoration project in the world, are ongoing. As of year end 2000, approximately DEM 6 700 million (USD 3 300 million) of the estimated DEM 13 000 million (USD 6 200 million) required to complete all decommissioning and remediation needs had been spent.

Portugal

Environmental activities consist of site rehabilitation (about 70 mines to reclaim), effluent management and monitoring.

Spain

Decommissioning of La Haba production centre, Badajoz province, was completed in 1997. A five-year supervision programme for verification of the decommissioning design criteria for the site was approved in January 1998. Restoration of 12 closed uranium mines in the Extremadura region and six additional mines in the Andalucia region was completed in 2000. The costs for these remediation activities totalled ESP 1 330 million (EUR 8 million) through 1998. The decommissioning plan for the Elefante heap leaching plant (Fe mine) was approved by the Regulatory Authorities in 2001. A plan for decommissioning of the Quercus plant, Salamanca Province, was submitted to Regulatory Authorities in 2003.

Sweden

Decommissioning and rehabilitation of the Ranstad mine was completed in the 1990s. An environmental monitoring programme is now being carried out, which is estimated to have cost a total of SEK 150 million.

Central, Eastern and Southeast Europe

Czech Republic/Czechoslovakia

Uranium mining and milling in the Czech Republic led to serious environmental impacts that will require significant resources over several decades to mitigate. Remediation of the tailings impoundments at the Pribram, Stráz pod Ralskem and Mydlovary processing plants have been the top environmental priority. Efforts have also focused on decommissioning of the Hamr, Olsi, Jasenice-Pucov, Zadni Chodov, Okrouhla Radoun and Licomerice-Brezinka mines, and remediation of the Stráz ISL operation, and also on long-abandoned production facilities including Jachymov and Horni Slavkov where mining took place in the 1940s and 1950s. Reclamation projects will continue until approximately 2040 and are projected to cost more than CZK 60 000 million (USD 2 600 million).

Hungary

A feasibility study for stabilising and remediation of tailings ponds in the Mecsek region was finalised following closure of the mines in 1998. Demolition of the ore processing plant began in 1999 and site remediation was expected to continue at least through the end of 2006. Costs of environmental management of the Mecsek site totalled HUF 2 948 million through year end 2002.

Poland

Only a limited number of serious environment impacts related to uranium production have been identified in Poland, the most important of which being the tailings pound in Kowary. A remediation plan has been developed to construct drainage systems and to cover the Kowary tailings ponds. A remediation programme has also been prepared for uranium production facilities in the Lower Silesia region.

Romania

Romania's environmental protection programme is focused on increasing water treatment capabilities in the Eastern Carpathians, Apuseni and Banat Mountains, increasing tailings impoundment capacities, processing and closure of ore storage areas at various mines, and long-term stabilisation, reclamation and revegetation of waste dumps and surrounding areas.

Slovenia

Environmental protection measures initiated after closure of the Zirovski vrh production facility in 1990, included geo-mechanical stabilisation of mine sites, protection of surface and groundwater, remediation of the mill, rehabilitation of waste dumps and tailings pounds and protection against radioactive contamination (radon exhalation).

Ukraine

Although no mines have yet been decommissioned in Ukraine, a programme is being undertaken by VostGOK to clean up and rehabilitate sites in Zheltiye Vody that are contaminated by uranium mill tailings. A State programme with a budget of USD 360 million has also been established for improving radiation protection at all facilities of the atomic industry. This programme includes remediation of contaminated lands, environmental monitoring, installation of personnel monitoring systems and improvement of effluent treatment.

Africa

Gabon

Following the March 1999 termination of uranium mining in Gabon, the government initiated a programme to rehabilitate seven sites (covering a total surface area of 60 hectares) comprising the Mounana mining and milling area. Rehabilitation of the Mounana site involved dismantling the mill and related facilities, closure of tailings impoundments, site clean-up and development of a lateritic cover over the tailings and re-vegetation. The objective was to assure a residual radiological impact that is as low as reasonably achievable, while insuring the physical stability of the impoundments, and to the extent possible, provide for the future utilisation of the affected area. A long-term programme for monitoring and surveillance of the tailings impoundment will be implemented after completion of the site rehabilitation.

Namibia

Application for a mining licence in Namibia requires that an environmental assessment study be carried out. Environmental damage due to mining must be minimised, and mining companies are obliged to rehabilitate the land after mining. For the Rössing mine, located in the Namib Desert, the principal environmental consideration is water management. The system used is aimed at reducing water consumption and minimising groundwater contamination. Cumulative environmental management costs have totalled over NAD 44 million (USD 8.5 million), including NAD 1.8 million (USD 252 000) in 2000.

South Africa

Environmental issues related to gold/uranium mining in the Witwatersrand district include dust pollution, surface and groundwater contamination and residual radioactivity. Closed gold/uranium plants are being decommissioned and former mines and areas surrounding uranium plants are surveyed for radioactivity and remediated before reuse.

Niger

Both producers in Niger, Cominak and Somaïr, committed to implementing an environmental management system that complies with ISO 14001. Cominak environment-related expenditures for impact assessment, monitoring, waste management and regulatory activities averaged USD 171 290 in 1999 and 2000. Expenditures for social and cultural programmes during the same time period averaged USD 4 million per year.

Middle East, Central and South Asia

India

Radiation, radon and dust impacts at uranium production facilities are monitored by the Health Physics Group of the Bhabha Atomic Research Centre in Bombay.

Jordan

A systematic study and evaluation of natural uranium concentration in Jordan's phosphate deposits was conducted to assess current and future environmental and safety effects of the uranium.

Kazakhstan

Environmental efforts in Kazakhstan are focused on wastes associated with closed uranium production facilities (uranium production over 40 years resulted in the accumulation of about 2 000 million tonnes of low-level waste dumps and mill tailings), as well as on environmental impacts of ISL mining. All uranium mine and mill sites were inventoried in 1997 and 1998, and it was determined that out of 100 waste storage sites, only five or six are of significant environmental concern, mainly related to the uncontrolled use of waste materials for construction by local inhabitants. A study of the long-term impact of ISL operations on aquifers, and an investigation of self-remediation of aquifers has been conducted.

Uzbekistan

The focus of environmental efforts in Uzbekistan has been on areas affected by past conventional mining and milling, as well as the environmental impacts associated with ISL operations. At Navoi hydrometallurgical plant, a system of wells has been installed to monitor and control potential ground-water contamination from the tailings impoundments, and research is underway to develop a tailings impoundment burial system for implementation by 2005.

Pacific Area

Australia

The Nabarlek mine ceased production in 1988; rehabilitation and decommissioning was completed in 1995 and the site is under environmental monitoring. Construction of the Jabiluka mine commenced in June 1988 following a comprehensive joint Commonwealth-Northern Territory environmental impact assessment process. Jabiluka is currently in care and maintenance status. In 2000-2001 the Beverley and Honeymoon ISL uranium mines were the subject of a joint Commonwealth-South Australia environmental assessment process. Beverley began commercial operations in 2000. Pilot leaching has been completed at Honeymoon, but no plans have been announced for commencement of commercial operations.

East Asia

China

China has used its many years of experience in uranium production to develop new regulations to control, monitor and reduce the environmental impacts of uranium production. These regulations have led to backfilling of waste rock and tailings into mined areas, treatment of mine water and used process water, and covering waste and tailings piles to reduce radon release. Extra high voltage electrostatic filters were installed at the Fuzhou and Hengyang ore processing plants to reduce release of fly dust. In addition to the environmental measures introduced at operating production centres, five small uranium mines have been completely decommissioned; seven other mines or mine-mill complexes are in various stages of decommissioning.

13. EPILOGUE – LESSONS LEARNT

Since its inception after World War II, the modern uranium industry has evolved from one exclusively satisfying military requirements to the current emphasis on satisfying fuel requirements for civilian nuclear reactors generating electricity. As the industry changed, so too did the Red Book. Resource terminology was expanded to include more definitive resource confidence levels in order to provide industry and government planners with better tools to assess the adequacy of uranium resources to meet future requirements. Resource production cost categories were periodically adjusted in response to changing market price and sections were added or deleted from successive Red Books as the industry matured and responded to changing market and regulatory requirements, as well as societal expectations. As the civilian nuclear industry grew, concerns about the adequacy of resources to meet future requirements emerged. That concern led to Red Book projections of nuclear power and related uranium requirements well into the future. A look back on these projections gives sobering lessons as to the impact of world and industry-specific events on the accuracy of these projections. The oil crisis of 1973 propelled nuclear power into the spotlight as an alternative to fossil fuels, which in turn led to overly optimistic projections of growth in generating capacity and uranium requirements. Subsequently, the Three Mile Island and Chernobyl accidents in 1979 and 1986, respectively, had a chilling affect on nuclear power that lasted for decades.

Today, with "nuclear renaissance" a common refrain, nuclear energy is once again in the forefront of discussion as a potential energy source in response to global warming threats in addition to continued energy security concerns. Uranium prices have reached levels not seen for decades and as a result exploration is increasing both in expenditure levels and geographic diversity. After decades in the doldrums, the nuclear industry has something to be optimistic about. This optimism, however, matches the mood that prevailed in the period preceding the Three Mile Island accident. As this Red Book Retrospective shows, projections made during such periods need to be put into perspective and tempered to avoid overstating future requirements. There is as yet no hard evidence that the perceived nuclear renaissance has led to concrete actions to build large numbers of new reactors outside of China and India, where expansion plans preceded talk of an international rebirth of nuclear power. This report shows that until there is hard evidence to support the current optimism; contributors to the Red Book would be well advised to temper projections for growth in nuclear generating capacity and uranium requirements. Such projections also need to include adequate timeframes for licensing and construction of reactors and fuel cycle facilities, as well as equally important time to fully assess the perceived maturing of public acceptance of nuclear power.

As country after country announces its intent to reconsider nuclear energy in its energy supply mix questions are again being raised about the adequacy of resources. Past experience shows that exploration responds positively to uranium price increases. Recent dramatic price rises, after a period of over 20 years of low prices and low levels of exploration, are similar to the rapid price increases experienced in the early 1970s. And just as then, exploration has rapidly responded to this market signal. The data compiled here show that this increased exploration activity can be expected to yield new deposits at reasonable costs and ultimately lead to increases in recoverable resources. Moreover, despite several decades of low-level exploration and the cumulative production of over 2 200 000 tU since 1945, reported uranium resources have steadily increased since the mid-1980s; maintaining a forward looking reserves to annual requirements ratio that has averaged about 45 over the past twenty years, despite steadily increasing requirements.

However, reliance on secondary sources and decades of low prices for uranium have had dramatic impacts on the industry in terms of consolidation of producers and significant reductions in primary production capability. The transition to a market fundamentally supplied by primary production after decades of reliance on secondary sources is the challenge now being faced. Given the extent of Known Conventional Resources and the recent increase in global exploration activity, the challenge in coming years is likely to be less one of adequacy of resources than adequacy of production capacity. Currently, the issue facing the uranium industry is obtaining the investment necessary to expand and construct production facilities that will be needed to fuel new reactors at the same time that secondary sources, so long relied on, become less readily available. Yet, as the historical data presented here shows, the amount of time needed to develop a new mine and bring new resources to market has steadily increased over time, in part due to increased regulatory requirements. Additionally, history shows that planned production capacity is seldom if ever reached in practice. These factors will have to be accounted for as the world transitions away from secondary sources and relies more and more on primary production to meet increasing reactor requirements. History shows that these challenges can be successfully met and already plans to increase production at existing centres and to open new mines are being announced and implemented.

What is clear from this review of Red Book history is the complexity and resiliency of the uranium industry. Undoubtedly, another look back in forty years time will provide additional insights to those presented in this volume. Until then, this work has attempted to provide some understanding of the industry's past as an aid to users as we look forward to a dynamic and vibrant period in the continuing story of uranium.

REFERENCES

CHAPTER 1

[1-1]   Willliams, R.M. (1993), Joint NEA/IAEA Uranium Activities – Background to the NEA Uranium Group, paper prepared for NEA Uranium Group Meeting, 22-23 June 1993.

CHAPTER 7

[7-1]   Euratom Supply Agency (2005), *Annual Report 2004*, CEC, Luxembourg.

[7-2]   OECD Nuclear Energy Agency (2001), *Management of Depleted Uranium*, Paris, France.

CHAPTER 8

[8-1]   OECD Nuclear Energy Agency (2001), *Management of Depleted Uranium*, Paris, France.

CHAPTER 10

[10-1]   International Atomic Energy Agency (2006), *Fissile Material Management Strategies for Sustainable Nuclear Energy*, Proceedings of a Technical Meeting held in Vienna, 12-15 September 2005, Proceedings Series, IAEA, Vienna.

[10-2]   Barthel, F.H. and F.J. Dahlkamp (1991), "Thorium", Supplement Vol. A1b, *Gemelin Handbook of Inorganic and Organometallic Chemistry*, Springer, Heidelberg, pp. 345-432.

[10-3]   Barthel, F.H. and F.J. Dahlkamp (1991), "Thorium Deposits and their Availability", IAEA-TECDOC-650, IAEA, Vienna, pp. 104-115.

[10-4]   Kotova, V.M. and J.I. Skorovarov (1997), "Thorium Deposits in the Commonwealth of Independent States and their Prospective Characteristics", IAEA-TECDOC-961, IAEA, Vienna, pp. 213-220.

[10-5]   World Nuclear Association (2003), Thorium, November.

[10-6]   Chung, T. (1996), "The Role of Thorium in Nuclear Energy", *Uranium Industry Annual*, Energy Information Administration, Washington, pp. ix-xvii.

[10-7]   World Nuclear Association (2005), Advanced Nuclear Power Reactors, April.

[10-8]   US Geological Survey (2005), Thorium, Mineral Commodity Summaries, 172-173.

[10-9]   US Geological Survey (2003), Thorium, Minerals Yearbook, 76.1-76.3.

[10-10] US Geological Survey (2005), Rare Earth, Mineral Commodity Summaries, 132-133.

[10-11] US Geological Survey (2003), Rare Earth, Minerals Yearbook, 60.1-60.9.

CHAPTER 12

[12-1]   International Atomic Energy Agency (1982), *Management of Wastes from Uranium Mining and Milling*, Proceedings of a Symposium, Albuquerque, 10-14 May 1982, jointly organised by IAEA and OECD/NEA, Vienna, Austria.

[12-2]   OECD Nuclear Energy Agency (1999), *Environmental Activities in Uranium Mining and Milling*, Joint NEA/IAEA Report, Paris, France.

[12-3]   OECD Nuclear Energy Agency (2002), *Environmental Remediation of Uranium Production Facilities*, Joint NEA/IAEA Report, Paris, France.

Further information available

Listed below are some publications that can provide more in-depth information on specific countries.

Australia

- McKay, A.D. and Y. Miezitis (2001), *Australia's Uranium Resources, Geology and Development of Deposits*, Commonwealth of Australia 2001, available at: www.ga.gov.au/pdf/RR0030.pdf.

Canada

- Williams, R.M. and H.W. Little (1973), *Canadian Uranium Resource and Production Capability,* Mineral Bulletin MR 140, Department of Energy, Mines and Resources, Catalogue no. M38-2/140.
- Cranstone, D.A. and R.T. Whillans (1989), *An Analysis of Uranium Discovery in Canada 1930-1982*, paper presented at Uranium Resources and Geology of North America, International Atomic Energy Agency, Saskatoon, Canada, 1-3 September 1987, published in IAEA-TECDOC-500, pp. 29-47.
- Cranstone, D.A. (2002), *A History of Mining and Mineral Exploration in Canada and Outlook for the Future*, Minister of Public Works and Government Services, Canada, Catalogue no. M37-51/2002, ISBN 0-662-32680-6.
- Uranium Resources Appraisal Group, 1976-1991 (biennial publication), *Uranium in Canada: Assessment of Supply and Requirements*, Department of Energy, Mines and Resources, Canada.
- Uranium Chapter in Canadian Minerals Yearbook, 1952-2004, Department of Natural Resources (former Department of Energy, Mines and Resources), Canada.

Estonia

- E. Lippmaa and E. Maremae (1999), *Dictyonema Shale and Uranium Processing at Sillamae*, Oil Shale 1999, vol. 16, no. 4.

France

- Paucard, A. (1996), *La mine et les mineurs de l'uranium français*, (Uranium exploration and production in France and abroad by French Companies), COGEMA (in French).
- Bavoux B. and P.C. Guiollard (1998), *L'uranium de la Crouzille*, COGEMA (in French).
- Bavoux B. and P.C. Guiollard (1999), *L'uranium du Lodévois*, COGEMA (in French).
- Chapot *et al.* (1996), *L'uranium vendéen*, Cahiers du Patrimoine, Nantes (in French).

United States

- *Uranium Industry Annual,* Energy Information Administration, US Department of Energy. Reports are published annually and provide historical information on US exploration statistics including exploration expenditures and drilling activities. This information is available at www.eia.doe.gov/fuelnuclear.html.
- Ringholz, R.C. (2002), *Uranium Frenzy: Saga of the Nuclear West*, Logan, Utah State University Press (lively account of the 1950s uranium boom on the Colorado Plateau).
- Amundson, M.A. (2002), *Yellowcake Towns: Uranium Mining Communities in the American West*, University Press of Colorado, Boulder, Colorado (this publication has an extensive bibliography on uranium exploration and mining history).

Czech Republic

- Rudné A Uranové Hornictvi Ceske Republiky, 2003 (in Czech).

Germany

- Barthel, F. (1977), Uran-Prospektion in der Bundesrepublik Deutschland, Naturwissenschaften 64, pp. 499-506, Springer, Heidelberg (in German).
- Barthel, F. and L. Lahner (1984), Verlauf und Ergebnisse des Explorationsförderprogramms der Bundesregierung, GEOL. JB. A75, pp. 447-463, Hannover (in German).
- Barthel, F., K. Busch *et al.* (1991), Zwanzig Jahre Explorationsförderung für mineralische Rohstoffe durch das Bundesministerium für Wirtschaft, GEOL. JB A127, pp. 271-288, Hannover (in German).
- Hagen, M., R. Scheid and W. Runge (1999), Chronik der Wismut, WISMUT GmbH, Chemnitz (in German).
- Müller, F. *et al.* (1990), Seilfahrt: Auf den Spuren des sächsischen Uranbergbaus, BODE Verlag Haltern (in German).
- Richter, H. and P. Mühlstedt (1991), Uranium Mining in Thuringia and Saxony, SDAG WISMUT, Chemnitz.

USSR/Russian Federation

- Naumov, S. and Y.L. Bastrikov (2005), *The History of Uranium Exploration in the Context of the "Atomic Project of the USSR"*, Razvedka I ohrana nedr. no. 10, pp. 2-7 (in Russian).
- Laverov, N.P., V.I. Velichkin, V.I. Vetrov *et al., Uranium Resources of the Union of Soviet Socialist Republic*, IAEA-TECDOC 650, pp. 172-187.
- Karimov, K.K., N.S. Bobonorov, K.G. Brovin *et al.* (1996), *Uranium Deposits of the Uchkuduk Type in the Republic of Uzbekistan*, Tashkent, Fan (in Russian).
- Petrov, N.N., V.G. Yazikov, B.R. Berikbolov *et al.* (2000), *Uranium Deposits in Kazakhstan*, Almaty, Gylym (in Russian).

Appendix 1.1.

COUNTRY PARTICIPATION IN THE RED BOOK (1965-2004)

The number of reports to individual Red Books is given in parenthesis after the country name. Countries followed by "*" are members of the OECD.

1. North America (3 countries)

Canada* (20), Mexico* (13), the United States of America* (20).

2. Central and South America (19 countries)

Argentina (19), Bolivia (5), Brazil (15), Chile (13), Colombia (9), Costa Rica (5), Cuba (4), Dominican Republic (1), Ecuador (6), El Salvador (2), Guatemala (2), Guayana (4), Jamaica (2), Panama (2), Paraguay (2), Peru (12), Surinam (2), Uruguay (6), Venezuela (2).

3. Western Europe and Scandinavia (15 countries)

Austria* (1), Belgium* (12), Denmark* (16), Finland* (16), France* (20), Germany* (15), Ireland* (6), Italy* (15), Netherlands* (10), Norway* (6), Portugal* (20), Spain* (20), Sweden* (20), Switzerland* (15), United Kingdom* (15).

4. Central Eastern and South-eastern Europe (14 countries)

Armenia (3), Bulgaria (5), Czech Republic* (6), Czech and Slovak Republic, prior to separation (1), Estonia (2), Greece* (11), Hungary* (7), Lithuania (6), Poland* (2), Romania (6), Russian Federation (5), Slovak Republic* (6), Slovenia (5), Turkey* (16), Ukraine (6), USSR, prior to separation (1), Yugoslavia, prior to separation (1).

5. Africa (31 countries)

Algeria (6), Benin (1), Botswana (4), Cameroon (3), Central African Republic (5), Egypt (11), Ethiopia (3), Gabon (11), Ghana (2), Ivory Coast (1), Lesotho (1), Liberia (2), Libyan Arab Jamahirya (1), Madagascar (7), Malawi (1), Mali (2), Mauritania (1), Morocco (11), Namibia (11), Niger (14), Nigeria (1), Rwanda (1), Senegal (1), Somalia (2), South Africa (18), Sudan (1), Tanzania (1), Togo (1), Zaire (4), Zambia (7), Zimbabwe (6).

6. Middle East, Central and Southern Asia (12 countries)

Bangladesh (2), India (18), Iran, Islamic Republic (5), Jordan (11), Kazakhstan (6), Kyrgyzstan (2), Pakistan (4), Sri Lanka (5), Syrian Arab Republic (6), Tajikistan (1), Turkmenistan (1), Uzbekistan (6).

7. South-eastern Asia (5 countries)

Indonesia (11), Malaysia (10), Philippines (11), Thailand (13), Vietnam (7).

8. Pacific (2 countries)

Australia* (19), New Zealand* (3).

9. East Asia (4 countries)

China (8), Japan* (19), Korea, Republic of* (15), Mongolia (3).

10. OECD

The reports refer to the OECD area, in which the following countries are members (accession dates in parenthesis):

Members since its foundation on 30 September 1961: Austria, Belgium, Canada, Denmark, France, Germany, Greece, Iceland, Ireland, Italy, Luxembourg, Netherlands, Norway, Portugal, Spain, Sweden, Switzerland, Turkey, United Kingdom and United States of America.

OECD membership was enlarged by: Japan (28 April 1964), Finland (28 January 1969), Australia (7 June 1971), New Zealand (29 May 1973), Mexico (18 May 1994), Czech Republic (21 December 1995), Hungary (7 May 1996), Poland (22 November 1996), Republic of Korea (12 December 1996) and Slovak Republic (14 December 2000).

11. Commonwealth of Independent States (CIS) or Newly Independent States (NIS)

Armenia, Kazakhstan, Tajikistan, Azerbaijan, Kyrgyzstan, Turkmenistan, Belarus, Moldavia, Ukraine, Georgia, Russian Federation, Uzbekistan.

12. European Union

Austria, Cyprus, Czech Republic, Estonia, France, Hungary, Italy, Latvia, Lithuania, Malta, Poland, Slovak Republic, Slovenia, Spain, Belgium, Germany, Luxemburg, Sweden, Denmark, Greece, Netherlands, United Kingdom, Finland, Ireland, Portugal.

13. International Organisations

The international organisations that have participated in the Uranium Group and in the production of the Red Book are:

- European Commission (Euratom Supply Agency).
- European Nuclear Energy Agency (ENEA), which evolved into the OECD Nuclear Energy Agency (NEA) on 20 April 1972.
- International Atomic Energy Agency (IAEA).

INDEX OF NATIONAL REPORTS IN RED BOOKS

The following index lists all national reports by the year in which these reports were published in the Red Books. A listing of all Red Book editions is shown at the end of this Index.

	1965	1967	1969	1970	1973	1975	1977	1979	1982	1983	1986	1988	1990	1992	1994	1996	1998	2000	2002	2004
Algeria						1975	1977	1979	1982										2002	2004
Argentina		1967	1969	1970	1973	1975	1977	1979	1982	1983	1986	1988	1990	1992	1994	1996	1998	2000	2002	2004
Armenia																		2000	2002	2004
Australia		1967	1969	1970	1973	1975	1977	1979	1982	1983	1986	1988	1990	1992	1994	1996	1998	2000	2002	2004
Austria							1977													
Bangladesh											1986	1988								
Belgium									1982	1983	1986	1988	1990	1992	1994	1996	1998	2000	2002	2004
Benin													1990							
Bolivia							1977	1979	1982	1983	1986									
Botswana								1979		1983	1986	1988								
Brazil				1970	1973	1975	1977	1979	1982	1983	1986			1992	1994	1996	1998	2000	2002	2004
Bulgaria													1990	1992	1994	1996	1998			
Cameroon							1977		1982	1983										
Canada	1965	1967	1969	1970	1973	1975	1977	1979	1982	1983	1986	1988	1990	1992	1994	1996	1998	2000	2002	2004
Central African Republic				1970	1973		1977	1979			1986									
Chile							1977	1979	1982	1983	1986	1988		1992	1994	1996	1998	2000	2002	2004
China													1990	1992	1994	1996	1998	2000	2002	2004
Colombia							1977	1979	1982	1983	1986	1988	1990			1996	1998			
Costa Rica									1982	1983	1986	1988	1990							
Cuba												1988		1992		1996	1998			
Czech Republic															1994	1996	1998	2000	2002	2004
Czech and Slovak Republic													1990							
Denmark (Greenland)	1965	1967	1969	1970	1973	1975	1977	1979	1982	1983	1986		1990	1992		1996	1998			2004
Dominican Republic									1982											
Ecuador							1977		1982	1983	1986	1988								
Egypt							1977	1979			1986	1988	1990	1992	1994	1996	1998	2000		2004
El Salvador										1983	1986									
Estonia																	1998			2004
Ethiopia								1979		1983	1986									
Finland					1973	1975	1977	1979	1982	1983	1986	1988	1990	1992	1994	1996	1998	2000	2002	2004
France	1965	1967	1969	1970	1973	1975	1977	1979	1982	1983	1986	1988	1990	1992	1994	1996	1998	2000	2002	2004
Gabon		1967		1970	1973				1982	1983	1986					1996	1998	2000	2002	2004
Germany				1970		1975	1977	1979	1982	1983	1986	1988	1990	1992	1994	1996	1998	2000	2002	
Ghana							1977			1983										
Greece							1977	1979	1982	1983	1986	1988	1990	1992	1994	1996	1998			
Guatemala											1986	1988								
Guyana								1979	1982	1983	1986									
Hungary														1992	1994	1996	1998	2000	2002	2004
India	1965	1967		1970	1973	1975	1977	1979	1982	1983	1986		1990	1992	1994	1996	1998	2000	2002	2004
Indonesia							1977				1986	1988	1990	1992	1994	1996	1998	2000	2002	2004
Iran, Islamic Republic of							1977										1998	2000	2002	2004
Ireland								1979	1982	1983	1986			1992			1998			
Italy		1967		1970	1973	1975	1977	1979	1982	1983	1986	1988		1992	1994	1996	1998	2000		
Ivory Coast									1982											
Jamaica									1982	1983										
Japan	1965	1967		1970	1973	1975	1977	1979	1982	1983	1986	1988	1990	1992	1994	1996	1998	2000	2002	2004
Jordan							1977				1986	1988	1990	1992	1994	1996	1998	2000	2002	2004

	1965	1967	1969	1970	1973	1975	1977	1979	1982	1983	1986	1988	1990	1992	1994	1996	1998	2000	2002	2004
Kazakhstan															1994	1996	1998	2000	2002	2004
Korea, Republic of						1975	1977	1979	1982	1983	1986	1988	1990	1992	1994	1996	1998	2000	2002	2004
Kyrgyzstan																1996			2002	
Lesotho												1988								
Liberia							1977			1983										
Libyan Arab Jamahirya										1983										
Lithuania															1994	1996	1998	2000	2002	2004
Madagascar						1975	1977	1979	1982	1983	1986	1988								
Malawi																		2000		
Malaysia									1982	1983	1986	1988	1990	1992	1994	1996	1998	2000	2002	
Mali											1986	1988								
Mauritania													1990							
Mexico				1970	1973	1975	1977	1979	1982		1986		1990	1992	1994	1996	1998	2000		
Mongolia															1994	1996	1998			
Morocco	1965	1967				1975	1977	1979	1982	1983	1986	1988	1990				1998			
Namibia								1979	1982	1983	1986	1988	1990			1996	1998	2000	2002	2004
Netherlands									1982	1983	1986		1990	1992	1994	1996	1998	2000	2002	
New Zealand		1967					1977	1979												
Niger		1967		1970	1973		1977				1986	1988	1990	1992	1994	1996	1998	2000	2002	2004
Nigeria								1979												
Norway								1979	1982	1983				1992		1996	1998			
Pakistan		1967															1998	2000	2002	
Panama										1983		1988								
Paraguay										1983	1986									
Peru							1977	1979		1983	1986	1988	1990	1992	1994	1996	1998	2000		2004
Philippines							1977		1982	1983	1986		1990		1994	1996	1998	2000	2002	2004
Poland																		2000	2002	
Portugal	1965	1967	1969	1970	1973	1975	1977	1979	1982	1983	1986	1988	1990	1992	1994	1996	1998	2000	2002	2004
Romania														1992	1994	1996	1998	2000	2002	
Russian Federation															1994		1998	2000	2002	2004
Rwanda											1986									
Senegal									1982											
Slovak Republic															1994	1996	1998	2000	2002	2004
Slovenia															1994	1996	1998		2002	2004
Somalia							1977	1979												
South Africa	1965	1967	1969	1970	1973	1975	1977	1979	1982	1983	1986			1992	1994	1996	1998	2000	2002	2004
Spain	1965	1967	1969	1970	1973	1975	1977	1979	1982	1983	1986	1988	1990	1992	1994	1996	1998	2000	2002	2004
Sri Lanka							1977		1982	1983	1986	1988								
Sudan							1977													
Surinam									1982	1983										
Sweden	1965	1967	1969	1970	1973	1975	1977	1979	1982	1983	1986	1988	1990	1992	1994	1996	1998	2000	2002	2004
Switzerland						1975	1977	1979	1982	1983	1986	1988	1990	1992	1994	1996	1998	2000	2002	2004
Syrian Arab Republic									1982	1983	1986	1988	1990		1994					
Tajikistan																			2002	
Tanzania														1990						
Thailand							1977	1979	1982	1983	1986	1988	1990	1992	1994	1996	1998	2000	2002	
Togo								1979												
Turkey					1973	1975	1977	1979	1982	1983	1986	1988	1990	1992	1994	1996	1998	2000	2002	2004
Turkmenistan																				2004
Ukraine															1994	1996	1998	2000	2002	2004
United Kingdom						1975	1977	1979	1982	1983	1986	1988	1990	1992	1994	1996	1998	2000	2002	2004
United States	1965	1967	1969	1970	1973	1975	1977	1979	1982	1983	1986	1988	1990	1992	1994	1996	1998	2000	2002	2004
Uruguay							1977		1982	1983	1986	1988	1990							
USSR														1992						
Uzbekistan															1994	1996	1998	2000	2002	2004
Venezuela											1986	1988								
Vietnam														1992	1994	1996	1998	2000	2002	2004
Yugoslavia														1992						
Zaire		1967			1973		1977					1988								
Zambia											1986	1988	1990	1992	1994	1996	1998			
Zimbabwe									1982			1988		1992	1994	1996	1998			

Appendix 2.1

SUMMARY OF NUCLEAR GENERATING HISTORY

The first commercial nuclear power plant was put into operation in 1957 in the United Kingdom. The Calder Hall-1 nuclear power plant had a generating capacity of 50 Megawatts electric (MWe). As of 1 January 2004, there were 435 operating reactors in 31 different countries (with six reactors operating in Taiwan) with an installed capacity of about 359 GWe. Two additional countries, Italy and Kazakhstan, operated nuclear power plants, which have now been closed. In defining the number of operating reactors and the installed capacity for each country, certain test reactors were not counted, i.e. Phénix in France, Fugen and Monju in Japan.

Argentina: Nuclear operations started in 1974 with the Atucha-1 reactor (321 MWe), with a second reactor, Embalse (600 MWe) connected to the grid in 1983.

Armenia: Nuclear operations started in 1976 (376 MWe), with a second reactor of the same size connected to the grid in 1980. The first reactor was shut down in 1990.

Belgium: The first nuclear reactor (11 MWe) began operating in 1962 and operated until 1987. Six other reactors of larger capacity (350 to 981 MWe) were connected to the grid between 1974 and 1985. In 2002, the government decided to phase out nuclear energy by limiting the operational lives of its reactors to 40 years. In 2005, over 55% of Belgium's electricity is generated with nuclear energy.

Brazil: The first nuclear reactor (626 MWe) was put into operation in 1982. A second reactor (1 245 MWe) was connected to the grid in 2000.

Bulgaria: From 1974 to 1991, six nuclear reactors were constructed at Kozloduy. The first two power plants were shut down in December 2002.

Canada: Twenty-five nuclear reactors were constructed in Canada between 1962 and 1993. The first reactor, NPD (22 MWe), was operated until 1987. Two other small size reactors (220 and 250 MWe) were shut down in 1977 and 1984. Following an independent performance assessment in 1997, eight reactors were shut down for refurbishment (four at Pickering-A site and four at Bruce-A site). During 2003, three of these reactors (two at Bruce and one at Pickering) were restarted. A second Pickering reactor was returned to service in September 2005. Plans to restart two of the remaining four reactors are being developed.

China: The first nuclear power plant in China was connected to the grid in 1993. As of 1 January 2005 there were nine reactors in operation (about 6.7 GWe net) and two under construction (about 2.0 GWe net). The government of China has announced plans to increase installed nuclear capacity to about 36 GWe by 2020 that will be accomplished through the construction of 27 reactors of at least 1.0 GWe net each beginning in 2005.

Czech Republic: The first nuclear power plant (396 MWe) was put into operation at Dukovany in 1985, followed by three other plants, of the same size and on the same site, in 1986 and 1987. The Temelin-1 plant (912 MWe) was connected to the grid in 2000, followed by Temelin-2 (950 MWe) in December 2002, which brings the total to seven operating reactors.

Finland: Four nuclear power plants were put in operation from 1977 to 1980. In May 2002 Finland approved the construction of a new nuclear power plant. Construction of Olkiluoto-3, an advanced European Pressurised Reactor, began in December 2005 and the reactor is expected to be operational by 2010.

France: A total of 69 nuclear reactors have been constructed in France. The first nuclear reactor (G338 MWe) was put in operation in 1959, and operated until 1984. Several generations of nuclear reactors were successively constructed in the 1960s (300 to 500 MWe each), 1970s to the early 1980s (about 900 MWe), the end of 1980s (1 300 MWe) and the end of 1990s (1 450 MWe). The Creys Malville (Super Phénix) fast breeder reactor (1 200 MWe) operated from 1986 to 1996. In 2005, 58 commercial reactors were operating in France, producing about 78% of the country's electricity. Also in 2005, France announced that it is planning to construct a new advanced reactor at Flammanville in the Normandy region. Construction of the 1.6 GWe net European Pressurised Reactor is expected to begin in 2007 with operation expected in 2012.

Germany: A total of 35 nuclear reactors were constructed in Germany between 1966 and 1989, but three of them (Greifswald 1, HDR and Niederaichbach) operated for only one or two years. Six nuclear reactors of Soviet design were built in the former GDR, all of which were decommissioned after German unification in 1990. In April 2002, the government enacted a new nuclear law that codifies the long-term phase-out of nuclear energy. For each plant operating as of 1 January 2000, a residual operating life was calculated based on a standard operating life of 32 years from the commencement of commercial operation. As of 1 January 2006, 17 reactors were still in operation, producing about 30% of Germany's electricity.

Hungary: Four nuclear reactors, about 415 MWe each, were constructed at Paks between 1982 and 1987.

India: The first two nuclear reactors (210 MWe each) were connected to the grid in 1969. Since then 13 additional reactors of similar size have constructed. In 2006, eight nuclear power plants with a total capacity of about 3.6 GWe were under construction, and plans are to increase the country's nuclear generation capacity to 20 GWe by 2020.

Iran: As of 1 January 2006, one reactor, Bushehr-1, was under construction, with start up of the first reactor expected in late 2006.

Italy: Three nuclear reactors were connected to the grid in 1963 and 1964. A fourth reactor was put into operation in 1978. The Latina reactor that began operating in 1963 was shut down in 1982. In 1987, a public referendum led to the shutdown of all nuclear power plants by 1990.

Japan: Since 1965, 57 commercial nuclear power reactors have been constructed in Japan. One, (Tokai-1, 137 MWe) was shut down in 1998. In 2005, nuclear reactors produced about 29% of Japan's electricity. Government plans continue to call for 59 power plants to be operational by 2010.

Kazakhstan: A small fast breeder nuclear reactor (135 MWe) operated from 1973 to 1999.

Korea, Republic of: The first nuclear reactor in South Korea was connected to the grid in 1977; as of 1 January 2006, 20 reactors were operating. Plans call for 28 nuclear reactors to be operational by 2015.

Lithuania: Two RBMK nuclear reactors were connected to the grid in 1983 and 1987. One of the plants was shut down in 2004 as part of Lithuania's accession into the European Union and plans are to shut down the remaining reactor at Ignalina by 2009.

Mexico: Two nuclear reactors (about 650 MWe each) were connected to the grid in 1989 and 1994.

Netherlands: Two nuclear reactors were connected to the grid in 1968 and 1973. The Dodewaard reactor (55 MWe) was shut down in 1997. The shutdown of the Borssele nuclear reactor (450 MWe), originally planned for 2005, has been delayed to at least 2013 with plans to extend the life of the plant to 2033 announced in September 2005.

Pakistan: Two nuclear reactors (128 MWe each) were connected to the grid in 1971. In 2002, the government announced plans to construct two new reactors.

Romania: The only operating nuclear reactor in Romania (655 MWe) was connected to the grid in 1996. Construction of the Cernavoda-2 plant was resumed in December 2002, after having been suspended in 1990. The plant is expected to begin operating in 2006.

Russian Federation: The first two nuclear reactors (102 and 197 MWe) were connected to the grid in 1964. As of 1 January 2006, there were 31 nuclear reactors operating in Russia, with capacities varying from 10 to 950 MWe and four reactors were under construction. Announced plans are to have at least 32 GWe net of installed capacity in operation by 2020.

Slovak Republic: The first nuclear reactor (A1 Bohunice, 110 MWe) was connected to the grid in 1972, but was closed in 1976. Between 1978 and 1985, four nuclear reactors were connected to the grid. Two additional reactors were connected in 1998 and 1999. In 2003, six reactors were operating.

Slovenia: The only operating nuclear reactor (632 MWe), which is jointly operated with Croatia, was connected to the grid in 1981.

South Africa: Two nuclear reactors (920 MWe each) were connected to the grid in 1984 and 1985.

Spain: The first nuclear reactor (153 MWe) was connected to the grid in 1968. Nine additional reactors were constructed in Spain between 1971 and 1987. One reactor (Vandellós-1, 480 MWe) was shut down in 1990.

Sweden: The first nuclear reactor (Agesta, 10 MWe) was connected to the grid in 1964 and was shut down in 1974. Twelve additional nuclear reactors were constructed from 1971 to 1985. Closure of a second reactor (Barseback-1, 615 MWe) resulted from the decision to phase out nuclear energy and occurred in 1999.

Switzerland: Five nuclear reactors were constructed in Switzerland between 1969 and 1984 and none have been shut down.

Taiwan: Six nuclear reactors were constructed in Taiwan between 1977 and 1985. As of 1 January 2006, two additional reactors were under construction.

Ukraine: The first nuclear reactor was connected to the grid in 1977. Sixteen additional reactors were constructed between 1978 and 1989 with two additional reactors starting operations in 2004. The Chernobyl-4 reactor exploded in 1986, the world's worst reactor accident and led to the closure of the three other reactors at the Chernobyl site in 1991, 1996 and 2000. As of 1 January 2006, two reactors were under construction.

United Kingdom: The first nuclear reactor (Calder Hall-1, 50 MWe) was connected to the grid in 1957. A total of 44 reactors were constructed in the United Kingdom between 1957 and 1995. One of the first reactors to come on line (Windscale, 24 MWe) was shut down in 1980 with four other reactors at Berkeley, Hunterston and Winfrith shut down in 1989. The reactors at the Trawsfynydd and Dounreay sites (three reactors total) were shut down in 1992 and 1993. In the early 2000s, low wholesale electricity prices placed pressure on operators to shut down some nuclear plants. As a consequence the four reactors at Calder Hall were shut down in March 2003 and the four reactors at Chapelcross were shut down in June 2004. The eight remaining MAGNOX reactors are all scheduled to be shut down by the end of 2010. As of 1 January 2006, 23 reactors were still in operation, producing over 20% of the United Kingdom's electricity.

United States: The first US nuclear reactor (Shippingport, 60 MWe) was connected to the grid in 1957. A total of 127 reactors were constructed in the United States between 1957 and 1996. Most of the first reactors were of small capacity (less than 100 MWe) and closed after a few years of operation. The first large reactor to be shut down was the Three Mile Island-2 (800 MWe) in 1979, following an accident. As of 1 January 2006 there were 104 operating reactors, with an average generating capacity of 945 MWe, producing about 20% of US electricity.*

As of 1 January 2006, a total of 24 reactors were under construction in Argentina (1), China (2), Finland (1), India (8), Islamic Republic of Iran (1), Japan (1), Pakistan (1), Romania (1), Russian Federation (4), Taiwan (2) and Ukraine (2).

---

* The Browns Ferry 1 reactor (1 065 MWe net) is listed as operational in the IAEA Power Reactor Information System though it has been shut down since June 1985.

OPERATING NUCLEAR REACTORS AND GENERATING CAPACITY (1957-2003)
(GWe net installed at end of year and number of operating reactors)

		1957	1958	1959	1960	1961	1962	1963	1964	1965	1966	1967	1968
Argentina	Capacity												
	Operating reactors												
Armenia	Capacity												
	Operating reactors												
Belgium	Capacity						11	11	11	11	11	11	11
	Operating reactors						1	1	1	1	1	1	1
Brazil	Capacity												
	Operating reactors												
Bulgaria	Capacity												
	Operating reactors												
Canada	Capacity						22	22	22	22	25	245	245
	Operating reactors						1	1	1	1	1	2	2
China	Capacity												
	Operating reactors												
Czech Republic	Capacity												
	Operating reactors												
Finland	Capacity												
	Operating reactors												
France	Capacity			38	38	38	38	38	108	318	798	1 178	1 178
	Operating reactors			1	1	1	1	1	2	3	4	6	6
Germany	Capacity					15	15	15	15	15	366	379	969
	Operating reactors					1	1	1	1	1	4	5	7
Hungary	Capacity												
	Operating reactors												
India	Capacity												
	Operating reactors												
Italy	Capacity							153	563	563	563	563	563
	Operating reactors							1	3	3	3	3	3
Japan	Capacity									166	166	166	166
	Operating reactors									1	1	1	1
Kazakhstan	Capacity												
	Operating reactors												
Korea, Republic of	Capacity												
	Operating reactors												
Lithuania	Capacity												
	Operating reactors												
Mexico	Capacity												
	Operating reactors												
Netherlands	Capacity												55
	Operating reactors												1
Pakistan	Capacity												
	Operating reactors												
Romania	Capacity												
	Operating reactors												
Russian Federation	Capacity								299	299	299	445	445
	Operating reactors								2	2	2	3	3
Slovak Republic	Capacity												
	Operating reactors												
Slovenia	Capacity												
	Operating reactors												
South Africa	Capacity												
	Operating reactors												
Spain	Capacity												153
	Operating reactors												1
Sweden	Capacity								10	10	10	10	10
	Operating reactors								1	1	1	1	1
Switzerland	Capacity												
	Operating reactors												
Taiwan	Capacity												
	Operating reactors												
Ukraine	Capacity												
	Operating reactors												
United Kingdom	Capacity	50	100	249	348	348	774	956	956	2 256	2 806	2 996	3 088
	Operating reactors	1	2	5	7	7	10	12	12	18	21	22	23
United States	Capacity	60	60	60	424	424	681	850	867	867	991	2 027	1 999
	Operating reactors	1	1	1	3	3	4	8	9	9	11	14	13
Total	Capacity	110	160	347	810	825	1 541	2 045	2 851	4 527	6 035	8 020	8 882
	Growth Rate		45.5	116.9	133.4	1.9	86.8	32.7	39.4	58.8	33.3	32.9	10.7
	Operating reactors	2	3	7	11	12	18	25	32	40	49	58	62

OPERATING NUCLEAR REACTORS AND GENERATING CAPACITY (1957-2003)
(GWe net installed at end of year and number of operating reactors)

		1969	1970	1971	1972	1973	1974	1975	1976	1977	1978	1979	1980
Argentina	Capacity						321	319	319	336	345	335	335
	Operating reactors						1	1	1	1	1	1	1
Armenia	Capacity								376	376	376	376	752
	Operating reactors								1	1	1	1	2
Belgium	Capacity	11	11	11	11	11	361	1 680	1 670	1 670	1 670	1 670	1 670
	Operating reactors	1	1	1	1	1	2	4	4	4	4	4	4
Brazil	Capacity												
	Operating reactors												
Bulgaria	Capacity						420	811	814	816	816	816	816
	Operating reactors						1	2	2	2	2	2	2
Canada	Capacity	245	245	1 299	2 122	2 536	2 535	2 531	2 534	4 014	4 508	5 248	5 172
	Operating reactors	2	2	4	6	7	7	7	7	9	9	10	10
China	Capacity												
	Operating reactors												
Czech Republic	Capacity												
	Operating reactors												
Finland	Capacity									431	1 176	1 065	2 217
	Operating reactors									1	2	2	4
France	Capacity	1 658	1 690	2 109	2 752	2 650	2 653	2 658	2 580	4 349	6 204	7 984	14 100
	Operating reactors	7	7	8	9	8	8	8	9	11	13	15	21
Germany	Capacity	973	1 019	1 669	2 211	2 177	3 769	4 068	6 871	7 692	9 371	10 242	10 240
	Operating reactors	7	8	9	9	10	12	12	15	16	18	19	18
Hungary	Capacity												
	Operating reactors												
India	Capacity	420	420	420	420	565	557	606	626	626	626	626	816
	Operating reactors	2	2	2	2	3	3	3	3	3	3	3	4
Italy	Capacity	563	571	576	576	547	550	563	551	564	1 216	1 112	1 112
	Operating reactors	3	3	3	3	3	3	3	3	3	4	4	4
Japan	Capacity	507	1 311	1 323	1 795	2 185	4 937	6 246	7 099	9 474	13 211	14 249	14 770
	Operating reactors	2	4	4	5	6	10	12	13	16	20	21	22
Kazakhstan	Capacity						135	135	135	135	135	135	135
	Operating reactors						1	1	1	1	1	1	1
Korea, Republic of	Capacity									555	555	564	564
	Operating reactors									1	1	1	1
Lithuania	Capacity												
	Operating reactors												
Mexico	Capacity												
	Operating reactors												
Netherlands	Capacity	55	55	54	54	499	528	498	501	501	496	496	498
	Operating reactors	1	1	1	1	2	2	2	2	2	2	2	2
Pakistan	Capacity			128	137	126	126	126	137	126	125	125	125
	Operating reactors			1	1	1	1	1	1	1	1	1	1
Romania	Capacity												
	Operating reactors												
Russian Federation	Capacity	781	781	1 166	1 166	2 887	3 318	4 253	5 188	5 188	5 188	7 038	8 548
	Operating reactors	4	4	5	5	8	11	13	15	15	15	17	19
Slovak Republic	Capacity					110	90	100	93	0	405	386	784
	Operating reactors					1	1	1	1	0	1	1	2
Slovenia	Capacity												
	Operating reactors												
South Africa	Capacity												
	Operating reactors												
Spain	Capacity	153	160	616	1 113	1 073	1 073	1 073	1 073	1 073	1 073	1 073	1 073
	Operating reactors	1	1	2	3	3	3	3	3	3	3	3	3
Sweden	Capacity	10	12	463	470	450	2 713	3 186	3 172	3 757	3 700	3 700	5 543
	Operating reactors	1	1	2	2	2	5	5	5	6	6	6	8
Switzerland	Capacity	364	364	1 027	967	1 006	1 006	1 006	1 006	1 020	1 020	1 944	1 940
	Operating reactors	1	1	3	3	3	3	3	3	3	3	4	4
Taiwan	Capacity									604	1 208	1 208	1 208
	Operating reactors									1	2	2	2
Ukraine	Capacity									723	1 850	1 850	2 211
	Operating reactors									1	2	2	3
United Kingdom	Capacity	3 410	3 360	3 628	4 056	3 964	3 964	4 147	5 312	5 853	6 220	6 223	6 398
	Operating reactors	24	24	25	26	26	26	27	31	32	33	33	33
United States	Capacity	3 666	6 295	8 341	13 893	21 341	31 581	35 144	42 457	45 029	49 738	49 719	50 822
	Operating reactors	14	18	21	29	38	50	53	62	64	69	68	69
Total	Capacity	12 816	16 294	22 830	31 743	42 262	60 637	69 150	82 514	94 912	111 232	118 184	131 849
	Growth Rate	44.3	27.1	40.1	39.0	33.1	43.5	14.0	19.3	15.0	17.2	6.3	11.6
	Operating reactors	70	77	91	105	123	150	161	182	197	216	223	240

OPERATING NUCLEAR REACTORS AND GENERATING CAPACITY (1957-2003)
(GWe net installed at end of year and number of operating reactors)

		1981	1982	1983	1984	1985	1986	1987	1988	1989	1990	1991	1992
Argentina	Capacity	335	335	935	935	935	935	935	935	935	935	935	935
	Operating reactors	1	1	2	2	2	2	2	2	2	2	2	2
Armenia	Capacity	752	752	752	752	752	752	752	752	752	376	376	376
	Operating reactors	2	2	2	2	2	2	2	2	2	1	1	1
Belgium	Capacity	1 670	3 574	3 467	3 467	4 445	5 487	5 503	5 480	5 501	5 501	5 501	5 501
	Operating reactors	4	6	6	6	7	8	8	7	7	7	7	7
Brazil	Capacity		626	626	626	626	626	626	626	626	626	626	626
	Operating reactors		1	1	1	1	1	1	1	1	1	1	1
Bulgaria	Capacity	1 224	1 632	1 632	1 632	1 632	1 632	2 579	2 519	2 585	2 585	3 538	3 538
	Operating reactors	3	4	4	4	4	4	5	5	5	5	6	6
Canada	Capacity	5 248	6 399	7 565	9 752	9 646	11 195	12 166	12 185	12 185	13 086	13 993	13 993
	Operating reactors	10	12	14	17	16	18	19	18	18	19	20	20
China	Capacity												
	Operating reactors												
Czech Republic	Capacity					396	1 219	1 632	1 632	1 632	1 632	1 632	1 632
	Operating reactors					1	3	4	4	4	4	4	4
Finland	Capacity	2 200	2 196	2 204	2 246	2 300	2 310	2 310	2 310	2 310	2 310	2 310	2 310
	Operating reactors	4	4	4	4	4	4	4	4	4	4	4	4
France	Capacity	21 345	23 054	26 603	32 789	37 536	44 542	49 347	52 300	52 335	55 995	56 910	57 959
	Operating reactors	29	31	35	41	44	48	52	54	54	56	56	56
Germany	Capacity	10 299	11 507	12 788	18 004	17 808	20 567	20 562	23 220	23 588	22 874	21 226	21 315
	Operating reactors	18	19	20	24	23	25	25	27	27	25	20	20
Hungary	Capacity		82	410	835	825	1 252	1 670	1 660	1 660	1 710	1 645	1 729
	Operating reactors		1	1	2	2	3	4	4	4	4	4	4
India	Capacity	860	832	959	975	1 163	1 154	1 154	1 154	1 364	1 324	1 324	1 551
	Operating reactors	4	4	5	5	6	6	6	6	7	7	7	9
Italy	Capacity	1 404	1 403	1 286	1 273	1 273	1 273	1 273	1 120	1 120	1 120		
	Operating reactors	4	4	3	3	3	3	3	2	2	2		
Japan	Capacity	16 438	16 438	18 849	21 536	23 464	25 646	26 709	28 153	29 139	30 716	31 843	34 090
	Operating reactors	24	24	27	30	32	34	35	37	38	40	41	43
Kazakhstan	Capacity	135	135	135	135	135	135	135	135	135	135	135	135
	Operating reactors	1	1	1	1	1	1	1	1	1	1	1	1
Korea, Republic of	Capacity	564	1 184	1 789	1 790	3 580	5 380	5 380	6 300	7 220	7 220	7 220	7 220
	Operating reactors	1	2	3	3	5	7	7	8	9	9	9	9
Lithuania	Capacity			1 380	1 380	1 380	1 380	2 816	2 760	2 760	2 760	2 760	2 370
	Operating reactors			1	1	1	1	2	2	2	2	2	2
Mexico	Capacity									654	640	640	654
	Operating reactors									1	1	1	1
Netherlands	Capacity	498	503	504	507	508	507	507	508	539	539	510	507
	Operating reactors	2	2	2	2	2	2	2	2	2	2	2	2
Pakistan	Capacity	125	125	125	137	137	125	125	125	125	125	125	125
	Operating reactors	1	1	1	1	1	1	1	1	1	1	1	1
Romania	Capacity												
	Operating reactors												
Russian Federation	Capacity	9 884	10 809	11 734	12 993	15 793	16 743	18 897	18 647	18 647	19 375	18 893	18 893
	Operating reactors	21	22	23	24	27	28	29	30	30	30	28	28
Slovak Republic	Capacity	780	796	796	816	1 632	1 632	1 632	1 632	1 632	1 632	1 632	1 632
	Operating reactors	2	2	2	2	4	4	4	4	4	4	4	4
Slovenia	Capacity	632	632	632	632	632	620	620	620	620	620	632	632
	Operating reactors	1	1	1	1	1	1	1	1	1	1	1	1
South Africa	Capacity				920	1 840	1 840	1 840	1 840	1 840	1 840	1 840	1 840
	Operating reactors				1	2	2	2	2	2	2	2	2
Spain	Capacity	2 007	2 003	3 891	4 731	5 636	5 608	6 562	7 526	7 595	7 595	7 121	7 115
	Operating reactors	4	4	6	7	8	8	8	9	10	10	9	9
Sweden	Capacity	6 420	7 330	7 307	7 355	9 491	9 602	9 652	9 678	9 843	9 824	9 999	10 041
	Operating reactors	9	10	10	10	12	12	12	12	12	12	12	12
Switzerland	Capacity	1 944	1 946	1 946	2 956	2 887	2 924	2 951	2 952	2 944	2 943	2 938	2 947
	Operating reactors	4	4	4	5	5	5	5	5	5	5	5	5
Taiwan	Capacity	2 156	3 104	3 104	3 994	4 884	4 884	4 884	4 884	4 890	4 828	4 890	4 890
	Operating reactors	3	4	4	5	6	6	6	6	6	6	6	6
Ukraine	Capacity	3 520	4 470	5 395	6 345	8 249	10 145	11 726	11 120	13 020	13 020	13 020	12 419
	Operating reactors	5	6	7	8	10	12	13	13	15	15	15	14
United Kingdom	Capacity	6 490	6 516	7 146	7 804	10 061	10 141	10 181	11 734	11 530	11 360	11 484	12 001
	Operating reactors	32	32	33	34	38	38	38	41	41	37	37	37
United States	Capacity	54 972	59 117	62 135	68 798	77 485	84 406	92 595	94 835	98 270	99 388	99 377	99 129
	Operating reactors	73	77	79	85	93	99	107	108	111	111	111	110
Total	Capacity	151 902	167 500	186 095	216 115	247 131	274 662	297 721	309 342	317 996	324 634	325 075	328 105
	Growth Rate	15.2	10.3	11.1	16.1	14.4	11.1	8.4	3.9	2.8	2.1	0.1	0.9
	Operating reactors	262	281	301	331	363	388	409	419	428	426	419	421

OPERATING NUCLEAR REACTORS AND GENERATING CAPACITY (1957-2003)
(GWe net installed at end of year and number of operating reactors)

		1993	1994	1995	1996	1997	1998	1999	2000	2001	2002	2003
Argentina	Capacity	935	935	935	935	935	935	933	978	935	935	935
	Operating reactors	2	2	2	2	2	2	2	2	2	2	2
Armenia	Capacity	376	376	376	376	376	376	376	376	376	376	376
	Operating reactors	1	1	1	1	1	1	1	1	1	1	1
Belgium	Capacity	5 501	5 535	5 573	5 667	5 728	5 712	5 753	5 753	5 753	5 801	5 801
	Operating reactors	7	7	7	7	7	7	7	7	7	7	7
Brazil	Capacity	626	626	626	626	626	626	626	1 976	1 976	1 901	1 901
	Operating reactors	1	1	1	1	1	1	2	2	2	2	2
Bulgaria	Capacity	3 538	3 538	3 538	3 538	3 538	3 538	3 538	3 666	3 538	3 538	2 722
	Operating reactors	6	6	6	6	6	6	6	6	6	6	4
Canada	Capacity	15 800	15 750	15 750	14 902	14 902	11 694	9 998	9 998	10 018	10 018	10 018
	Operating reactors	22	22	22	21	21	16	14	14	14	14	14
China	Capacity	288	2 167	2 188	2 167	2 188	2 167	2 167	2 188	2 167	3 715	5 312
	Operating reactors	1	3	3	3	3	3	3	3	3	5	7
Czech Republic	Capacity	1 782	1 782	1 782	1 648	1 760	1 648	1 648	1 648	1 648	2 560	3 510
	Operating reactors	4	4	4	4	4	4	4	4	4	5	6
Finland	Capacity	2 310	2 310	2 310	2 310	2 398	2 656	2 656	2 656	2 656	2 656	2 656
	Operating reactors	4	4	4	4	4	4	4	4	4	4	4
France	Capacity	58 775	58 880	58 340	58 340	60 050	60 050	61 500	62 950	62 950	62 840	63 180
	Operating reactors	56	56	55	55	56	56	57	58	58	58	58
Germany	Capacity	21 398	21 450	20 942	21 010	21 049	21 063	21 122	21 177	21 283	21 283	21 283
	Operating reactors	20	20	19	19	19	19	19	19	19	19	19
Hungary	Capacity	1 726	1 729	1 729	1 729	1 729	1 729	1 729	1 729	1 755	1 755	1 755
	Operating reactors	4	4	4	4	4	4	4	4	4	4	4
India	Capacity	1 561	1 550	1 747	1 680	1 658	1 708	1 708	2 508	2 508	2 547	2 548
	Operating reactors	9	9	10	10	10	10	10	14	14	14	14
Italy	Capacity											
	Operating reactors											
Japan	Capacity	37 881	39 808	40 604	43 235	43 234	43 234	43 097	43 097	43 893	43 893	43 893
	Operating reactors	47	49	50	52	52	52	51	51	52	52	52
Kazakhstan	Capacity	70	64	50	52	52	52	52	0	0	0	0
	Operating reactors	1	1	1	1	1	1	1	0	0	0	0
Korea, Republic of	Capacity	7 220	8 088	9 115	9 120	9 770	12 340	12 990	12 990	13 940	14 890	15 850
	Operating reactors	9	10	11	11	12	15	16	16	17	18	19
Lithuania	Capacity	2 370	2 370	2 370	2 370	2 370	2 370	2 370	2 370	2 370	2 370	2 370
	Operating reactors	2	2	2	2	2	2	2	2	2	2	2
Mexico	Capacity	654	1 256	1 256	1 274	1 242	1 310	1 338	1 290	1 290	1 360	1 360
	Operating reactors	1	2	2	2	2	2	2	2	2	2	2
Netherlands	Capacity	510	510	510	510	505	450	449	449	449	450	450
	Operating reactors	2	2	2	2	2	1	1	1	1	1	1
Pakistan	Capacity	125	125	125	125	125	125	125	425	425	425	425
	Operating reactors	1	1	1	1	1	1	1	2	2	2	2
Romania	Capacity				655	655	655	655	655	655	655	655
	Operating reactors				1	1	1	1	1	1	1	1
Russian Federation	Capacity	19 843	19 843	19 843	19 843	19 843	19 843	19 843	19 843	20 793	20 793	20 793
	Operating reactors	29	29	29	29	29	29	29	29	30	30	30
Slovak Republic	Capacity	1 632	1 620	1 632	1 744	1 696	2 040	2 440	2 440	2 440	2 442	2 442
	Operating reactors	4	4	4	4	4	5	6	6	6	6	6
Slovenia	Capacity	620	620	620	620	620	620	620	646	656	676	676
	Operating reactors	1	1	1	1	1	1	1	1	1	1	1
South Africa	Capacity	1 840	1 840	1 840	1 840	1 840	1 840	1 840	1 840	1 840	1 800	1 800
	Operating reactors	2	2	2	2	2	2	2	2	2	2	2
Spain	Capacity	7 158	7 158	7 098	7 208	7 113	7 350	7 369	7 457	7 469	7 574	7 588
	Operating reactors	9	9	9	9	9	9	9	9	9	9	9
Sweden	Capacity	10 002	10 037	10 058	10 050	10 048	10 070	10 038	9 419	9 416	9 440	9 462
	Operating reactors	12	12	12	12	12	12	12	11	11	11	11
Switzerland	Capacity	2 978	3 005	3 056	3 075	3 093	3 093	3 127	3 170	3 200	3 220	3 220
	Operating reactors	5	5	5	5	5	5	5	5	5	5	5
Taiwan	Capacity	4 890	4 888	4 884	4 884	4 884	4 884	4 884	4 884	4 884	4 884	4 884
	Operating reactors	6	6	6	6	6	6	6	6	6	6	6
Ukraine	Capacity	12 162	12 072	13 045	12 860	12 120	12 120	12 120	12 120	11 207	11 207	11 207
	Operating reactors	14	14	15	15	14	14	14	14	13	13	13
United Kingdom	Capacity	11 726	11 695	12 914	12 914	12 920	12 920	12 920	12 450	12 450	12 004	12 004
	Operating reactors	35	34	35	35	35	35	35	33	33	27	27
United States	Capacity	98 878	98 784	99 004	100 378	100 181	97 378	96 987	97 221	97 493	97 964	98 357
	Operating reactors	109	109	109	110	109	105	104	104	104	104	104
Total	Capacity	335 175	340 411	343 860	347 685	349 248	346 596	347 018	350 369	352 433	355 972	359 433
	Growth Rate	2,2	1.6	1.0	1.1	0.4	-0,8	0.1	1 0	0.6	1.0	1.0
	Operating reactors	426	431	434	437	437	431	431	433	435	433	435

Source: IAEA Power Reactor Information System as modified using other sources.

Appendix 3.1

ANNUAL URANIUM REQUIREMENTS (1956-2003)
(tU)

	1956	1957	1958	1959	1960	1961	1962	1963	1964	1965	1966	1967	1968	1969	1970	1971	1972	1973	1974
Argentina																			
Armenia																			
Belgium																			
Brazil																			100
Bulgaria																			
Canada																			
China																			
CSFR																			
Czech Republic																			
Finland																			
France																			
Germany																			
Hungary																			
India															84	84	84	113	111
Italy																			
Japan																		70	70
Kazakhstan																			
Korea																			
Lithuania																			
Mexico																			
Netherlands																			
Pakistan																15	15	15	15
Romania																			
Russian Federation														164	164	245	245	606	697
Slovak Republic																		40	32
Slovenia																			
South Africa																			
Spain																			
Sweden																			
Switzerland																			
Taiwan																			
Ukraine																			
United Kingdom																			
United States																			
WORLD	770	516	732	716	1 655	1 463	2 079	3 696	4 543	4 928	5 082	7 084	9 625	12 200	13 964	15 245	16 245	17 716	18 499

ANNUAL URANIUM REQUIREMENTS (1956-2003)
(tU)

	1975	1976	1977	1978	1979	1980	1981	1982	1983	1984	1985	1986	1987	1988	1989	1990	1991	1992	1993
Argentina						165	126	156	170	175	160	114	126	109	100	147	150	148	150
Armenia					90					180	180	180	180	180	180	90	90	90	90
Belgium						1 050	1 050	1 450	700	990	960	950	950	950	950	950	950	950	950
Brazil						106	103	93	106	112	112	110	110	110	110	110	140	110	110
Bulgaria	195	195	195	195	195	195	213	390	390	390	390	390	390	600	700	700	700	844	844
Canada						1 000	1 100	1 200	1 400	1 300	960	1 400	1 600	1 700	1 800	1 900	1 900	1 900	1 900
China																	50	48	197
CSFR											247	761	863	863	730	730	916		
Czech Republic																		380	390
Finland						270	270	270	270	350	350	480	480	480	480	480	572	460	460
France						3 700	4 400	5 100	6 500	6 000	6 000	6 200	6 200	6 050	7 200	7 200	7 300	7 800	7 900
Germany						2 760	1 949	2 033	2 700	3 400	3 300	3 300	3 300	3 300	3 300	3 300	3 300	3 200	3 200
Hungary								21	104	211	209	317	423	420	420	420	420	415	434
India	121	125	125	125	125	163	172	166	192	195	233	230	230	230	220	220	255	207	207
Italy						46	38	46	232	185	185	191	0	0					
Japan						4 000	4 700	4 900	3 500	4 900	4 500	4 800	6 000	6 400	6 600	6 900	7 100	7 600	8 100
Kazakhstan	70	70	70	70	70	70	70	70	70	70	70	70	70	70	70	70	70	70	70
Korea						191	191	309	470	540	920	1 190	1 390	840	1 040	1 158	1 233	1 551	2 144
Lithuania												516	516	516	516	516	516	516	516
Mexico															145	111	120	116	220
Netherlands						95	95	95	95	95	95	95	95	95	95	95	95	76	85
Pakistan	15	15	15	15	15	15	15	15	15	15	15	15	15	15	15	15	15	16	16
Romania																			
Russian Federation	893	1 089	1 089	1 089	1 478	1 795	2 076	2 270	2 464	2 729	3 317	3 516	3 968	3 916	3 916	4 069	4 000	4 000	4 210
Slovak Republic	36	33	0	97	92	187	186	180	180	184	360	360	360	360	360	360	360	360	360
Slovenia							104	104	104	104	102	102	102	102	102	102	102	102	102
South Africa						576	424	374	145	289	289	270	270	270	270	270	270	200	200
Spain						797	391	564	707	510	543	1 019	1 300	1 254	1 100	1 124	1 233	1 015	1 335
Sweden						311	702	1 024	1 300	1 400	1 200	1 148	1 325	1 500	1 400	1 400	1 500	1 600	1 600
Switzerland						400	533	483	350	570	570	566	570	529	529	570	570	532	532
Taiwan			111	223	223	223	397	572	572	736	900	900	900	900	940	900	850	810	810
Ukraine			147	376	376	449	715	907	1 095	1 288	1 675	2 059	2 380	2 257	2 643	2 643	2 643	2 526	2 060
United Kingdom						1 900	1 300	1 500	1 500	2 150	2 200	1 500	1 550	1 700	1 900	2 000	2 000	2 330	2 035
United States						7 400	8 300	7 400	8 900	12 900	12 600	11 850	11 150	14 190	14 500	11 400	16 000	17 300	17 300
WORLD	19 944	23 137	25 251	27 117	29 301	28 044	29 800	31 872	34 711	42 268	42 942	44 599	46 813	49 906	52 331	49 950	55 420	57 272	58 527

Appendix 3.1 (cont'd 3/3)

ANNUAL URANIUM REQUIREMENTS (1956-2003) (tU)

	1994	1995	1996	1997	1998	1999	2000	2001	2002	2003	Total (1956-2003)
Argentina	150	150	150	150	150	150	120	120	120	120	3 376
Armenia	90	89	89	89	89	89	89	89	70	90	3 213
Belgium	1 030	1 030	1 050	1 050	1 050	1 050	1 050	1 050	1 150	1 150	24 460
Brazil	120	120	120	120	120	310	450	450	450	450	4 252
Bulgaria	844	844	844	844	844	844	840	840	840	840	16 635
Canada	1 900	1 900	1 800	1 800	1 200	1 300	1 800	1 800	1 400	1 700	37 660
China	300	300	300	300	380	380	380	380	790	1 100	4 905
CSFR											5 110
Czech Republic	370	335	370	370	440	516	722	512	745	745	5 895
Finland	450	450	495	496	550	550	500	500	500	500	10 663
France	8 900	8 900	8 900	8 600	8 200	8 200	8 879	8 568	8 570	8 570	173 837
Germany	3 000	3 300	3 200	2 900	3 100	3 100	3 350	3 200	3 150	3 200	73 842
Hungary	420	420	415	415	400	400	368	311	370	370	7 703
India	194	220	220	220	376	433	397	397	430	465	7 369
Italy											923
Japan	7 700	7 700	8 700	7 500	7 810	9 290	7 500	11 100	7 840	8 380	163 520
Kazakhstan	50	50	50	50	50						1 720
Korea	1 580	1 850	1 810	2 760	2 400	2 500	3 400	2 900	2 780	2 780	37 927
Lithuania	470	450	385	415	480	640	240	240	360	310	9 018
Mexico	169	323	325	170	360	190	187	189	330	230	3 185
Netherlands	95	99	93	74	74	74	84	10	95	95	2 094
Pakistan	16	16	16	16	16	16	65	65	65	65	703
Romania		100	100	100	100	100	100	100	100	100	900
Russian Federation	4 210	4 210	3 800	3 800	3 600	3 600	4 000	4 000	4 600	5 100	94 925
Slovak Republic	440	440	440	770	566	618	63	63	500	500	8 887
Slovenia	102	102	102	102	110	110	120	120	190	230	2 622
South Africa	200	200	200	200	200	200	200	200	280	280	6 277
Spain	1 320	1 375	1 155	1 075	1 500	1 150	1 400	1 700	1 470	1 500	26 537
Sweden	1 500	1 500	1 500	1 500	1 600	1 600	1 500	1 500	1 600	1 600	32 810
Switzerland	532	532	537	499	570	480	360	375	360	375	11 924
Taiwan	810	810	810	810	810	810	830	830	830	830	19 147
Ukraine	2 015	2 650	2 490	2 640	2 350	2 433	2 200	2 050	2 200	2 200	49 467
United Kingdom	2 432	2 512	2 622	2 622	2 356	2 356	2 250	2 250	1 930	1 760	48 655
United States	15 600	18 400	17 400	21 300	17 700	18 100	20 570	18 420	22 700	22 800	364 180
World total	57 009	61 377	60 488	63 757	59 551	61 589	64 014	64 329	66 815	68 435	1 513 327

Notes:

Individual country data prior to 1980 are estimated using capacity information provided by the IAEA Power Reactor Information System and/or member countries.

Data after 1980 are taken from a Red Book unless otherwise indicated.

Data in *bold grey* from a Red Book that was a Secretariat estimate.

Data in **bold red** are estimates made using a ratio of tonnes U per megawatt net of installed capacity using the nearest applicable data and based on data from the IAEA Power Reactor Information System.

Armenia	376 MWe net = 90 tU = 0.239 tU/MWe
Bulgaria	1 632 MWe net = 390 tU = 0.239 tU/MWe
Hungary	1 660 MWe net = 420 tU = 0.253 tU/MWe
India	1 154 MWe net = 230 tU = 0.199 tU/MWe
Kazakhstan	135 MWe net = 70 tU = 0.51 tU/MWe
Korea	For 1981, same as 1980. For 1982, 1981 plus ratio for a Canadian PHWR ca 1980 of 0.19 tU/MWe net based on 5 172 MWe net = 1 000 tU.
Russia	18 893 MWe net = 4 000 tU = 0.21 tU/MWe (Domestic requirements in 1992).
Slovak Republic	1 632 MWe net = 360 tU = 0.221 tU/MWe
Slovenia	620 MWe net = 102 tU = 0.165 tU/MWe
Ukraine	12 420 MWe net = 2 520 tU = 0.203 tU/MWe

CSFR = Czech-Slovak Federated Republic.

ACTUAL AND PROJECTED URANIUM REQUIREMENTS BY RED BOOK EDITION (tU)

	1956	1960	1965	1970	1975	1980	1985	1990	1995	2000	2005	2010	2015	2020	2025	2030
actual	770	1 650	4 920	14 230	18 750	26 000	36 100	49 290	61 380	64 010						
Red Book edition																
1969																
forecast low		1 650	4 920	14 230	33 840	56 140										
forecast high					45 370	81 510										
1970																
actual				9 000												
forecast					29 000	56 000	100 000									
1973																
low					23 000	51 000	79 000	100 000								
high					26 000	66 000	127 000	224 000								
1976																
actual					18 000											
low					18 000	48 000	82 000	130 000	188 000	236 000						
high					18 000	53 000	101 000	168 000	249 000	313 000						
1977																
low						41 000	65 000	85 000	104 000	125 000						
high						43 000	88 000	156 000	234 000	338 000						
1979																
actual						28 000	NA	56 000	NA	96 000	NA	NA	88 000	NA	48 000	
low						32 000	NA	88 000	NA	199 000	NA	NA	433 000	NA	593 000	
high																
1982																
low							44 000	53 000	64 000	81 000	59 000	54 000	45 000	55 000	41 000	
high							45 000	65 000	93 000	129 000	164 000	212 000	267 000	317 000	374 000	
1983																
low							43 000		58 000	67 000	50 000	45 000	39 000	45 000	38 000	
high							NA	54 000	70 000	90 000	119 000	165 000	214 000	265 000	321 000	
1986								55 000								
actual							37 000									
low								48 000	55 000	62 000	NA	NA	NA	NA	106 000	
high								NA	NA	NA	NA	NA	NA	NA		
1988																
reference								44 800	47 200	52 400	NA	NA	NA	NA	NA	
1989																
actual								41 600								
low									NA	52 000	NA	56 500	NA	60 400	NA	63 700
high									NA	58 400	NA	76 900	NA	90 600	NA	108 700
1991																
reference								49 290	60 920	67 820	71 260	77 670	NA	NA	NA	NA
1993																
reference									60 810	63 500	68 420	75 670	NA	NA	NA	NA
1995																
reference									61 380	64 350	69 290	NA	NA	NA	NA	NA
low										NA	NA	69 130	NA	NA	NA	NA
high										NA	NA	74 680	NA	NA	NA	NA
1997																
reference										64 520						
low											66 810	70 980	62 540	NA	NA	NA
high											69 430	77 050	82 800	NA	NA	NA
1999																
reference										64 200						
low											57 230	59 290	54 460	NA	NA	NA
high											70 380	73 160	79 800	NA	NA	NA
2001																
reference										64 010	65 920					
low												64 920	58 040	58 010	NA	NA
high												71 790	72 540	80 250	NA	NA
2003																
reference											70 600					
low												70 610	76 710	73 500	NA	NA
high												73 280	84 410	86 070	NA	NA

NA = Not available.

178

URANIUM MARKET PRICES (1970-2002)
(USD/kgU))

	1970	1971	1972	1973	1974	1975	1976	1977	1978	1979	1980
Current dollars											
Red Book Dat											
US (Domestic Uranium)								51.35	56.16	62.01	73.19
US (Imported Uranium)											
Australia											
Euratom (Multi-annual Contracts)											93.40
Euratom (Spot Contracts)											90.82
Niger											
Commercial Data (used with permission)											
NUEXCO "EV" (Spot)	16.22	15.76	15.21	16.67	28.88	61.62	103.22	109.72	112.37	110.66	82.68
Constant dollars (2003)											
PPI Inflation Index	3.646	3.538	3.428	3.143	2.724	2.462	2.357	2.215	2.053	1.847	1.628
NUEXCO "EV" (Spot)	59.15	55.74	52.14	52.39	78.67	151.71	243.29	243.03	230.70	204.39	134.60

	1981	1982	1983	1984	1985	1986	1987	1988	1989	1990	1991
Current dollars											
Red Book Dat											
US (Domestic Uranium)	96.00	102.00	99.35	84.90	81.72	78.03	71.16	67.99	50.86	40.82	35.52
US (Imported Uranium)		67.00	68.00	58.84	52.20	52.20	49.76	49.46	43.55	32.66	40.43
Australi			93.00	92.00	75.00	70.00	75.00	79.00	74.00	47.60	55.10
Euratom (Multi-annual Contracts)	86.74	83.16	80.55	77.42	75.82	80.25	84.53	82.60	76.18	76.20	67.89
Euratom (Spot Contracts)	73.05	62.38	60.42	50.09	38.83	45.95	44.85	41.89	31.63	25.08	23.56
Niger											
Commercial Data (used with permission)											
NUEXCO "EV" (Spot)	62.89	51.74	59.75	44.90	40.58	44.20	43.60	37.83	26.00	25.38	22.60
Constant dollars (2003)											
PPI Inflation Index	1.491	1.433	1.41	1.382	1.369	1.389	1.36	1.327	1.261	1.202	1.177
NUEXCO "EV" (Spot)	93.77	74.14	84.25	62.05	55.55	61.39	59.30	50.20	32.79	30.51	26.60

	1992	1993	1994	1995	1996	1997	1998	1999	2000	2001	2002
Current dollars											
Red Book Data											
US (Domestic Uranium)	34.96	34.17	26.79	28.89	35.91	33.46	31.99	30.90	29.77	27.17	26.91
US (Imported Uranium)	29.48	27.37	23.27	26.52	34.19	30.69	29.08	27.42	25.58	24.74	26.14
Australia	43.51	42.15	39.01	40.10	42.83	37.35	35.58	35.36	33.16	30.34	31.71
Euratom (Multi-annual Contracts)	64.35	54.99	52.66	45.52	40.64	39.26	38.08	37.18	34.04	34.43	32.30
Euratom (Spot Contracts)	25.03	23.99	22.31	19.98	22.54	33.90	28.00	26.48	20.93	18.90	24.23
Niger		66.59	54.96	58.53	53.55	47.17	44.05	42.36	36.53	32.81	34.24
Commercial Data (used with permission)											
NUEXCO "EV" (Spot)	20.70	18.24	18.33	21.97	33.49	27.39	23.61	21.45	18.04	21.06	25.69
Constant dollars (2003)											
PPI Inflation Index	1.163	1.149	1.142	1.12	1.091	1.087	1.096	1.077	1.038	1.018	1.032
NUEXCO "EV" (Spot)	24.07	20.96	20.93	24.61	36.54	29.77	25.88	23.10	18.73	21.44	26.51

Source: NUEXCO EV spot price data courtesy of TradeTech (www.uranium.info).

Appendix 4.2

URANIUM, GOLD AND OIL MARKET PRICES (1970-2004)
(USD current)

		1970	1971	1972	1973	1974	1975	1976	1977	1978	1979	1980	1981
Gold	USD/troy oz	36.02	40.62	58.42	97.39	154.00	160.86	124.74	147.84	193.40	306.00	615.00	460.00
Uranium	USD/lb U_3O_8	6.24	6.06	5.95	6.41	11.03	23.68	39.70	42.20	43.23	42.57	31.79	24.19
Crude Oil	USD/bbl	3.18	3.39	3.39	3.89	6.87	7.67	8.19	8.57	9.00	12.64	21.59	31.77

		1982	1983	1984	1985	1986	1987	1988	1989	1990	1991	1992	1993
Gold	USD/troy oz	376.00	424.00	361.00	317.00	368.00	447.00	437.00	381.00	383.50	382.10	343.80	359.80
Uranium	USD/lb U_3O_8	19.90	22.98	17.27	15.60	17.00	16.78	14.55	10.00	9.76	8.70	7.95	7.12
Crude Oil	USD/bbl	28.52	26.19	25.88	24.09	12.51	15.40	12.58	15.86	20.03	16.54	15.99	14.25

		1994	1995	1996	1997	1998	1999	2000	2001	2002	2003	2004
Gold	USD/troy oz	384.00	384.20	387.70	331.00	294.20	279.00	279.10	271.00	309.70	363.40	409.70
Uranium	USD/lb U_3O_8	7.05	8.45	14.19	10.27	9.08	8.25	6.94	8.10	9.88	11.56	18.55
Crude Oil	USD/bbl	13.19	14.62	18.46	17.23	10.87	15.56	26.72	21.84	22.51	27.54	37.66

Sources:
Oil: US DOE Energy Information Administration (Domestic Crude Oil price).
Gold: National Mining Association.
Uranium: NUEXCO EV spot price.

Appendix 5.1

ANNUAL EXPLORATION EXPENDITURES (1945-2003) (USD million)

Country	1945	1946	1947	1948	1949	1950	1951	1952	1953	1954	1955	1956	1957	1958	1959	1960
Argentina																
Australia																
Bangladesh																
Belgium																
Bolivia (c)																
Botswana (c)																
Brazil (c)																
Bulgaria																
Cameroon																
Canada																
Central African Republic (c)																
Chile																
China																
Colombia																
Congo, Democratic Republic																
Costa Rica																
Cuba																
Czech Republic																
Denmark (c)																
Ecuador (c)																
Egypt (c)																
Finland (c)																
France (c)																
Gabon (c)																
Germany, Federal Republic																
Germany, Democratic Republic (a)		17.860	17.860	17.860	17.860	17.860	17.860	17.860	17.860	30.480	30.480	30.480	30.480	30.480	30.620	30.620
Ghana																
Greece																
Guatemala																
Hungary																
India (c)																
Indonesia (c)																
Iran																
Ireland																
Italy (c)																
Jamaica																
Japan (c)																
Jordan																
Kazakhstan																
Korea, Republic of																
Lesotho																
Madagascar																
Malaysia																
Mali (c)																

Appendix 5.1 (cont'd 2/8)

ANNUAL EXPLORATION EXPENDITURES (1945-2003) (USD million)

Country	1945	1946	1947	1948	1949	1950	1951	1952	1953	1954	1955	1956	1957	1958	1959	1960
Mexico																
Mongolia																
Morocco																
Namibia (c)																
Niger																
Nigeria (c)																
Norway																
Pakistan																
Paraguay (c)																
Peru (c)																
Philippines (c)																
Poland																
Portugal (c)																
Romania																
Russian Federation																
Rwanda																
Slovenia																
Somalia																
South Africa (c)																
Spain (c)																
Sri Lanka (c)																
Sudan																
Sweden (c)																
Switzerland (c)																
Syria (c)																
Thailand																
Turkey (c)																
Ukraine																
United Kingdom (c)																
United States				0.500	1.000	1.000	1.000	1.000	1.500	1.500	2.000	2.500	2.500	2.500	3.000	3.000
Uruguay																
USSR (b)	2.600	13.740	13.740	13.740	13.740	13.740	27.300	27.300	27.300	27.300	27.300	51.000	51.000	51.000	51.000	51.000
Uzbekistan																
Vietnam																
Yugoslavia																
Zambia																
Zimbabwe (c)																
Total (d)	2.600	31.600	31.600	32.100	32.600	32.600	46.160	46.160	46.660	59.280	59.780	83.980	83.980	83.980	84.620	84.620

Appendix 5.1 (cont'd 3/8)

ANNUAL EXPLORATION EXPENDITURES (1945-2003) (USD million)

	1961	1962	1963	1964	1965	1966	1967	1968	1969	1970	1971	1972	1973	1974	1975	1976
Argentina																0.000
Australia							1.000	4.000	7.000	9.000	10.000	12.800	16.300	14.600	10.000	14.000
Bangladesh																0.000
Belgium																
Bolivia (c)											0.200	0.060	0.070	1.150	2.000	3.300
Botswana (c)															0.060	0.070
Brazil (c)											10.000	1.500	8.500	10.000	12.000	24.400
Bulgaria																
Cameroon																
Canada											NR	24.900			23.500	43.100
Central African Republic											15.000	0.000	0.000		1.700	1.500
Chile											NR				0.145	0.570
China																
Colombia													0.000	0.500	0.500	1.500
Congo, Democratic Republic																
Costa Rica																
Cuba																
Czech Republic											13.845	13.845	13.845	13.845	13.845	13.845
Denmark (c)											0.450	0.000	0.170	0.170	-0.350	
Ecuador (c)																0.040
Egypt (c)																
Finland (c)											0.700	0.500	0.600	0.900	1.400	1.440
France (c)											87.000	4.800	5.400	8.200	14.600	19.600
Gabon (c)																33.700
Germany, Federal Republic (a)												1.200	1.700	2.000	2.700	3.100
Germany, Democratic Republic (a)	32.000	32.000	32.110	32.200	32.000	32.000	32.100	32.000	32.620	35.100	36.190	40.140	48.140	49.280	51.960	50.850
Ghana											NR	0.000	0.000	0.000	0.000	0.000
Greece											0.000	0.090	0.090	0.122	0.224	0.432
Guatemala																
Hungary																
India (c)											39.146	2.584	2.402	3.214	3.400	3.400
Indonesia (c)											NR					
Iran											NR	0.000	0.000			
Ireland													NR		2.000------	
Italy (c)																
Jamaica																
Japan (c)											10.617	0.630	0.533	0.507	0.360	0.460
Jordan											0.000		0.028	0.020		
Kazakhstan																
Korea, Republic of											0.000	0.065	0.043	0.030	0.058	0.246
Lesotho																
Madagascar											NR	0.000	0.000	0.000	0.000	0.050
Malaysia																
Mali (c)																
Mexico											NR	0.488	1.554	1.709	2.152	2.766
Mongolia																
Morocco																NR
Namibia (c)														0.600	1.882	3.874

Appendix 5.1 (cont'd 4/8)

ANNUAL EXPLORATION EXPENDITURES (1945-2003) (USD million)

Country	1961	1962	1963	1964	1965	1966	1967	1968	1969	1970	1971	1972	1973	1974	1975	1976
Niger										NR			34.000			17.023
Nigeria (c)														2.250	0.900	0.600
Norway																NR
Pakistan																
Paraguay (c)																
Peru (c)														0.097	0.033	0.069
Philippines (c)											0.300	0.001	0.001	0.020	0.040	0.046
Poland																
Portugal (c)											5.594	0.135	0.207	0.246	0.243	0.276
Romania																
Russian Federation																
Rwanda																
Slovenia																
Somalia										NR	------	-3.750-		6.250	------	0.000
South Africa (c)											0.841		0.376	0.822	3.442	7.257
Spain (c)											13.336	1.787	1.929	2.576	6.483	7.100
Sri Lanka (c)																0.005
Sudan													NR	0.000	0.000	0.200
Sweden (c)											1.000	1.200	1.480	2.070	2.900	5.050
Switzerland (c)											0.742	0.040	0.166	0.142	0.126	0.069
Syria (c)																0.800
Thailand																NR
Turkey (c)											3.500	0.400	0.500	0.650	0.800	0.900
Ukraine																
United Kingdom (c)											0.637	0.110	0.068	0.040	0.010	0.015
United States	3.000	3.000	3.000	3.000	4.000	7.037	18.644	34.913	33.896	31.099	41.146	32.400	49.470	80.950	128.290	195.900
Uruguay																NR
USSR (b)	65.000	65.000	65.000	65.000	65.000	73.160	73.160	73.160	73.160	73.160	94.380	94.380	94.380	94.380	94.380	127.040
Uzbekistan																
Vietnam																
Yugoslavia																
Zambia																
Zimbabwe (c)																0.400
Total (d)	100.000	100.000	100.110	100.200	101.000	112.197	124.904	144.073	146.676	148.359	203.611	242.105	290.835	303.676	389.531	550.623

185

Appendix 5.1 (cont'd 5/8)

ANNUAL EXPLORATION EXPENDITURES (1945-2003) (USD million)

Country	1977	1978	1979	1980	1981	1982	1983	1984	1985	1986	1987	1988	1989	1990	1991	1992
Argentina	4.784	7.210	5.605	7.405	5.200	4.138	5.700	2.015	0.069	0.965	0.087	0.484	0.642	0.340	0.588	1.330
Australia	18.000	29.000	32.000	45.000	43.000	30.000	13.000	11.000	9.000	12.000	18.000	20.360	17.460	11.835	10.803	10.273
Bangladesh			NR	0.203	0.046	0.034	0.034	0.040	0.043	0.040	0.013					
Belgium	0.021	0.061	1.200	0.526	0.094				0.065	0.085	0.330	0.105		0.000	0.000	0.000
Bolivia (c)			1.595		0.968											
Botswana (c)	0.105	0.070	0.100	0.210				0.210								
Brazil (c)	25.600	29.300	12.500	8.600	18.830	17.890	4.038	2.970								0.000
Bulgaria			1.282													
Cameroon																
Canada	68.000	80.000	111.900	111.550	85.543	56.900	33.950	27.970	25.310	25.170	29.720	49.720	48.000	39.380	39.250	38.420
Central African Republic (c)	1.200	1.200	1.200													
Chile			0.840	0.840	0.840	0.840	0.120	0.115	0.123	0.107	0.102	0.120	0.129	0.082	0.099	0.117
China																
Colombia	0.250	4.401	2.500	3.000	6.000	0.740	0.150	0.150	0.063	0.079	0.037	0.036	0.040			
Congo, Democratic Republic																
Costa Rica			NR	0.002	0.003	0.008	0.001		0.250	0.100						
Cuba																0.236
Czech Republic	13.845	13.845	13.845	13.845	21.540	20.700	18.820	15.800	15.490	17.810	19.080	16.200	18.600	7.370	1.540	0.660
Denmark (c)	1.100	0.100	0.500	0.350	0.200	0.400	0.200	0.100	0.050							
Ecuador (c)	0.008	0.160	0.050	0.180	0.654	0.203	0.258	0.192	0.200	0.200						
Egypt (c)						8.571	1.574	1.250	1.050	1.072	1.213	3.123	2.688	4.373	3.614	4.505
Finland (c)	0.958	1.084	1.135	1.128	1.029	0.971	0.665	0.544	0.336	0.304	0.109	0.136	0.045	0.000	0.000	0.000
France (c)	18.200	26.200	61.700	89.500	70.450	55.730	49.320	48.580	47.740	64.026	55.143	47.020	33.046	32.472	23.725	14.984
Gabon (c)	4.800	8.500	9.600	8.600	8.253	7.741	4.737	3.982	5.000	4.600	5.800	3.000	0.000			2.011
Germany, Federal Republic	9.292	11.462	10.000	11.000	13.265	7.350	5.400	2.200	2.800							0.000
Germany, Democratic Republic (a)	55.240	63.730	69.830	70.490	56.640	52.700	70.450	66.770	64.570	80.640	92.870	97.530	70.740	36.510	0.000	0.000
Ghana					0.030	0.060										
Greece	0.833	0.781	1.151	1.448	1.500	1.650	1.723	1.484	1.095	0.810	0.510	0.354	0.540	0.658	0.395	0.389
Guatemala			NR					0.580		0.030						
Hungary												2.500	1.200	0.000	0.000	0.000
India (c)	4.400	4.900	7.660				9.850	8.670	10.190	11.580	13.430	13.664	12.803	15.420	13.230	9.010
Indonesia (c)		2.449	0.392	1.163	1.600	1.300	0.874	0.170	0.187	0.052	0.229	0.206	0.222	0.368	0.886	1.230
Iran																
Ireland	------															
Italy (c)		26.000	13.000	13.000	9.000	7.000	4.000	1.000	0.600	0.700	0.760					
Jamaica						NR	0.030	0.010								
Japan (c)	0.390	0.550	0.660	0.700	0.770	0.620	0.580	0.560	0.480	0.760	0.520	0.000	0.000	0.000	0.000	0.000
Jordan												0.163	0.120	0.108	0.042	0.036
Kazakhstan																2.500
Korea, Republic of			1.949	5.741	3.961	3.688	0.477	0.302	0.213	0.081	0.315	0.329	0.325	0.038	0.025	0.000
Lesotho					NR			0.010	0.006	0.005						
Madagascar	0.112	0.152	0.398	1.049	1.489	1.087	0.914	0.042								
Malaysia		NR	0.050	3.064	0.039	0.041	0.635	0.705	0.738	0.725	0.665	0.680	0.684	0.246	0.281	0.310
Mali (c)		19.700	10.775	10.775	9.535	7.506	0.072	0.330								
Mexico	1.368	1.192	1.677	3.400	7.000	7.000			0.000	0.000	0.000	0.000	0.000	0.000	0.000	0.000
Mongolia					1.500	1.000										0.048
Morocco (c)									0.200	0.028	0.024					
Namibia (c)	2.903	3.718	2.139	3.689	2.456	1.962										0.364

ANNUAL EXPLORATION EXPENDITURES (1945-2003) (USD million)

Country	1977	1978	1979	1980	1981	1982	1983	1984	1985	1986	1987	1988	1989	1990	1991	1992
Niger	11.112	4.989	28.257	20.132	13.715	5.025	4.890	0.879	2.130	5.497	4.870	4.788	4.697	1.432	1.128	1.343
Nigeria (c)	1.000	1.200	1.000													
Norway	0.350	0.350	0.480	0.500	0.450	0.650	0.400									
Pakistan																
Paraguay (c)		2.870	4.140	7.250	6.750	4.500	0.850									
Peru (c)	0.156	0.990	0.625	0.460	0.900	0.600	0.021	0.054	0.112	0.189	0.300	0.046	0.015	0.060	0.036	0.009
Philippines (c)	0.180	0.360	0.421	0.507	0.662	0.562	0.062	0.025	0.025	0.025	0.016	0.016	0.016	0.010	0.010	0.010
Poland																
Portugal (c)	0.279	0.285	0.465	0.763	0.668	0.597	0.705	0.705	0.693	0.744	0.798	1.210	0.964	0.736	0.269	0.277
Romania																
Russian Federation																9.710
Rwanda					1.505											
Slovenia																
Somalia	0.000	0.000												0.000	0.000	0.000
South Africa (c)	12.350	24.567	31.271	28.699	19.150	6.794	5 277									
Spain (c)	6.800	8.000	12.000	11.000	16.000	16.000	10.000	7.000	0.382	0.680	0.990	1.224	1.861	2.485	3.552	4.119
Sri Lanka (c)			0.008	0.008	0.012	0.008	0.007	0.002	0.001							
Sudan																
Sweden (c)	5.400	5.900	5.000	5.530	5.100	3.400	2.100	1.200	0.570					0.000	0.000	0.000
Switzerland (c)	0.275	0.360	0.365	0.396	0.275	0.145	0.126	0.112	0.020							
Syria (c)														0.089		
Thailand	0.110	0.244	0.310	0.401	0.242	0.094	0.050	0.065	1.966	4.477	2.716	0.035	0.027	0.063	0.179	0.083
Turkey (c)	2.000	1.200	0.730	0.860	1.000	1.400	2.000	1.000	1.000	1.000	1.000	0.403	0.161	0.077	0.063	0.063
Ukraine																
United Kingdom (c)	0.020	0.400	0.525	0.625	0.635	0.730										0.000
United States	293.500	371.500	316.070	267.066	145.035	74.024	44.200	31.300	25.290	25.100	22.000	22.100	16.900	19.200	19.700	16.000
Uruguay							0.031	0.052	0.034	0.029	0.027	0.028	0.030			
USSR (b)	127.040	127.040	127.040	127.040	138.640	138.640	138.640	138.640	138.640	147.690	147.690	147.690	147.690	147.690		
Uzbekistan																
Vietnam								NR								
Yugoslavia						NR		0.750	0.175	0.295	0.040	0.070	0.215	0.462	0.353	0.252
Zambia		NR														0.021
Zimbabwe (c)		0.035						0.367	0.583	1.261		1.429	1.064	0.719	0.526	0.518
Total (d)	692.381	835.436	907.932	889.695	722.274	542.528	436.931	379.892	357.489	408.756	419.404	434.769	380.924	322.259	120.294	118.828

Appendix 5.1 (cont'd 7/8)

ANNUAL EXPLORATION EXPENDITURES (1945-2003) (USD million)

Country	1993	1994	1995	1996	1997	1998	1999	2000	2001	2002	2003	Total 1945-2003 (e)
Argentina	1.242	0.700	0.950	0.000	0.000	0.000		0.791	0.777	0.265	0.976	52.263
Australia	5.790	4.904	5.942	11.842	18.754	12.030	6.260	4.390	2.470	3.020	4.116	508.949
Bangladesh												0.453
Belgium	0.000	0.000	0.000	0.000	0.000	0.000	0.000	0.000	0.000	0.000	0.000	2.487
Bolivia (c)												9.343
Botswana (c)												0.825
Brazil (c)	0.000	0.000	0.000	0.000	0.000	0.000	0.000	0.000				186 128
Bulgaria												NR
Cameroon												1.282
Canada	31.820	26.090	32.350	28.470	42.000	41.000	33.000	30.700	16.200	22.900	21.687	1 288.500
Central African Republic (c)												21.800
Chile	0.115	0.094	0.218	0.143	0.154	0.196	0.178	0.214	0.126	0.154	0.115	6.896
China								4.200	6.000	7.200	6.000	23.400
Colombia		0.000	0.000	0.000	0.000	0.000	0.000	0.000				19.946
Congo, Democratic Republic												NR
Costa Rica												0.364
Cuba	0.230	0.228	0.142	0.086	0.050	0.090	0.060	0.040	0.050	0.020	0.000	0.972
Czech Republic	0.579	0.468	0.282	0.201	0.163							314.013
Denmark (c)	0.000	0.000	0.000	0.000	0.000	0.000	0.000	0.000	0.000	0.000	0.000	4.140
Ecuador (c)												1.945
Egypt (c)	6.647	3.245	3.264	6.528	7.418	7.976	7.976	10.499	9.404	7.186	5.631	108.807
Finland (c)	0.000	0.000	0.000	0.000	0.000	0.000	0.000	0.000	0.000	0.000	0.000	13.984
France (c)	9.963	6.217	2.882	7.960	1.742	1.040	0.000	0.000	0.000	0.000	0.000	907.240
Gabon (c)	1.839	1.050	0.939	1.338	0.343	0.000	0.000	0.000	0.000	0.000	0.000	102.433
Germany, Federal Republic	0.000	0.000	0.000	0.000	0.000	0.000	0.000	0.000	0.000	0.000	0.000	96.869
Germany, Democratic Republic (a)	0.000	0.000	0.000	0.000	0.000	0.000	0.000	0.000	0.000	0.000	0.000	1 905.920
Ghana												0.090
Greece	0.403	0.154	0.148	0.273	0.290							17.547
Guatemala												0.610
Hungary	0.000	0.000	0.000	0.000	0.000	0.000	0.000	0.000	0.000	0.000	0.000	3.700
India (c)	9.519	9.363	9.536	9.250	11.183	12.812	12.090	14.368	12.060	11.922	14.172	315.228
Indonesia (c)	1.523	0.648	0.574	0.695	0.632	0.114	0.217	0.061	0.023	0.030	0.033	15.878
Iran						0.857	1.000	1.700	1.004	1.389	3.781	9.731
Ireland (c)		0.000	0.000	0.000	0.000	0.000	0.000	0.000				6.200
Italy (c)												75.060
Jamaica												0.030
Japan (c)	0.000	0.000	0.000	0.000	0.000	0.000	0.000	0.000	0.000	0.000	0.000	19.697
Jordan	0.013	0.010	0.030	0.100	0.100	0.150	0.000	0.000	0.000	0.000	0.000	0.920
Kazakhstan	2.525	1.290	0.113	0.242	0.160	0.000	0.000	11.035	13.175	11.836	4.372	47.248
Korea, Republic of	0.000	0.000	0.000	0.000	0.000	0.000	0.000	0.000	0.000	0.000	0.000	17.886
Lesotho												0.021
Madagascar												5.293
Malaysia	0.368	0.399	0.163	0.000	0.245	0.188	0.186	0.660				11.072
Mali (c)												58.693
Mexico	0.000	0.000	0.000	0.000	0.000	0.000	0.000	0.000	0.000	0.000	0.000	30.306
Mongolia	0.060	0.700	1.650	2.560	3.135							8.153
Morocco												2.752

ANNUAL EXPLORATION EXPENDITURES (1945-2003) (USD million)

Country	1993	1994	1995	1996	1997	1998	1999	2000	2001	2002	2003	Total 1945-2003 (e)
Namibia (c)	0.000	0.000	2.044	0.000	0.000	0.000	0.000	0.000	0.000	0.000	0.110	25.741
Niger	0.440	1.481	1.665	0,427	1 653	0.754	0.471	0.633	1.088	3.126	4.545	216.121
Nigeria (c)												6.950
Norway		0.000	0.000		0.000	0.000	0.000	0.000	0.000	0.000	0.000	3.180
Pakistan												NR
Paraguay (c)												26.360
Peru (c)	0.000	0.004	0.000	0.000	0.000	0.000	0.000	0.000	0.000	0.000	0.000	4.776
Philippines (c)	0.010	0.030	0.030	0.019	0.019	0.013	0.011	0.005	0.004	0.004	0.000	3.460
Poland												NR
Portugal (c)	0.135	0.106	0.130	0.114	0.154	0.102	0.018	0.019	0.000	0.000	0.000	17.637
Romania		2.998	2.448	1.776	1.198	0.934	0.549	0.157				10.060
Russian Federation	2.828	4.197	5.581	4.281	10.052	8.650	6.870	13.300	11.470	10.420	7.241	94.600
Rwanda												1.505
Slovenia												NR
Somalia												10.000
South Africa (c)				0.000	0.000	0.000	0.000	0.000	0.000	0.000	0.073	140.919
Spain (c)	2.872	0.891	0.000	1.388	0.000	0.000	0.000	0.000	0.000	0.000	0.000	140.455
Sri Lanka (c)												0.043
Sudan												0.200
Sweden (c)	0.000	0.000	0.000	0.000	0.000	0.000	0.000	0.000	0.000	0.000	0.000	47.900
Switzerland (c)	0.000	0.000	0.000	0.000	0.000	0.000	0.000	0.000	0.000	0.000	0.000	3.359
Syria (c)	0.000											1.151
Thailand	0.138	0.116	0.119	0.000	0.000	0.000	0.000					11.299
Turkey (c)					0.200	1.200	0.000	0.000				21.988
Ukraine				1.376	1.611	1.940	1.606	2.107	1.701	1.898	0.007	15.654
United Kingdom (c)	0.000	0.000	0.000	0.000	0.000	0.000	0.000	0.000	0.000	0.000		3.815
United States	12.000	4.329	6.009	10.054	30.426	21.724	8.968	6.694	4.827	0.352		2 507.113
Uruguay												0.231
USSR (b)												3 692.350
Uzbekistan		0.472	6.197	22.067	21.954	19.652	19.392	14.152	8.516	13.255	13.923	139.580
Vietnam	0.324	0.137	0.161	0.208	0.227	0.120	0.120	0.104	0.104	0.130	0.196	2.898
Yugoslavia												1.581
Zambia		0.004										0.025
Zimbabwe (c)	0.000	0.000	0.000	0.000	0.000	0.000						6.902
Total (d)	91.383	70.325	83.567	111.398	153.863	131.542	98.972	115.829	88.999	95.107	90.393	13 382.927

Total Spending

Notes:

Exploration expenditures reported in national currencies were converted to USD by the Secretariat using an annual average exchange rate provided by the United Nations Development Program's Department of Finance.

--- Reported numbers refer to the total expenditures over the period indicated by the horizontal line.

NR Exploration activities were reported but no expenditures were provided.

a) Total uranium exploration expenditures in GDR amounts to about GDR Mark 5 500 mil.. Expenditures from 1946 to 1953 were about GDR Mark 600 mil.. Expenditures from 1954 to 1982 were about GDR Mark 3 900 million. Expenditures have been divided equally per year for each period. The exchange rate for GDR Mark and USD is the same as DM/USD.

b) Expenditures for the USSR were given in 5-year intervals. These have been equally divided within each period. The exchange rate used was USD 1 = RUB 0.75.

c) Expenditures in first year of reported data represent spending for all previous exploration activity.

d) Total does not include multi-year expenditures reported in the first year that data was provided. See note (c).

e) Includes all exploration expenditures reported by the country.

Appendix 5.2

SUMMARY OF EXPLORATION HISTORIES BY RED BOOK

1965 and 1967 Red Books

In 1965, uranium exploration activity (and production) was mainly focused on three geological deposit types:

- sandstone deposits;
- quartz-pebble conglomerate deposits;
- vein deposits.

In the United States, although uranium deposits occurred in a wide variety of rocks and in many types of deposits, about 95% were associated with coarse clastic sediments as irregularly shaped, but generally tabular disseminated deposits. Uranium associated with sandstone deposits was also reported in Spain and Japan.

Resources in the conglomeratic deposits in the Blind River, Elliot Lake area of Ontario, constituted about 93% of the total resources in Canada. Uranium in South Africa was associated with conglomerates of the Witwatersrand system, and was recovered as a by-product of gold.

Vein-type deposits were explored and mined in the Beaverlodge area of Saskatchewan in Canada (and totalled about 6% of total resources of Canada), in France, United States and Portugal.

Uranium was also reported in other formations such as pegmatite (Bancroft area, Canada), lignite (Spain), bituminous black shale (Sweden), and intrusive formations (Ilimaussaq deposit in Greenland).

In addition to the conventional deposits, uranium resources were reported in phosphate rock deposits (United States and Morocco).

In 1966-1967, exploration was directed preferentially towards finding low-cost uranium, and in developing extensions of known deposits. Therefore, most of the activity emphasised discovery of sedimentary, shale and phosphate deposits in the United States, gold conglomerate ores in South Africa, shale deposits in Sweden and monazite sand deposits in India.

Argentina reported uranium resources associated with sandstones (92% of the estimated resources) and vein-type deposits.

In **Australia**, most of the resources were associated with vein-type deposits. Exploration was also directed towards the discovery of large conglomeratic deposits, and other types of low-grade mineralisation, but grades appeared disappointingly low.

In **Canada**, exploration continued for conglomeratic deposits (mainly in Ontario, but also in Northwest Territories, Labrador, Nova Scotia, and Québec). Other types of deposits, such as Tertiary and Mesozoic basins and lignites, were also explored as possible sources of low-grade uranium. Vein-type deposits were exploration targets in Northern Territories, Nova Scotia, Ontario and Labrador.

In **Democratic Republic of Congo (formally Zaire)**, vein deposits in Precambrian formations, similar to the Shinkolobwe deposit, were explored.

Resources in **France** were reported in vein-type deposits in crystalline rocks (77% of the RAR, 87% of the EAR), in stratiform type deposits in sandstone (17% of the RAR, 13% of the EAR) and black shale (6% of the RAR).

In **Gabon**, uranium was found in coarse sandstones in the lower part of the Francevillian Formation (upper Precambrian), the deposit being related to sedimentary and structural features.

In **Niger** uranium sedimentary type deposits, mainly in sandstone formations, were explored.

Geological distribution of uranium resources (December 1967)

Deposit type	Reasonably Assured Resources						Estimated Additional Resources					
	<USD 26/kgU		USD 26-39/kgU		USD 39-78/kgU		<USD 26/kgU		USD 26-39/kgU		USD 39-78/kgU	
	10^3 tU	%	10^3 tU	%	10^3 tU	%	10^3 tU	%	10^3 tU	%	10^3 tU	%
Sandstones	160	25.4	97	16.1	85	25.4	262	47.4	146	33.5	153	17.2
Quartz-pebble conglomerates	289	45.9	146	24.3	126	37.7	223	40.3	154	35.3	287	32.2
Vein and related types	60	9.5	12	2.0	6	1.8	45	8.1	23	5.3	21	2.4
Shales	3	0.5	274	45.6	0	0	0	0	40	9.2	0	0
Phosphates	75	11.9	49	8.2	77	23.1	0	0	0	0	231	26.0
Others*	43	6.8	23	3.8	40	12.0	23	4.2	73	16.7	197	22.2
Total	630	100.0	601	100.0	334	100.0	553	100.0	436	100.0	889	100.0

* Others include pegmatite, schists, uranothorianite, copper leach by-products, hyper-alkaline silicates, lignites, monazites, volcanics.

1970 Red Book

As a result of the upswing in orders for nuclear power stations by utilities, and of an increase in the price of uranium, prospecting restarted in most of the producing countries, and unprecedented amount of drilling was completed.

In **Argentina**, virtually all the uranium resources were contained in sedimentary rocks in continental or transitional facies from the Tertiary, Cretaceous or Permian periods. Exploration in the previous years had started in the Sierra Pintada district, leading to the discovery of several deposits.

In 1969-70, exploration in **Australia** was conducted in Queensland, South Australia and Northern Territory. In addition to prospecting for finding deposits of the types already known (hydrothermal vein type), exploration was directed towards the discovery of conglomerate, dolomite and phosphate type deposits.

In **Brazil**, several types of deposits were explored, including uranium in mineralised volcanic breccias, veins of tinguaite (Poços de Caldos), a by-product of uranium in niobium-tantalum deposits (Araxá), uraniferous conglomerates (Jacobina) and phosphate type deposits (Olinda).

In **Canada** more than 80% of the RAR were contained in quartz-pebble conglomerates of Hudsonian age in the Elliot Lake and Agnew Lake areas of Ontario. Most of the remaining RAR in the USD 10/lb U_3O_8 category were in pitchblende bearing vein-type related deposits, which occurred primarily in Saskatchewan, British Columbia, North West Territories and Labrador.

Several areas of Canada were also explored for the occurrence of Colorado Plateau type deposits, uranium bearing lignites, radioactive pegmatites and uranium as a by-product of niobium. With the discovery of the Rabbit Lake deposit (deposit described as hydrothermal in origin, occurring in highly altered meta-sediments, but not yet referred as unconformity-related deposit), in the Wollaston area of North Saskatchewan, the uranium potential of this new province seemed promising.

In **Central African Republic**, three deposits geologically associated with an Eocene uranium-bearing phosphatic formation with unusually high uranium content for a formation of this type were discovered.

In **France**, exploration continued both inside and outside the producing regions. Resources were distributed between vein-type deposits (80%) and stratiform deposits in sandstones (20%).

Extensive exploration between Mounana and Franceville in **Gabon** lead to the discovery of the Oklo deposit and revealed additional mineralised areas associated with coarse sandstones.

In **India**, exploration in focused mainly in Pre-Cambrian terrains, but also on uraniferous copper ores, monazites and sedimentary formations.

In **Niger**, uranium resources were associated with sandstone deposits or continental pelitic formations, belonging to the Carboniferous and to a lesser extend the Jurassic systems. These deposit types were the main exploration targets.

In **Portugal**, uranium occurrences were revealed in highly tectonised zones, either in monzonitic Hercynian granites, or in metamorphic schists of the ante-Ordovician complex.

The conglomerates of the Witwatersrand system continued to be the principal source of uranium in **South Africa** and the main exploration targets.

In **Namibia**, a uranium prospect located at Rössing had been under investigation for the past two years. Drilling indicated the existence of an extensive low-grade deposit of uranium, which could be mineable by open-pit if proven to be economic.

Exploration activity in **Spain** emphasised research of pegmatite deposits, vein-type deposits in granite and meta-sedimentary formations.

In the **United States**, the principal areas of drilling activity continued to be previously known uranium provinces. About 94% of the resources were in coarse clastic sediments, more than 84% being located in well established districts (Colorado Plateau, Wyoming Basins and Gulf Coastal Plains).

1973 Red Book

In **Argentina**, exploration continued to be oriented towards discovery of uranium deposits associated with sedimentary formations (Sierra de Pichinan area, Sierra Pintada district), but also toward granite bearing areas which could be favourable for uranium deposits.

During the previous two years, drilling programmes outlined substantial new resources in **Australia**, associated with typical vein-type deposits, but also large disseminated-type ore-bodies (in Northern Territories). Numerous deposits (Westmoreland, Mt. Isa Belt, Mary Kathleen) were also been found in volcanics and in sediments of Carpentaria age near the Northern Territories/Queensland border and in metasomatites (Mary Kathleen). In South Australia, in the Mt Painter area, substantial resources were discovered in Archean granites and gneissic rocks, as well as in hematitic breccia masses. Deposits were also discovered in sedimentary formations in the Lake Frome Basin. A drilling programme was conducted to evaluate discovery of uranium mineralisation near Yeelirrie (Western Australia). It appears that there is an extensive, flat-lying shallow deposit of sedimentary uranium (predominantly carnotite) in calcareous rock (calcrete) infilling old drainage channels. No significant economically exploitable uraniferous conglomerates were discovered, but preliminary drilling indicated that large low-grade resources of Witwatersrand/Blind River type deposits may be present in the Kimberley and Pilbara areas of Western Australia.

In **Brazil**, ten of the most promising occurrences were evaluated by extensive drilling. Two small size deposits were evaluated in the Poços de Caldas area (vein-type deposits in an intrusive pipe of alkaline rocks), and promising occurrences were found in the Cambai-Figueira district (State of Paraná) and in the iron Quadrangle (State of Minas Gerais).

Quartz-pebble conglomerate and vein-type deposits continued to be the most promising subjects of exploration in **Canada**, but also new areas of potential were recognised such as the Carswell Dome and the Wollaston Lake Belt in Northern Saskatchewan, Baker Lake, Amer Lake, Great Slave Lake and Nonacho Lake in the Northwest Territories. Exploration of the lignitic uranium occurrences in Southern Saskatchewan was abandoned mainly because of the growing emphasis on preservation of the environment.

Exploration in **France** continued in each of the three mining districts (vein-type deposits), but also on sedimentary formations, primarily those of Permian period; encouraging results were recorded in the basin of the Hérault.

In **India**, new discoveries were made in amphibole schist exposed on the hanging wall side of copper mineralisation in Jhunjhunu district, and an uraniferous syenite was reported in the Sarguja district.

After 1971, prospecting work in **Portugal** increased significantly in the overseas provinces of Angola and Mozambique, mainly on phosphate type deposits (Angola).

In **Namibia**, extensive diamond drilling confirmed the existence of a low-grade deposit at Rössing.

In **South Africa**, uranium mineralisation was discovered near the surface in sandstone and conglomeratic mudstone of the Lower Beaufort Member of the Karoo Formation, which covers more than 40% of South Africa, and also extends into Namibia and Angola. Drilling of uraniferous conglomerates (down to more than 1 500 m) continued. Fairly extensive deposits of uranium of good grade were proved, but associated gold values were very low.

Exploration continued in **Spain** but emphasis shifted from granitic terrains and surrounding sediments, towards continental sedimentary formations of Triassic, Lower Cretaceous and Tertiary age.

Since 1970, uranium exploration in the **United States** declined. Resources (RAR and EAR were distributed among sandstone (94%), vein-type deposits (3%) and others (limestone, conglomerate, lignite) (3%).

World resources by geological ore types

Deposit type	Reasonably Assured Resources				Estimated Additional Resources			
	< USD 26/kgU		USD 26-39/kgU		< USD 26/kgU		USD 26-39/kgU	
	10^3 tU	%	10^3 tU	%	10^3 tU	%	10^3 tU	%
Sandstones	336	39	161	24	555	61	256	41
Quartz-pebble conglomerates	302	35	172	26	133	15	150	24
Vein and related types	152	17	45	7	188	20	125	20
Others	75	9	292	43	35	4	96	15
Total	855	100	670	100	911	100	627	100

Exploration was also conducted in countries without known major uranium resources (Austria, Egypt, Germany, Greece, Iran, Italy, Korea, Switzerland and the United Kingdom). Most of the investigations were essentially at a broad reconnaissance level (geology, some radiometric surveys), and therefore it was not possible to appraise potential resources.

In **Egypt** uranium was discovered in both sedimentary rocks (shale and sandstone) and basement rocks (granite). In the **Federal Republic of Germany**, exploration indicated the existence of small deposits in Bavaria and in the Black Forest. In **Korea**, radiometric surveys were conducted over meta-sedimentary formations, and additional work (detailed surveys, drilling) was conducted in the most promising areas. In **Switzerland**, prospecting was directed toward Permian age continental detrital sediments and volcanic rocks, as well as to the Hercynian granitic massifs and their surrounding metamorphic formations. In the **United Kingdom**, geological terrain considered to be favourable for uranium, were investigated by radiometric and chemical techniques. Numerous anomalies were found in Devonian sediments resting on granite and metamorphic rocks in Caithness and Orkney. Uranium concentrations were discovered associated with wrench faults in proximity to Hercynian granites in the South West of England.

1977 Red Book

Exploration programmes were reported in more than 50 countries, many of which had no previous uranium exploration activity. All geological types of deposits were explored, but exploration focused on three types of deposits. In Precambrian areas of the world, considerable attention focused on exploration for Lower Proterozoic, unconformity-related deposits. Since 1970, large tonnages of uranium of this type of deposit had been discovered in Canada (Saskatchewan) and Australia (Northern Territory). In desert areas increasing attention was given to the search for calcrete deposits, which are localised near the water table under the present cycle of erosion. Significant discoveries were made in Australia, Namibia and Somalia. In all continents, continuing work was devoted to the discovery of sandstone-type ore-bodies, in particular in Africa, partially due to the success of uranium exploration in Niger, Gabon and South Africa.

1979 Red Book

Exploration in sandstones in all continents accounted for a large part of world uranium exploration expenditures between 1977 and 1979. In Precambrian areas worldwide, (particularly Australia, Canada), considerable attention was focused on exploration for Lower Proterozoic, unconformity-related deposits. In desert areas attention was given to the search for younger surficial calcrete deposits.

The search for other types of deposits in igneous and metamorphic rocks continued in many areas of the world.

Over 50% of **Canada's** resources were contained in quartz-pebble conglomerates, but in more recent years exploration focused on unconformity-related deposits, mainly in Saskatchewan and Northern territories, with new discoveries in the Collins Bay, Midwest Lake, Cluff Lake and Schultz Lake areas.

In **Europe** exploration activity increased. In **France**, exploration focused on sedimentary formations in small intragranite basins and sedimendary rocks bordering the Massif Central. In the **Federal Republic of Germany**, uranium mineralisation was identified in Upper Carboniferous sandstones and shales in the northern part of the Black Forest. **Denmark** evaluated the Kvanefjeld deposit in Greenland and conducted a regional programme in east Greenland.

In **Africa**, there was also an increase in activity. Partially due to the success of exploration in Niger and Gabon, interest in sandstone basins in north and central Africa increased, with activities taking place in **Algeria**, **Central African Republic**, **Chad**, **Libya**, **Mali**, **Mauritania**, **Nigeria** and **Sudan**. Discoveries in the Karoo in South Africa and in other younger surficial sediments inspired interest in similar geological environments in **Lesotho**, **Botswana**, **Zambia**, **Malawi** and **Madagascar**. Extensive prospecting operations also took place in the dominion reef and Witwatersrand basins in **South Africa** and in granitic rocks in **Namibia**.

In the **United States**, drilling was undertaken to develop deposits in sandstones of Triassic and Jurassic ages in the Colorado Plateau region of Arizona, Colorado, Utah and New Mexico, and sandstones of Tertiary age in the Wyoming basins and Gulf Coastal Plain of Texas. A National Uranium Resource Evaluation (NURE) programme was initiated to develop an authoritative, comprehensive assessment of the country's uranium resources. A report which identified areas favourable for uranium occurrences was released.

1982 Red Book

In response to the uranium over-supply situation and to declining uranium prices, exploration expenditures in WOCA countries declined from their record level in 1979 in most of the major producing countries.

In **Brazil**, following the discovery of the Itataia deposit (metasomatic deposit), exploration expenditures decreased.

In **Canada** most of the drilling activity focused on exploring and delineating Proterozoic unconformity-related deposits in Saskatchewan. Significant drilling was also reported in Northern territories, Ontario, Newfoundland, Québec and Nova Scotia.

In **France** most of the activity was concentrated in the granitic areas of the Hercynian Massifs and their surrounding sedimentary basins.

Expenditures increased in **Mexico**, exploration focused on vein and disseminated type deposits, but also on sandstone type deposits in the Nuevo Leon area.

In the **United States**, exploration continued to focus on the traditional uranium areas (sandstones deposits), but also on the search for non-sandstone deposits (vein-type and breccia pipe deposits).

Distribution of uranium resources by deposit type

Deposit type	Reasonably Assured Resources (<USD 130/kgU)		Estimated Additional Resources (<USD 130/kgU)	
	10^3 tU	%	10^3 tU	%
Quartz-pebble conglomerate deposits	390	17	544	20
Unconformity-related deposits	390	17	272	10
Disseminated deposits	298	13	408	15
Vein deposits	115	5	218	8
Sandstone deposits	917	40	1 115	41
Other	183	8	163	6
Total	2 293	100	2 720	100

1983 Red Book

Expenditures on uranium exploration declined rapidly in the previous three years following the record levels in the late 1970s. A reduction in expenditures was reported by nearly all countries. The areas attracting the greatest interest continued to be those with major uranium resources. In spite of the cut-backs in exploration, there were several successes during the period.

In **Australia**, drilling confirmed the importance of the Ranger deposit (unconformity-related deposit). A major effort was made to delineate the copper-uranium deposit at Olympic Dam. Copper-uranium mineralisation was also intersected in the Stuart Shelf region at the Acropolis project.

Brazil continued exploration and delineation of the Itataia deposit.

In Northern Saskatchewan, **Canada**, unconformity-related mineralisation was identified at Peter River (Cluff Lake), Eagle Point, Dawn Lake and Waterbury Lake (future Cigar Lake deposit).

In **Greenland**, regional reconnaissance programmes were conducted around the Ilimanssaq intrusive complex, where high grade occurrences were found in granites, and in the alkaline Igaliko intrusive complex where low-grade mineralisation may exist.

In **Italy**, exploration led to the discovery of mineralisation in a mylonite at the tectonic contact between metamorphic basement and late Paleozoic continental sediments.

In **Peru**, uranium was discovered in Tertiary acidic tuffs.

In **Turkey** , uranium was found in Eocene sediments in Central Anatolia.

In the **United States**, significant new deposits were found. Uranium was discovered at Crow Butte, Nebraska, in a channel fill sandstone of Oligocene age. The Swanson deposit, which occurs in highly fractured Early Proterozoic gneiss, was also delineated in South Cental Virginia. In Arizona, there was an increase in drilling activity, where the search is for collapsed breccia pipe structures containing high grade ore.

1986 Red Book

Expenditures on uranium exploration continued to decline between 1983 and 1985. However, some successes were reported during this period.

Argentina maintained a quite stable level of expenditures, concentrated in the Puna, Sierras de Cordoba, Sierra Pintana areas, as well as in Patagonia. At Sierra de Pichinan, an ore-body was delineated in shallow Cretaceous conglomerates. Exploration continued in Aguilini area (acid volcanics).

In **Australia**, activities were concentrated in the Stuart Shelf area to define the extent of the Olympic Dam copper-gold-uranium deposit, and to search for similar mineralisation in the area. In addition, exploration focused on areas along the margins of sedimentary basins overlying Archean and Proterozoic basements, such as in the Carnavon, Eucla, Georgina and Officer basins.

In **Brazil** exploration activities resulted in discovery of new uranium occurrences in the Rio Cristina area (State of Pará).

In **Canada**, activities concentrated mainly in Saskatchewan and Northern territories, with Proterozoic unconformity-type deposits being the primary targets. Discovery of the Cigar Lake deposit was announced in early 1983.

In **Ecuador**, exploration was conducted at the Puyango prospect, underlain by cretaceous phosphatic sediments.

In **France**, exploration targets including mainly the areas surrounding producing mines, but also the Alps and the Aquitaine basin.

In **Peru**, following the discovery of uranium in Tertiary acidic tuffs, exploration revealed new occurrences at Cerro Calvario, Cerro Concharrumio, Chapi and Pinocho.

In **Portugal**, targets included Hercynian granites and their contacts with Cambrian and Paleozoic host rocks.

In **South Africa**, exploration for uranium associated with gold continued in the Witwatersrand Basin. Outside the Witwatersrand Basin, exploration continued in the Karoo basin.

In the **United States**, activities concentrated mainly in the vicinity of the major producing areas, including the Colorado Plateau, Wyoming Sedimentary basins and the Texas Coastal Plain. Exploration for collapsed breccia pipes in Arizona resulted in the discovery of several high-grade deposits.

1988 Red Book

No significant discoveries of new uranium deposits were reported in 1986-1987. The largest share of expenditures for regional ground reconnaissance programmes occurred in areas with known potential, such as the Athabasca Basin (Canada), the Stuart Shelf area (Australia) and northwestern Arizona (United States). The main focus of exploration during that period was the search for high-grade deposits in established uranium districts. Exploration in areas without known deposits received relatively little attention, except in Canada where there was an increase in grass-roots exploration.

In **Argentina**, exploration confirmed the uranium potential of Cretaceous sediments of the Chubut Basin.

In **Australia**, delineation of the Olympic Dam deposit continued. Exploration was concentrated in Western Australia, where work continued at Mulga Rock, Manyingee and Kintyre.

In **Canada**, over 95% of exploration efforts remained concentrated in Saskatchewan, where the Eagle Point deposit was confirmed, and in the Northwest Territories. New grass-roots exploration projects started in the Northwest Territories and in Eastern Canada.

In **Egypt**, exploration targets included vein-type occurrences related to granite, and sandstone uranium projects in mid-Carboniferous sediments.

In **Federal Republic of Germany**, exploration focused around the Menzenschwand, (Black Forest) and Grossschloppen (Northern Bavaria) deposits.

In the **United States**, most of the exploration was concentrated in the Grand Canyon area in northwest Arizona, where the targets were collapse breccia pipe deposits.

In **Zambia**, foreign companies explored for uranium in Karoo sediments and in contact areas between the Precambrian Katanga Formation and the basement.

Similarly, the Karoo sediments were explored in the Zambezi valley of **Zimbabwe**.

1989 Red Book

The main focus of exploration efforts continued to be on the search for high-grade uranium deposits in established districts, and the development of known deposits. During the last period, no major discovery had been reported in a WOCA country, since discovery of the Kintyre deposit in Australia in 1985.

Areas of investigation in **Argentina** were located in the Sierra of Cordoba (Paleozoic granites), the Puna (acid volcanics of Tertiary age), Sierra Pintada (Paleozoic volcagenic sediments) and Patagonia (Cretaceous sandstones).

The main exploration projects in **Australia** continued to be in the Stuart Shelf region of South Australia and the Rudall area, northern Paterson Province. The objectives of these programmes were better definition of Olympic Dam mineralisation, and further exploration and evaluation of the Kintyre deposit and its surrounding area. In addition, exploration for unconformity-related deposits continued in the Pine Creek geosyncline in the Myra Falls, Rum Jungle-Waterhouse and the Ranger project areas.

In **Canada** the main concentration of exploration projects was in the Athabasca basin, although grass-roots projects continued in Eastern Canada and in the Northwest Territories. Discovery of ore grade mineralisation at the Wholly project in Saskatchewan was announced early 1989.

In **Finland**, evidence for the presence of an unconformity-related uranium occurrence was found at Riutta, Koni area, eastern Finland.

India carried out extensive exploration, mainly in the Singhbhum thrust belt, as well in new areas with potential for uranium deposits in Cretaceous sandstone and metamorphic rocks.

In **Indonesia** exploration continued in the Sibolga area (North Sumatra) and in Kalan (West Kalimatan). These areas are underlain by Paleogene sediments of granitic provenance and by Permo-Carboniferous meta-sediments, respectively.

Exploration targets in the **United States** were located mainly in Arizona (collapse breccia pipes), Texas and Wyoming (sandstone uranium deposits amenable to ISL extraction).

1991 Red Book

In **Australia**, in the Pine creek geosyncline, exploration for unconformity-related deposits continued, mainly in Arnheim land, Rum Jungle and in the Allander area. High-grade mineralisation was encountered in Kintyre vicinity.

In **Canada**, exploration for unconformity-related deposits continued in the Athabasca basin. Discovery of the P2 North deposit at McArthur River was announced in May 1990. Grass-roots exploration near Great Bear Lake, Northern Territories, was started to test a conceptual model developed by the Geological Survey of Canada, indicating that Olympic Dam type mineralisation could be found in the area.

In **Egypt**, emphasis was directed towards investigation of vein-type deposits related to Pan-African granites and mineralisation hosted in Carbonaceous sediments.

In **France**, grass-roots activities were reduced, and exploration concentrated on mining properties in the north-western and southern part of the Massif Central.

In 1991, after **Germany's** re-unification, exploration activities in the former German Democratic Republic were reported for the first time. Exploration since 1945 resulted in confirming mineable uranium deposits associated with known multi-element vein deposits of the Erzgebirge and polygenetic deposits of lens shape and stockwork type in black shales, limestones and diabase of the Thuringian Ronneburg district. Deposits of the sedimentary type were found and exploited in Cretaceous sandstones of Königstein/Saxony, in Zechstein sediments in Culmitzsch/Thuringia and in Lower Permian coal of Freital/Saxony.

India reported discoveries of deposits in heretofore unexplored geological environments, including upper Cretaceous Mahadek sandstones in the state of Meghalaya, middle Proterozoic limestone of the Cuddapah basin, Andhra Pradesh State, as well as shear zones in metamorphosed lower Proterozoic basalts and rhyolites in Badal, Madhya Pradesh state.

Korea explored the Gongju area (uraniferous graphite layers in hornfelses) and the Munkyeang area (uraniferous granites and potassium rich volcanics).

In **Niger**, exploration continued in the Tin Mersoi basin, the main project being the evaluation of the Akola deposit.

In **Spain**, exploration was concentrated in the vicinity of the Mina Fe deposit, where a number of anomalies and deposits were found. In Caceres and Badajoz Provinces exploration focused on Precambrian-Cambrian schists intruded by Hercynian granites.

In **Zimbabwe** exploration was undertaken in the western and northern parts of the country underlain by Karoo sediments. The results were positive with the discovery of the Kanyemba deposit.

1993 Red Book

Exploration in countries with substantial exploration programmes were targeted on resources with low production costs, including high-grade deposits (unconformity-related deposits) and sandstone-hosted deposits amenable to ISL mining.

In **Australia**, exploration focused on targets for high-grade deposits in the Paterson province (Western Australia) in the vicinity of the Kintyre deposit, in the Pine Creek Geosyncline (Northern Territory) the Westmoreland area (Queensland) and on Olympic Dam-type mineralisation on the Stuart Shelf (South Australia).

In **Canada** two new ore-bodies were discovered in the Sue zone (McClean project) and mineralisation was found in the area adjacent of McArthur River. Exploration continued in the Baker Lake area (Northwest Territories).

In **China**, new areas of exploration included the Erlian Basin, Inner Mongolia Autonomous Region, and the Yili Basin, Xinjiang Autonomous Region, where sandstone hosted deposits amenable to ISL mining were the main targets.

Exploration in **Czechoslovakia** was concentrated in the vicinity of known deposits, mainly around the Hamr and Straz deposits.

India reported the discovery of a new sandstone hosted deposit at Domiasiat, in the Meghalaya plateau, hosted in Upper cretaceous fluviatile sandstone.

In **Jordan**, initial reconnaissance of basement complexes and Precambrian sandstones was initiated.

The following five countries reported for the first time:

Exploration in **Kazakhstan** was concentrated on sandstone hosted deposits in the Chu-Saryssu and Syr-Darya basins of Mesozoic to Cenozoic age.

In **Mongolia**, uranium deposits occurring in veins and stockworks in volcanogenic complexes and sandstone-hosted deposits were discovered.

In the **Russian Federation**, exploration targets included valley-type deposits, where epigenetically enriched uranium occurs in young fluviatile sandstones in drainage systems.

In **Ukraine** exploration concentrated on metasomatite deposits in the central and western part of the crystalline shield.

In **Vietnam**, exploration focused on evaluation of the uranium potential of the Triassic Nongson sedimentary basin. The potential of the acidic volcanic rocks of the Da Lat zone was also evaluated.

1995 Red Book

In **Australia**, exploration focused on the Paterson province where mineralisation similar to that at the Kintyre deposit is expected to occur. In the Mount Painter area, low-grade uranium mineralisation within hematite-rich breccias was intersected. Limited work was done in the Arnheim area, Northern territory, for unconformity-related deposits.

In **Canada**, exploration work concentrated on targets favourable for unconformity-related deposits, mostly in the Athabasca Basin. Most of the spending was attributed to the development of advanced projects in Saskatchewan. In addition, detailed exploration was carried out in the Kiggavik area and the south-west part of the Thelon basin in the Northwest Territories. Initial exploration was also carried out in the western part of the Athabasca basin and in the Great Bear magmatic zone.

In **India**, exploration focused on Cretaceous sandstones in the Meghalaya area, but also included reconnaissance exploration for unconformity-related deposits and breccia complex deposits.

In **Kazakhstan**, exploration occurred mainly in the southern part of the country for sandstone-hosted deposits, and in the northern part for vein-type deposits in the Kosachinoe area.

Russia reported a geographically extensive exploration programme in diverse geological environments. Exploration focused on seven regions, Karelia, Zaural, Western Siberia, Zayan-Baikalak, Aldansk, Bureinsk and Chankaisk.

In the southern part of **Ukraine**, activities focused on the crystalline shield covered by sedimentary rocks.

In the **United States**, exploration continued in Wyoming (sandstone type deposits), and to a lesser degree in Arizona, Colorado, Nebraska and Texas.

1997 Red Book

In **Argentina**, exploration continued on the Cerro Solo sandstone deposit and in the Las Termas occurrence. The later consists of vein-type mineralisation in metamorphic rocks in the vicinity of a granitic intrusion.

In **Australia**, exploration expenditures increased and focused on three areas: the Paterson province and Arnheim Land (unconformity related deposits), and Westmoreland area (sandstone type deposits).

In **Canada** exploration emphasis was unchanged, with most of the effort having been concentrated on targets favourable for unconformity-related high-grade deposits, mostly in the Athabasca Basin. Most of the expenditures were attributed to advance underground exploration and deposit appraisal at such projects as Cigar Lake and McArthur River. Detailed exploration was also carried out in the Kiggavik area and in the extreme western and north-eastern part of the Thelon Basin. Some geological research and grass-roots exploration was carried out in the western part of the Athabasca basin, in Alberta and in the Great Bear magmatic zone.

China concentrated its exploration efforts on sandstone hosted deposits amenable to ISL mining in the Yili and Erlian basins.

Kazakhstan decreased its exploration efforts, relying on already discovered large resources associated with sandstone type deposits. The main grassroots activities were carried out in the northern part of the country, targeting unconformity-related deposits.

In **Russia**, expenditures focused on valley-type sandstone deposits amenable to ISL mining.

1999 Red Book

In **Argentina**, exploration involved drilling on the Cerro Solo sandstone deposit.

In **Australia**, the main focus of exploration was on unconformity-type deposits in Arnheim land and the Paterson Province, as well as for sandstone and calcrete-type deposits in South Australia and Western Australia.

In **Canada**, a significant portion of exploration expenditures continued to be attributed to projects awaiting production approvals. Most of the basic grassroots expenditures were spent in Saskatchewan, targeting unconformity-related deposits.

China continued exploration for sandstone-type deposits in Xinjiang, Inner Mongolia and Northern China, but also for volcanic and unconformity-related deposits (Liaoning Province).

The **Russian Federation** concentrated its activities on sandstone deposits amenable to ISL mining. Major drilling programmes continued in the Transural and West Siberian districts, and in the Vitim region.

2001 Red Book

Worldwide exploration continued to be unevenly distributed geographically, with the majority of exploration expenditures concentrated in areas considered to have the best likelihood for the discovery of economically attractive deposits.

In **Australia**, the main focus of exploration continued to be on unconformity-type deposits in Arnheim land and the Paterson Province, as well as for sandstone and calcrete-type deposits in South Australia and Western Australia.

As in previous years, uranium exploration in Canada remained concentrated in areas favourable for the occurrence of deposits associated with Proterozoic unconformities, most notably in the Athabasca basin of Saskatchewan, but also in the Thelon basin.

China continued exploration for sandstone-type deposits in Xinjiang Province, Inner Mongolia and Northern China, but also for volcanic and unconformity-related deposits (Liaoning Province).

India's exploration activities continued in Proterozoic basins, Cretaceous sandstones and other geologic environments.

The **Russian Federation** concentrated its exploration activities on sandstone deposits amenable to ISL mining and on unconformity-related deposits. Major drilling programmes were in the Transural and the West Siberian districts and in the Vitim region.

2003 Red Book

In **Australia**, areas explored included Arnhem Land for unconformity-related deposits, the Frome Embayment for sandstone deposits, and the Gawler Craton/Stuart Shelf region for hematite breccia complex deposits. In November 2001, the discovery at Prominent Hill of copper, gold, uranium and rare-earth mineralisation in hematite breccia (a similar setting to the Olympic Dam deposit) was announced.

Canada continued to be the world's leader in domestic exploration activities. However, a significant portion of the overall expenditures was attributed to advanced underground exploration, deposit appraisal activities, and care and maintenance expenditures associated with projects awaiting production approvals. Basic grass-roots exploration took place in Saskatchewan, but also in Alberta, Labrador, Nunavut and Quebec.

In **Egypt**, exploration and evaluation activities concentrated on the black sand deposits of the northern Nile River delta, that constitute a non-conventional resource for uranium and thorium, as well as on uranium deposits in the eastern and south-western desert regions of the Sinai.

In **Iran**, activities included exploration and evaluation of uranium resources associated with Precambrian magmatic and metasomatic complexes in the Bafq-Posht-e-Badam province, and exploration of sedimentary basins in central and north-western Iran.

In **Kazakhstan**, exploration was conducted in the Shu-Saryssu province, where three ISL leach test sites were completed and mining tests were initiated.

In **Niger**, activities focused on resource development in and around the existing mine sites in an effort to expand the resource base in the western Arlit area.

In the **Russian Federation**, exploration activities were concentrated on sandstone deposits amenable to ISL mining and unconformity-related deposits.

Ukraine continued exploration for vein-type and unconformity-related deposits.

2005 Red Book

After a break of 23 years, the 2005 Red Book again reported resources by deposit type. A summary is given in the table below.

Identified resources by deposit type (as of 1 January 2005)

Geological type of deposit	Reasonably Assured Resources						Inferred Resources*					
	<USD 40/kgU		<USD 80/kgU		<USD 130/kgU		<USD 40/kgU		<USD 80/kgU		<USD 130/kgU	
	10^3 tU	%	10^3 tU	%	10^3 tU	%	10^3 tU	%	10^3 tU	%	10^3 tU	%
Unconformity-related	433.2	22.2	492.2	18.6	498.5	15.1	151.6	19.0	169.6	14.6	171.3	11.8
Sandstone	552.5	28.4	716.5	27.1	986.6	29.9	172.9	21.6	256.3	22.1	301.6	20.9
Hematite breccia complex	513.3	26.4	513.3	19.4	522.4	15.8	281.9	35.3	286.9	24.7	288.5	20.0
Quartz-pebble conglomerate	85.6	4.4	153.3	5.8	229.3	7.0	50.5	6.3	72.0	6.2	84.8	5.9
Vein	0	0	84.0	3.2	258.8	7.9	14.8	1.9	136.1	11.7	231.8	16.0
Intrusive	63.7	3.3	150.6	5.7	202.9	6.2	60.6	7.6	81.1	7.0	109.6	7.6
Volcanic and caldera-related	49.9	2.6	135.5	5.1	140.3	4.3	1.5	0.2	5.7	0.5	7.1	0.5
Metasomatite	109.3	5.6	157.6	6.0	179.8	5.5	5.6	0.7	22.5	1.9	87.2	6.0
Other **	129.2	6.6	164.7	6.2	186.2	5.6	49.6	6.2	102.8	8.9	125.4	8.7
Unspecified	10.6	0.5	75.6	2.9	91.9	2.8	9.9	1.2	28.1	2.4	38.7	2.7
Total	1 947.3	100.0	2 643.3	100.0	3 296.7	100.0	799.0	100.0	1 161.0	100.0	1 446.2	100.0

* Formerly EAR-I with the name changed for the 2005 edition of the Red Book.

** Includes surficial, collapse breccia pipe, metamorphic, limestone and uranium coal deposits. Rock types with elevated uranium contents such as pegmatite, granites and black shale are not included.

Appendix 5.3

DEFINITIONS OF URANIUM DEPOSIT TYPES

These definitions of the geological types of uranium deposits were developed by the IAEA in 1988-1989 and updated for the 2003 edition of the Red book.

Uranium resources can be assigned on the basis of their geological setting to the following categories of uranium ore deposit types (arranged according to their approximate economic significance):

1. Unconformity-related deposits.
2. Sandstone deposits.
3. Hematite breccia complex deposits.
4. Quartz-pebble conglomerate deposits.
5. Vein deposits.
6. Intrusive deposits.
7. Volcanic and caldera-related deposits.
8. Metasomatite deposits.
9. Surficial deposits.
10. Collapse breccia pipe deposits.
11. Phosphorite deposits.
12. Other types of deposits.
13. Rock types with elevated uranium content.

1. Unconformity-related deposits

Unconformity-related deposits are associated with and occur immediately below and above an unconformable contact that separates a crystalline basement intensively altered from overlying clastic sediments of either Proterozoic or Phanerozoic age.The unconformity-related deposit type includes the following sub-types:

- *Unconformity contact*

  i) Fracture bound deposits occur in meta-sediments immediately below the unconformity. Mineralisation is monometallic and of medium grade. Examples include Rabbit Lake and Dominique Peter in the Athabasca Basin, Canada.

  ii) Clay-bound deposits occur associated with clay at the base of the sedimentary cover directly above the unconformity. Mineralisation is commonly polymetallic and of high to very high grade. An example is Cigar Lake in the Athabasca Basin, Canada.

- *Sub-unconformity-post-metamorphic deposits*

  i) Deposits are strata-structure bound in meta-sediments below the unconformity on which clastic sediments rest. These deposits can have large resources, at low to medium grade. Examples are Jabiluka and Ranger in Australia.

2. Sandstone deposits

Sandstone uranium deposits occur in medium to coarse-grained sandstones deposited in a continental fluvial or marginal marine sedimentary environment. Uranium is precipitated under reducing conditions caused by a variety of reducing agents within the sandstone, for example, carbonaceous material, sulphides (pyrite), hydrocarbons and ferro-magnesium minerals (chlorite), etc. Sandstone uranium deposits can be divided into four main sub-types:

- *Roll-front deposits:* The mineralised zones are convex down the hydrologic gradient. They display diffuse boundaries with reduced sandstone on the down-gradient side and sharp contacts with oxidised sandstone on the up-gradient side. The mineralised zones are elongate and sinuous approximately parallel to the strike, and perpendicular to the direction of deposition and groundwater flow. Resources can range from a few hundred tonnes to several thousands of tonnes of uranium, at grades averaging 0.05% to 0.25%. Examples are Moynkum, Inkay and Mynkuduk (Kazakhstan); Crow Butte and Smith Ranch (United States) and Bukinay, Sugraly and Uchkuduk (Uzbekistan).

- *Tabular deposits* consist of uranium matrix impregnations that form irregularly shaped lenticular masses within reduced sediments. The mineralised zones are largely oriented parallel to the depositional trend. Individual deposits can contain several hundreds of tonnes up to 150 000 tonnes of uranium, at average grades ranging from 0.05% to 0.5%, occasionally up to 1%. Examples of deposits include Westmoreland (Australia), Nuhetting (China), Hamr-Stráz (Czech Republic), Akouta, Arlit, Imouraren (Niger) and Colorado Plateau (United States).

- *Basal channel deposits*: Paleodrainage systems consist of several hundred metres wide channels filled with thick permeable alluvial-fluvial sediments. Here, the uranium is pre-dominantly associated with detrital plant debris in ore bodies that display, in a plan-view, an elongated lens or ribbon-like configuration and, in a section-view, a lenticular or, more rarely, a roll shape. Individual deposits can range from several hundreds to 20 000 tonnes uranium, at grades ranging from 0.01% to 3%. Examples are the deposits of Dalmatovskoye (Transural Region), Malinovskoye (West Siberia), Khiagdinskoye (Vitim district) in Russia and Beverley in Australia.

- *Tectonic/lithologic deposits* occur in sandstone related to a permeable zone. Uranium is precipitated in open zones related to tectonic extension. Individual deposits contain a few hundred tonnes up to 5 000 tonnes of uranium at average grades ranging from 0.1% to 0.5%. Examples include the deposits of Mas Laveyre (France) and Mikouloungou (Gabon).

3. Hematite breccia complex deposits

Deposits of this group occur in hematite-rich breccias and contain uranium in association with copper, gold, silver and rare earths. The main representative of this type of deposit is the Olympic Dam deposit in South Australia. Significant deposits and prospects of this type occur in the same region, including Prominent Hill, Wirrda Well, Acropolis and Oak Dam as well as some younger breccia-hosted deposits in the Mount Painter area.

4. Quartz-pebble conglomerate deposits

Detrital uranium oxide ores are found in quartz-pebble conglomerates deposited as basal units in fluvial to lacustrine braided stream systems older than 2.3-2.4 Ga. The conglomerate matrix is pyritiferous, and gold, as well as other oxide and sulphide detrital minerals are often present in minor amounts. Examples include deposits found in the Witwatersrand Basin where uranium is mined as a by-product of gold. Uranium deposits of this type were mined in the Blind River/Elliot Lake area of Canada.

5. Vein deposits

In vein deposits, the major part of the mineralisation fills fractures with highly variable thickness, but generally important extension along strike. The veins consist mainly of gangue material (e.g. carbonates, quartz) and ore material, mainly pitchblende. Typical examples range from the thick and massive pitchblende veins of Pribram (Czech Republic), Schlema-Alberoda (Germany) and Shinkolobwe (Democratic Republic of Congo), to the stockworks and episyenite columns of Bernardan (France) and Gunnar (Canada), to the narrow cracks in granite or metamorphic rocks, also filled with pitchblende of Mina Fe (Spain) and Singhbhum (India).

6. Intrusive deposits

Deposits included in this type are those associated with intrusive or anatectic rocks of different chemical composition (alaskite, granite, monzonite, peralkaline syenite, carbonatite and pegmatite). Examples include the Rossing and Trekkopje deposits (Namibia), the uranium occurrences in the porphyry copper deposits such as Bingham Canyon and Twin Butte (United States), the Ilimaussaq deposit (Greenland), Palabora (South Africa), as well as the deposits in the Bancroft area (Canada).

7. Volcanic and caldera-related deposits

Uranium deposits of this type are located within and nearby volcanic caldera filled by mafic to felsic volcanic complexes and intercalated clastic sediments. Mineralisation is largely controlled by structures (minor stratabound), occurs at several stratigraphic levels of the volcanic and sedimentary units and extends into the basement where it is found in fractured granite and in meta-morphites. Uranium minerals are commonly associated with molybdenum, other sulphides, violet fluorine and quartz. Most significant commercial deposits are located within Streltsovsk caldera in the Russian Federation. Examples are known in China, Mongolia (Dornot deposit), Canada (Michelin deposit) and Mexico (Nopal deposit).

8. Metasomatite deposits

Deposits of this type are confined to the areas of tectono-magmatic activity of the Precambrian shields and are related to near-fault alkali meta-somatites, developed upon different basement rocks: granites, migmatites, gneisses and ferruginous quartzites with production of albitites, aegirinites, alkali-amphibolic and carbonaceous-ferruginous rocks. Ore lenses and stocks are a few metres to tens of metres thick and a few hundred metres long. Vertical extent of ore mineralisation can be up to 1.5 km. Ores are uraninite-brannerite by composition and belong to ordinary grade. The reserves are usually medium scale or large. Examples include Michurinskoye, Vatutinskoye, Severinskoye, Zheltorechenskoye and Pervomayskoye deposits (Ukraine), Lagoa Real, Itataia and Espinharas (Brazil), the Valhalla deposit (Australia) and deposits of the Arjeplog region in the north of Sweden.

9. Surficial deposits

Surficial uranium deposits are broadly defined as young (Tertiary to Recent) near–surface uranium concentrations in sediments and soils. The largest of the surficial uranium deposits are in calcrete (calcium and magnesium carbonates), and they have been found in Australia (Yeelirrie deposit), Namibia (Langer Heinrich deposit) and Somalia. These calcrete-hosted deposits are associated with deeply weathered uranium-rich granites. They also can occur in valley-fill sediments along Tertiary drainage channels and in playa lake sediments (e.g. Lake Maitland, Australia). Surficial deposits also can occur in peat bogs and soils.

10. Collapse breccia pipe deposits

Deposits in this group occur in circular, vertical pipes filled with down-dropped fragments. The uranium is concentrated as primary uranium ore, generally uraninite, in the permeable breccia matrix, and in the arcuate, ring-fracture zone surrounding the pipe. Type examples are the deposits in the Arizona Strip north of the Grand Canyon and those immediately south of the Grand Canyon in the United States.

11. Phosphorite deposits

Phosphorite deposits consist of marine phosphorite of continental-shelf origin containing syn-sedimentary stratiform, disseminated uranium in fine-grained apatite. Phosphorite deposits constitute large uranium resources, but at a very low grade. Uranium can be recovered as a by-product of phosphate production. Examples include New Wales Florida (pebble phosphate) and Uncle Sam (United States), Gantour (Morocco) and Al-Abiad (Jordan). Other type of phosphorite deposits consists of organic phosphate, including argillaceous marine sediments enriched in fish remains that are uraniferous (Melovoe deposit, Kazakhstan).

12. Other deposits

Metamorphic deposits: In metamorphic uranium deposits, the uranium concentration directly results from metamorphic processes. The temperature and pressure conditions and age of the uranium deposition have to be similar to those of the metamorphism of the enclosing rocks. Examples include the Forstau deposit (Austria) and Mary Kathleen (Australia).

Limestone deposits: This includes uranium mineralisation in the Jurassic Todilto Limestone in the Grants district (United States). Uraninite occurs in intra-formational folds and fractures as introduced mineralisation.

Uranium coal deposits: Elevated uranium contents occur in lignite/coal, and in clay and sandstone immediately adjacent to lignite. Examples are uranium in the Serres Basin (Greece), in North and South Dakota (United States), Koldjat and Nizhne Iliyskoe (Kazakhstan) and Freital (Germany). Uranium grades are very low and average less than 50 ppm U.

13. Rock types with elevated uranium contents

Elevated uranium contents have been observed in different rock types such as pegmatite, granites and black shale. In the past no economic deposits have been mined commercially in these types of rocks. Their grades are very low, and it is unlikely that they will be economic in the foreseeable future.

Rare metal pegmatites: These pegmatites contain Sn, Ta, Nb and Li mineralisation. They have variable U, Th and rare earth elements contents. Examples include Greenbushes and Wodgina pegmatites (Western Australia). The Greenbushes pegmatites commonly have 6-20 ppm U and 3-25 ppm Th.

Granites: A small proportion of un-mineralised granitic rocks have elevated uranium contents. These "high heat producing" granites are potassium feldspar-rich. Roughly 1% of the total number of granitic rocks analysed in Australia have uranium-contents above 50 ppm.

Black Shale: Black shale-related uranium mineralisation consists of marine organic-rich shale or coal-rich pyritic shale, containing syn-sedimentary disseminated uranium adsorbed onto organic material. Examples include the uraniferous alum shale in Sweden and Estonia, the Chatanooga shale (United States), the Chanziping deposit (China), and the Gera-Ronneburg deposit (Germany).

Appendix 5.4

WORLD AND OECD EXPLORATION EXPENDITURES
AND DRILLING ACTIVITIES (1970-2003)

Year	Uranium spot price (NUEXCO) (USD/kgU)	Expenditures (USD thousand)				Surface + development drilling		
		OECD	Number of reporting countries	World	Number of reporting countries	Drilling (km)	Number of reporting countries	Number of holes reported
1970	16.22	31 039	1	148 359	4	7 422	NA	NA
1971	15.76	51 146	2	203 611	7	5 157	NA	NA
1972	15.47	80 362	13	242 105	27	5 150	22	NA
1973	16.67	78 050	12	290 835	27	5 590	25	NA
1974	28.68	113 066	15	303 676	29	7 330	26	NA
1975	61.57	191 851	14	389 531	34	8 492	30	78 362
1976	103.22	291 557	16	550 623	37	12 206	31	70 183
1977	109.72	425 428	14	692 381	38	15 837	34	137 554
1978	112.40	537 083	17	835 436	39	17 257	35	126 963
1979	110.68	570 221	18	907 932	47	15 353	41	95 066
1980	82.65	562 242	17	889 695	45	10 137	43	63 414
1981	62.89	393 344	17	722 274	48	7 273	44	31 744
1982	51.74	257 047	17	542 528	45	5 169	42	14 656
1983	59.75	167 789	15	436 931	43	2 601	37	10 298
1984	44.64	134 195	13	379 892	43	2 244	38	7 218
1985	40.56	114 951	15	357 489	41	1 846	37	5 172
1986	44.20	135 219	12	408 756	36	1 952	33	4 887
1987	43.63	135 160	12	419 404	32	1 895	29	4 792
1988	37.83	145 632	11	434 769	31	1 883	30	6 376
1989	26.00	118 977	8	380 924	29	1 592	27	4 677
1990	25.38	106 843	7	322 259	27	1 552	28	3 974
1991	22.62	97 694	7	120 294	23	1 320	25	4 156
1992	20.67	84 462	5	118 828	29	1 663	28	9 715
1993	18.51	62 983	6	91 383	24	1 459	24	7 879
1994	18.33	42 691	6	70 325	28	1 200	28	7 877
1995	21.97	47 461	6	83 567	26	688	26	5 678
1996	36.89	60 302	8	111 398	24	721	23	7 792
1997	26.70	93 729	6	153 863	25	868	25	11 046
1998	23.61	77 186	6	131 542	20	2 199	22	9 728
1999	21.45	48 306	4	98 972	18	1 543	18	6 687
2000	18.04	41 843	4	115 829	20	1 181	20	5 216
2001	21.06	23 547	3	88 999	18	1 258	17	4 782
2002	25.69	26 942	3	95 107	18	1 137	18	4 251
2003	30.06	25 803	2	90 393	18	1 117	17	3 737

Appendix 5.5

EXPLORATION ACTIVITIES IN SELECTED COUNTRIES

Australia

Year	1965	1966	1967*	1968*	1969*	1970*	1971*	1972	1973	1974
Expenditures (AUD million)			1.0	3.0	6.0	8.0	9.0	13.0	11.0	11.0
Expenditures (USD million)			1.0	4.0	7.0	9.0	10.0	12.8	16.3	14.6
Surface drilling (km)										

Year	1975	1976	1977	1978	1979	1980	1981	1982	1983	1984
Expenditures (AUD million)	8.0	13.0	17.0	25.0	29.0	35.0	38.0	29.0	14.0	13.0
Expenditures (USD million)	10.0	14.0	18.0	29.0	32.0	45.0	43.0	30.0	13.0	11.0
Surface drilling (km)	65.0	168.0	240.0	335.0	274.0	489.0	425.0	254.0	101.0	77.0

Year	1985	1986	1987	1988	1989	1990	1991	1992	1993	1994
Expenditures (AUD million)	13.00	18.00	24.00	26.00	22.04	15.74	14.26	13.56	8.28	6.67
Expenditures (USD million)	9.000	12.000	18.000	20.360	17.460	11.835	10.803	10.273	5.790	4.904
Surface drilling (km)	56.00	100.00	143.00	173.52	115.43	105.85	93.11	77.79	37.03	12.38

Year	1995	1996	1997	1998	1999	2000	2001	2002	2003	Total
Expenditures (AUD million)	8.26	14.92	23.63	19.37	9.61	7.59	4.80	5.34	6.38	544.45
Expenditures (USD million)	5.942	11.842	18.754	12.030	6.260	4.390	2.470	3.020	4.116	508.949
Surface drilling (km)	16.13	19.29	63.42	78.09	33.13	19.29	13.70	24.10	33.90	3 643.16

No data before 1967.

* Total expenditures for the period 1967 to 1971 of USD 31 million have been dispersed into annual expenditures based on consultations with Australia.

Canada

Year	1965	1966	1967	1968	1969	1970	1971*	1972*	1973*	1974*
Expenditures (USD million)								24.90		
Surface drilling (km)								NA		

Year	1975	1976	1977	1978	1979	1980	1981	1982	1983	1984
Expenditures (USD million)	23.50	43.10	68.00	80.00	111.90	111.55	85.54	56.90	33.95	27.97
Surface drilling (km)	NA	155.0	304.1	333.9	483.3	503.0	359.0	247.0	153.0	197.0

Year	1985	1986	1987	1988	1989	1990	1991	1992	1993	1994
Expenditures (USD million)	25.31	25.17	29.72	49.72	48.00	39.38	39.25	38.42	31.82	26.09
Surface drilling (km)	183.0	162.0	164.0	202.0	160.0	66.0	67.0	79.0	62.0	68.0

Year	1995	1996	1997	1998	1999	2000	2001	2002	2003	Total
Expenditures (USD million)	32.35	28.47	42.00	41.00	33.00	30.70	16.20	22.90	21.69	1 288.50
Surface drilling (km)	75.0	79.0	104.0	95.0	89.0	77.0	48.0	78.0	74.0	4 667.3

NA   Not available.

No data before 1971.

* Expenditures for the period 1971 to 1974 total USD 24.9 million and no breakdown per year is available.

EXPLORATION ACTIVITIES IN SELECTED COUNTRIES

Czechoslovakia/Czech Republic

Year	1965	1966	1967	1968	1969	1970	1971*	1972*	1973*	1974*
Expenditures (CZK million)							247.53	247.53	247.53	247.53
Expenditures (USD million)							13.85	13.85	13.85	13.85
Surface drilling (km)							121.55	121.55	121.55	121.55

Year	1975*	1976*	1977*	1978*	1979*	1980*	1981	1982	1983	1984
Expenditures (CZK million)	247.53	247.53	247.53	247.53	247.53	247.53	290.80	288.80	269.00	264.70
Expenditures (USD million)	13.85	13.85	13.85	13.85	13.85	13.85	21.54	20.70	18.82	15.80
Surface drilling (km)	121.55	121.55	121.55	121.55	121.55	121.55	112.54	105.93	90.53	92.45

Year	1985	1986	1987	1988	1989	1990	1991	1992	1993	1994
Expenditures (CZK million)	263.4	265.0	262.4	238.4	282.3	119.9	47.6	18.2		
Expenditures (USD million)	15.490	17.810	19.080	16.200	18.600	7.370	1.540	0.660	0.579	0.468
Surface drilling (km)	93.56	92.48	85.78	73.17	58.20	18.35	6.18	1.24	2.38	

Year	1995	1996	1997	1998	1999	2000	2001	2002	2003	Total
Expenditures (CZK million)										
Expenditures (USD million)	0.282	0.201	0.163	0.090	0.060	0.040	0.050	0.020	0.000	314.013
Surface drilling (km)										2 048.29

No data before 1971.

* Expenditures for the period 1971 to 1980 total CZK 2 475.3 million (USD 138.95 million) which were equally divided among the 10 years.

* Surface drilling for the period 1971 to 1980 totalled 1 215.54 km equally divided between the 10 years.

German Democratic Republic (Wismut)

Year	1945	1946	1947	1948	1949	1950	1951	1952	1953	1954
Expenditures (GDR Mark million)		75	75	75	75	75	75	75	75	128
Expenditures (USD million)		17.86	17.86	17.86	17.86	17.86	17.86	17.86	17.86	30.48
Surface drilling (km)										

Year	1975	1976	1977	1978	1979	1980	1981	1982	1983	1984
Expenditures (Mark million)	128	128	128	128	128	128	128	128	180	190
Expenditures (USD million)	51.96	50.85	55.24	63.73	69.83	70.49	56.64	52.70	70.45	66.77
Surface drilling (km)	367	357	326	304	297	31	222	191	134	163

Year	1985	1986	1987	1988	1989	1990	1991	1992	1993	1994
Expenditures (Mark million)	190	175	167	174	133	59	0	0	0	0
Expenditures (USD million)	64.57	80.64	92.87	97.53	70.74	36.51	0.00	0.00	0.00	0.00
Surface drilling (km)	164	119	116	120	68	17	0	0	0	0

Year	1995	1996	1997	1998	1999	2000	2001	2002	2003	Total
Expenditures (Mark million)	0	0	0	0	0	0	0	0	0	5 580
Expenditures (USD million)	0.00	0.00	0.00	0.00	0.00	0.00	0.00	0.00	0.00	1 905.92
Surface drilling (km)	0	0	0	0	0	0	0	0	0	27 814

Expenditures from 1946 to 1953 of about GDR Mark 600 million and expenditures from 1954 to 1982 of about GDR Mark 3 900 million have been equally divided between the years.

The exchange rate between GDR Mark/USD was assumed to be the same as for DM/USD.

EXPLORATION ACTIVITIES IN SELECTED COUNTRIES

United States

Year	1945	1946	1947	1948	1949	1950	1951	1952	1953	1954
Expenditures (USD million)				0.5	1.0	1.0	1.0	1.0	1.5	1.5
Surface drilling (km)				64.00	125.88	237.13	435.25	506.80	1 224.57	1 405.13

Year	1955	1956	1957	1958	1959	1960	1961	1962	1963	1964
Expenditures (USD million)	2.0	2.5	2.5	2.5	3.0	3.0	3.0	3.0	3.0	3.0
Surface drilling (km)	1 837.64	2 679.19	2 804.20	2 210.70	1 722.10	1 709.90	1 374.34	1 193.00	870.81	674.22

Year	1965	1966	1967	1968	1969	1970	1971	1972	1973	1974
Expenditures (USD million)	4.000	7.037	18.644	34.913	33.896	31.099	41.146	32.400	49.470	80.950
Surface drilling (km)	644.04	1 280.20	3 280.90	7 240.20	9 099.80	7 171.30	4 709.80	4 701.20	5 005.10	6 705.60

Year	1975	1976	1977	1978	1979	1980	1981	1982	1983	1984
Expenditures (USD million)	128.290	195.900	293.500	371.500	316.070	267.066	145.035	74.024	44.200	31.300
Surface drilling (km)	7 786.20	10 434.00	12 360.00	14 327.00	12 426.00	8 490.20	4 303.20	1 851.70	1 000.00	780.00

Year	1985	1986	1987	1988	1989	1990	1991	1992	1993	1994
Expenditures (USD million)	25.290	25.100	22.000	22.100	16.900	19.200	19.700	16.000	12.000	4.329
Surface drilling (km)	540.0	630.0	600.0	920.0	750.0	512.0	561.0	324.0	338.0	200.0

Year	1995	1996	1997	1998	1999	2000	2001	2002	2003	Total
Expenditures (USD million)	6.009	10.054	30.426	21.724	8.968	6.694	4.827	0.352	NA	2 507.113
Surface drilling (km)	411.0	928.0	1 488.0	1 415.0	763.0	312.0	201.0	NA	NA	155 564.3

NA   Not available.

Expenditures from 1948 to 1965 are estimated from non-Red Book sources.

USSR

Year	1945	1946	1947	1948	1949	1950	1951	1952	1953	1954
Expenditures (USD million)	2.60	13.74	13.74	13.74	13.74	13.74	27.30	27.30	27.30	27.30
Surface drilling (km)	2.0	80.2	80.2	80.2	80.2	80.2	233.6	233.6	233.6	233.6

Year	1955	1956	1957	1958	1959	1960	1961	1962	1963	1964
Expenditures (USD million)	27.30	51.00	51.00	51.00	51.00	51.00	65.00	65.00	65.00	65.00
Surface drilling (km)	233.6	1 323.2	1 323.2	1 323.2	1 323.2	1 323.2	1 848.8	1 848.8	1 848.8	1 848.8

Year	1965	1966	1967	1968	1969	1970	1971	1972	1973	1974
Expenditures (USD million)	65.00	73.16	73.16	73.16	73.16	73.16	94.38	94.38	94.38	94.38
Surface drilling (km)	1 848.8	2 206.6	2 206.6	2 206.6	2 206.6	2 206.6	3 074.0	3 074.0	3 074.0	3 074.0

Year	1975	1976	1977	1978	1979	1980	1981	1982	1983	1984
Expenditures (USD million)	94.38	127.04	127.04	127.04	127.04	127.04	138.64	138.64	138.64	138.64
Surface drilling (km)	3 074.0	4 322.4	4 322.4	4 322.4	4 322.4	4 322.4	5 135.8	5 135.8	5 135.8	5 135.8

Year	1985	1986	1987	1988	1989	1990	Total 1945-1990
Expenditures (USD million)	138.64	147.69	147.69	147.69	147.69	147.69	3 692.35
Surface drilling (km)	5 135.8	5 098.0	5 098.0	5 098.0	5 098.0	5 098.0	116 615.0

Expenditures and surface drilling for the USSR were given in 5-year intervals and are equally divided among the years. The exchange rate used for currency conversion was USD.
1 = RUB 0.75.

Appendix 5.6 NON-DOMESTIC URANIUM EXPLORATION EXPENDITURES
(USD thousand in year of expenditure)

Country	Pre-1972	1972	1973	1974	1975	1976	1977	1978	1979
Australia	215	0.0	0.0	0.0	0.0	0.0	0.0	NA	NA
Belgium	NA	NA	NA	NA	NA	NA	NA	1 100	1 200
Canada	NA	NA	NA	NA	NA	NA	NA	NA	NA
France	58 200	8 600	14 900	19 000	22 700	31 900	32 000	36 900	52 300
Germany	36 000	5 000	6 000	25 000	14 000	21 000	21 000	28 000	30 000
Italy	NA	NA	NA	NA	NA	NA	NA	NA	NA
Japan	8 310	1 210	1 230	1 440	2 880	8 350	24 030	24 430	24 540
Korea, Repulic of	NA	NA	NA	NA	NA	NA	NA	1 045	1 949
Nigeria	NA	NA	NA	NA	NA	NA	2 400	NA	NA
Spain	0.0	0.0	0.0	0.0	4 130	1 330	1 440	2 500	2 000
Switzerland	0.0	0.0	0.0	1 500	1 500	800	2 600	2 200	2 949
United Kingdom	NA	NA	NA	NA	NA	NA	NA	NA	NA
United States	15 000	2 000	3 000	5 000	5 000	19 000	31 160	35 890	43 240
Total	117 725	16 810	25 130	51 940	50 210	82 380	114 630	132 065	158 178

Country	1980	1981	1982	1983	1984	1985	1986	1987	1988
Australia	NA	NA	NA	NA	NA	NA	NA	NA	NA
Belgium	900	500	300	200	100	100	100	0.0	0.0
Canada	NA	NA	NA	NA	NA	NA	NA	NA	NA
France	68 200	66 230	43 460	28 700	18 764	15 013	17 595	14 525	7 748
Germany	30 000	26 000	26 000	15 500	11 500	15 800	20 000	19 000	14 000
Italy	NA	NA	NA	NA	NA	NA	NA	NA	NA
Japan	29 320	30 199	24 237	24 220	20 684	20 592	27 170	26 450	17 570
Korea, Republic of	5 741	3 961	3 688	3 295	1 177	257	43	5	48
Nigeria	NA	NA	NA	NA	NA	NA	NA	NA	NA
Spain	2 000	2 000	3 000	NA	NA	NA	NA	NA	0
Switzerland	3 222	2 753	1 273	294	340	2 570	2 400	1 005	1 000
United Kingdom	NA	NA	NA	2 875	7 402	11 854	6 705	6 761	6 406
United States	38 970	35 280	13 820	4 680	1 830	NA	NA	NA	NA
Total	178 353	166 923	115 778	79 764	61 797	> 66 186	> 74 013	> 67 746	> 46 772

Country	1989	1990	1991	1992	1993	1994	1995	1996	1997
Australia	NA	0.0	0.0	0.0	NA	NA	NA	NA	NA
Belgium	0.0	0.0	0.0	0.0	0.0	0.0	0.0	0.0	0.0
Canada	NA	NA	NA	NA	NA	1 449	1 471	3 650	3 986
France	9 128	5 726	11 076	19 438	32 619	30 959	10 245	6 808	8 972
Germany	9 000	6 766	4 853	2 898	3 107	2 646	2 951	3 137	4 000
Italy	NA	NA	200	NA	NA	NA	NA	NA	NA
Japan	11 990	10 990	11 210	12 010	11 620	12 923	14 771	7 533	4 752
Korea, Republic of	108	158	177	260	225	175	178	511	603
Nigeria	NA	NA	NA	NA	NA	NA	NA	NA	NA
Spain	0.0	0.0	0.0	0.0	0.0	0.0	0.0	0.0	0.0
Switzerland	500	600	540	482	502	627	0.0	0.0	0.0
United Kingdom	8 006	8 300	1 900	899	155	0.0	0.0	0.0	0.0
United States	0.0	0.0	0.0	0.0	0.0	NA	NA	422	3 050
Total	38 732	32 540	29 956	35 987	48 228	> 48 779	29 616	22 061	25 363

Country	1998	1999	2000	2001	2002	2003	Total
Australia	NA	NA	NA	NA	NA	NA	215
Belgium	0.0	0.0	0.0	0.0	0.0	0.0	4 500
Canada	3 000	3 000	3 667	2 597	2 549	2 547	27 916
France	8 777	7 120	7 330	7 690	14 370	16 701	753 694
Germany	0	0	0	0	0	0	403 158
Italy	NA	NA	NA	NA	NA	NA	200
Japan	2 280	1 390	NA	NA	NA	NA	418 331
Korea, Republic of	445	NA	NA	NA	NA	NA	24 049
Nigeria	NA	NA	NA	NA	NA	NA	2 400
Spain	0.0	0.0	0.0	0.0	0.0	0.0	18 400
Switzerland	0.0	0.0	0.0	0.0	0.0	0.0	29 657
United Kingdom	0.0	0.0	0.0	0.0	0.0	0.0	61 263
United States	3 616	NA	NA	NA	NA	NA	260 958
Total	18 118	11 510	10 997	10 287	16 919	19 248	1 701 245

NA = Not available

212

Appendix 5.7

DISCOVERY COST OF KNOWN CONVENTIONAL RESOURCES
AND PRODUCTION IN SELECTED COUNTRIES

Country	RAR (tU)	EAR-I (tU)	Known Conventional Resources (RAR + EAR-I) (tU)	Production (tU)	KCR + Production (tU)	Exploration expenditures (USD 1000)	Discovery cost (USD/kgU)
Argentina	7 080	8 560	15 640	2 631	18 271	52 263	2.86
Australia	735 000	323 000	1 058 000	113 304	1 171 304	508 949	0.43
Belgium	0	0	0	680	680	2 487	3.66
Brazil	86 190	57 140	143 330	1 645	144 975	186 128	1.28
Canada	333 834	104 710	438 544	374 548	813 092	1 288 500	1.58
Central African Republic	12 000	0	12 000	0	12 000	21 800	1.82
Chile	560	885	1 445	0	1 445	6 896	4.77
China	35 060	14 690	49 750	27 689	77 439	23 400	0.30
Czech Republic[a]	830	90	920	108 649	109 569	314 013	2.87
Denmark	20 250	12 000	32 250	0	32 250	4 140	0.13
Finland	1 125	0	1 125	41	1 166	13 984	11.99
France	0	9 510	9 510	75 965	85 475	907 240	10.61
Gabon	4 830	1 000	5 830	25 403	31 233	102 433	3.28
Germany[b]	3 000	4 000	7 000	219 239	226 239	2 002 789	8.85
Greece	1 000	6 000	7 000	0	7 000	17 547	2.51
Hungary	0	13 800	13 800	21 080	34 880	3 700	0.11
India	40 980	18 935	59 915	7 963	67 878	315 228	4.64
Indonesia	4 620	1 155	5 775	0	5 775	15 878	2.75
Iran	370	700	1 070	0	1 070	9 731	9.09
Italy	4 800	1 300	6 100	0	6 100	75 060	12.30
Japan	6 600	0	6 600	84	6 684	19 697	2.95
Madagascar	0	0	0	785	785	5 293	6.74
Mexico	1 275	525	1 800	49	1 849	30 306	16.39
Mongolia[c]	46 200	15 750	61 950	535	62 485	8 153	0.13
Namibia	170 532	87 085	257 617	78 794	336 411	25 741	0.08
Niger	102 227	125 377	227 604	91 186	318 790	216 121	0.68
Peru	1 215	1 265	2 480	0	2 480	4 776	1.93
Portugal	7 470	1 450	8 920	3 715	12 635	17 637	1.40
Romania[d]	3 325	3 608	6 933	17 989	24 922	10 060	0.40
Slovenia /Yugoslavia	2 200	10 000	12 200	380	12 580	1 581	0.13
Somalia	4 950	2 550	7 500	0	7 500	10 000	1.33
South Africa	315 330	80 340	395 670	157 618	553 288	140 919	0.25
Spain	4 925	6 380	11 305	6 156	17 461	140 455	8.04
Sweden	4 000	6 000	10 000	91	10 091	47 900	4.75
Thailand	5	5	10	0	10	11 299	1129.90
Turkey	6 845	0	6 845	0	6 845	21 988	3.21
United States	345 000	0	345 000	356 485	701 485	2 507 113	3.57
Vietnam	1 005	5 435	6 440		6 440	2 898	0.45
Zambia	0	0	0	102	102	25	0.25
Zimbabwe	1 350	0	1 350	0	1 350	6 902	5.11
Sub-total	2 315 983	923 245	3 239 228	1 692 806	4 932 034	9 101 030	1.85

USSR + Kazakhstan-Russian Federation-Ukraine-Uzbekistan (from 1945 to 2003)*							
	817 760	488 630	1 306 390	467 482	1 773 872	3 989 432	2.25

| World Total | 3 133 743 | 1 411 875 | 4 545 618 | 2 160 288 | 6 705 906 | 13 090 462 | 1.95 |

Notes:
*   Including USD 3 692 350 000 in USSR from 1945 to 1990.
a)   Exploration expenditures from 1971 to 2003.
b)   Including German Democratic Republic.
c)   Exploration expenditures from 1992 to 1997.
d)   Exploration expenditures since 1994.

Appendix 5.8

DISCOVERY COST OF TOTAL (KNOWN AND UNDISCOVERED) RESOURCES AND PRODUCTION IN SELECTED COUNTRIES

Country	Known Conventional Resources (tU)	Undiscovered Resources (tU)	Total Resources (tU)	Production (tU)	Total Resources plus production (tU)	Exploration expenditures (USD 1000)	Discovery cost (USD/kgU)
Argentina	15 640	1 400	17 040	2 631	19 671	52 263	2.66
Australia [a]	1 058 000	2 600 000	3 658 000	113 304	3 771 304	508 949	0.13
Belgium	0	0	0	680	680	2 487	3.66
Brazil	143 330	620 000	763 330	1 645	764 975	186 128	0.24
Canada	458 544	850 000	1 308 544	374 548	1 683 092	1 288 500	0.77
Central African Republic	12 000	0	12 000	0	12 000	21 800	1.82
Chile	1 445	4 700	6 145	0	6 145	6 896	1.12
China [b]	49 750	1 773 600	1 823 350	27 689	1 851 039	23 400	0.01
Colombia	0	228 000	228 000	0	228 000	19 946	0.09
Czech Republic [c]	920	179 200	180 120	108 649	288 769	314 013	1.09
Denmark	32 250	60 000	92 250	0	92 250	4 140	0.04
Egypt [d]	0	15 000	15 000	0	15 000	108 807	7.25
Finland	1 125	0	1 125	41	1 166	13 984	11.99
France	9 510	0	9 510	75 965	85 475	907 240	10.61
Gabon [e]	5 830	1 610	7 440	25 403	32 843	102 433	3.12
Germany [f]	7 000	74 000	81 000	219 239	300 239	2 002 789	6.67
Greece	7 000	6 000	13 000	0	13 000	17 547	1.35
Hungary	13 800	32 480	46 280	21 080	67 360	3 700	0.05
India	59 915	32 500	92 415	7 963	100 378	315 228	3.14
Indonesia	5 775	4 100	9 875	0	9 875	15 878	1.61
Iran	1 070	13 900	14 970	0	14 970	9 731	0.65
Italy	6 100	10 000	16 100	0	16 100	75 060	4.66
Japan	6 600	0	6 600	84	6 684	19 697	2.95
Madagascar	0	0	0	785	785	5 293	6.74
Mexico	1 800	13 000	14 800	49	14 849	30 306	2.04
Mongolia [g]	61 950	1 390 000	1 451 950	535	1 452 485	8 153	0.01
Namibia	257 617	0	257 617	78 794	336 411	25 741	0.08
Niger	227 604	9 500	237 104	91 186	328 290	216 121	0.66
Peru [h]	2 480	39 700	42 180	0	42 180	4 776	0.11
Portugal	8 920	6 500	15 420	3 715	19 135	17 637	0.92
Romania [i]	6 933	6 000	12 933	17 989	30 922	10 060	0.33
Slovenia /Yugoslavia	12 200	1 100	13 300	380	13 680	1 581	0.12
Somalia	7 500	0	7 500	0	7 500	10 000	1.33
South Africa	395 670	1 223 200	1 618 870	157 618	1 776 488	140 919	0.08
Spain	11 305	0	11 305	6 156	17 461	140 455	8.04
Sweden	10 000	0	10 000	91	10 091	47 900	4.75
Thailand	10	0	10	0	10	11 299	1 129.9
Turkey	6 845	0	6 845	0	6 845	21 988	3.21
United States	345 000	2 613 000	2 958 000	356 485	3 314 485	2 507 113	0.76
Vietnam	6 440	237 900	244 340	0	244 340	2 898	0.01
Zambia	0	22 000	22 000	102	22 102	25	0.00
Zimbabwe	1 350	25 000	26 350	0	26 350	6 902	0.26
Sub-Total	**3 259 228**	**12 093 390**	**15 352 618**	**1 692 806**	**17 045 424**	**9 229 783**	**0.54**

USSR + Kazakhstan-Russian Federation-Ukraine-Uzbekistan (from 1945 to 2003)*							
	1 306 390	1 947 700	3 254 090	467 482	3 721 572	3 989 432	1.07

| **World Total** | **4 565 618** | **14 041 090** | **18 606 708** | **2 160 288** | **20 766 996** | **13 219 215** | **0.64** |

Notes:
* Including USD 3 692 350 000 in USSR from 1945 to 1990.
a) Speculative Resources of 2 600 000 – 3 900 000 tU taken from Red Book 1993.
b) Speculative Resources of 1 770 000 tU taken from Red Book 1997.
c) Exploration expenditures from 1971 to 2003 only.
d) Speculative Resources of 15 000 tU taken from Red Book 1997.
e) EAR-II of 1 610 tU taken from Red Book 1997.
f) Includes data from the former German Democratic Republic.
g) Exploration expenditures from 1992 to 1997 only.
h) EAR-II of 20 000 tU and Speculative Resources of 19 700 tU taken from Red Book 1997.
i) Exploration expenditures since 1994 only.

Appendix 5.9

EXPLORATION EXPENDITURES FOR COUNTRIES WITHOUT REPORTED RESOURCES

Country	Exploration expenditures (USD 1000)
Bangladesh	453
Bolivia	9 343
Botswana	825
Cameroon	1 282
Costa Rica	364
Cuba	972
Ecuador	1 945
Ghana	90
Guatemala	610
Ireland	6 200
Jamaica	30
Jordan	920
Korea, Republic of	17 886
Lesotho	21
Malaysia	11 072
Mali	58 693
Morocco	2 752
Nigeria	6 950
Norway	3 180
Paraguay	26 360
Rwanda	1 505
Sri Lanka	43
Sudan	200
Switzerland	3 359
Syria	1 151
United Kingdom	3 815
Uruguay	231
Total	160 252

Appendix 6.1
EVOLUTION OF RESOURCES DEFINITIONS
IN SUCCESSIVE EDITIONS OF THE RED BOOK

To evaluate mineral resources, clear definitions of what is to be evaluated must be given. In addition to geological criteria, such definitions must contain elements for confidence in the estimates and of economics and feasibility of exploitation. In consideration of these parameters, resources estimates reported in the Red Books are divided into separate categories reflecting different levels of confidence in the quantities reported, and are further separated into categories based on the cost of production. As part of a continuing effort to develop illustrations of current and future uranium supply possibilities, resource definitions have been revised several times to improve their clarity, and to account for changing economic, technical and political conditions. This appendix provides a detailed chronology of changes in uranium resource categories in successive Red Books.

In 1965, in the first edition of the Red Book, estimates of uranium resources were made based on "prices". Beginning in 1976 and lasting through 2003 "costs of recovery" of uranium were used to define resource categories. Three price ranges were introduced in the first Red Book in 1965:

- USD 5 – 10 per pound U_3O_8 (USD 13 – 26 per kgU)
- USD 10 – 15 per pound U_3O_8 (USD 26 – 39 per kgU)
- USD 15 – 30 per pound U_3O_8 (USD 39 – 78 per kgU)

Within each price range, estimates were given for two different levels of confidence, Reasonably Assured Resources (RAR) and Possible Additional Resources.

Reasonably Assured Resources referred to material which occurred in known ore deposits of such grade, quantity and configuration that it could, within the given price range, be profitably mined and processed with presently known mining and processing technology. Estimates of tonnage and grade were based on specific sample data and measurements of the deposits and on knowledge of ore body characteristics. They were considered as roughly equivalent to Ore Reserves in the usual sense, particularly in the lower price ranges. Table 6.1A summarises the classification of resources used in the 1965 Red Book.

Possible Additional Resources referred to material surmised to occur in unexplored extensions of known deposits, or in undiscovered deposits, in known or postulated uranium districts, and which were expected to be economically discoverable and exploitable in the given price range. Estimates of tonnage and grade of Possible Additional Resources in known districts were based primarily on knowledge of known deposits within the same districts. The existence and size of undiscovered deposits in specific unexplored areas was inferred on the basis of comparisons with resource distribution in known areas.

Table 6.1A. Resource classification – 1965 Red Book

USD 13 – 26 per kgU	Reasonably Assured Resources	Possible Additional Resources
USD 26 – 39 per kgU	Reasonably Assured Resources	Possible Additional Resources
USD 39 – 78 per kgU	Reasonably Assured Resources	Possible Additional Resources

Decreasing economic attractiveness ↓

Decreasing confidence in estimates ⟶

In 1967, in response to current trends in uranium marketing, it was no longer realistic to consider USD 5 per pound U_3O_8 as an attainable lower price limit. Therefore the lower price range was modified to "less than USD 10 per pound". A change in resources terminology was also introduced. Estimated Additional Resources (EAR) replaced Possible Additional Resources, with the latter potentially being interpreted as representing a maximum resource total, which was not the intent.

Between 1973 and 1975, important events occurred in the energy field, namely OPEC's decision to raise oil prices dramatically by reducing the world oil supply. As a consequence, the uranium market was also affected and uranium prices followed the general upward trend of energy prices (average Nuexco spot price of uranium increased from USD 11.03/lb U_3O_8 in 1973, to USD 23.68/lb U_3O_8 in 1975). Furthermore, worldwide inflation considerably increased uranium production costs. It was therefore necessary to reconsider the price categories for uranium resources, which were used in the previous reports. In addition, to follow the international metric system, kilograms of uranium (kgU) instead of lb U_3O_8 was introduced.

In the December 1975 edition of the Red Book, the Working Group decided to abandon "price categories" in favour of "cost categories". This was done in order to maintain a certain stability in the resource categories employed, and to become independent of possible significant price changes in the future, which might be a reflection of the market situation and not necessarily be linked with actual production costs of a given type of resources. Costs in general included not only the direct costs of mining, milling and extraction, but also the cost of capital spent in providing and maintaining the production unit. Exploration costs (sunk costs) were not included. The new cost ranges that were adopted included:

- < USD 15 per pound U_3O_8 (< USD 39 per kgU)
- USD 15 – 30 per pound U_3O_8 (USD 39 – 78 per kgU)

This was a considerable change compared with the previous figures. However, the new low cost category corresponded to a large extend to the previous low price category. On the other hand, the new high category contained the bulk of the resources previously listed in the USD 10-15/lb U_3O_8 category, together with some new material not previously considered.

Between 1975 and 1976, uranium prices continued to increase, with inflation in the cost of uranium recovery continuing to be a significant factor. Therefore it was again necessary to revise the range of the cost categories used to report uranium resources in the next edition of the Red Book published in 1977. The adopted cost brackets were:

- < USD 30 per pound U_3O_8 (< USD 80 per kgU)
- USD 30 – 50 per pound U_3O_8 (USD 80 – 130 per kgU)

The new figures represented a considerable change from the previous ones. However the Working Group concluded that these differences did not significantly affect the comparability of resources totals with previous Red Books. In addition, revision of the cost categories provided a more accurate estimate of worldwide resources. A higher cost resources category (> USD 130/kgU) was added to take into account uranium that will not likely be recovered in the 20th century, but which in some areas has been investigated for possible production in the very long term. Table 6.2A shows the cost and confidence categories used in the 1977 Red Book.

Table 6.2A. Resource classification – 1977 Red Book

Sub Economic Resources	Exploitable at costs from USD 130 – X/kgU	Reasonably Assured Resources	Estimated Additional Resources
	Exploitable at costs from USD 80 – 130/kgU	Reasonably Assured Resources	Estimated Additional Resources
Economic Resources	Exploitable at costs up to USD 80/kgU	Reasonably Assured Resources	Estimated Additional Resources

Decreasing economic attractiveness (top to bottom)

Decreasing confidence in estimates ⟶

For the first time, the distinction between in place (*in situ*) and recoverable resources was introduced into the Red Book and it was requested that resource estimates should be expressed in terms of quantities of uranium recoverable from *in situ* resources, after accounting for ore dilution, and for mining and milling losses. Except where noted, resource estimates were expressed in terms of recoverable rather than in-situ quantities. However, in some countries the practice was to report estimates as quantities of uranium contained in mineable ore after allowing for ore dilution, particularly where processing losses were not considered to be significant.

The Working Group also considered that current estimates represented an incomplete appraisal, as additional potential resources were likely to exist. Regional exploration results supported establishment of another resource category with a lower degree of confidence than Estimated Additional Resources. Such a category could include uranium, in the same cost ranges, that is thought to exist, mostly on the basis of indirect indications and geological extrapolations, in deposits discoverable with existing exploration techniques. The location of deposits envisaged in this category could generally be specified only as being within a given region or geological trend.

Table 6.3A correlates resource categories used in the 1977 Red Book with terms used in other major resource classification systems:

**Table 6.3A. Comparison of 1977 resource nomenclature
with Canadian and US resource classifications**

NEA/IAEA 1976 Red Book	Reasonably Assured		Estimated Additional			
Energy Mines and Resources Canada	Demonstrated		Surmised		Speculative	
	Measured	Indicated	Inferred	Prognosticated	In areas with occurrences	In virgin areas
United States (ERDA) 1976	Reserves		Probable Potential Resources		Possible and Speculative Potential Resources	

In the 1979 Red Book, a third resource category, based on level of confidence, was introduced: Speculative Resources. These lower confidence resources were further separated into resources that can be produced at less than USD 80/kgU, between USD 80 and 130/kgU, and higher than USD 130/kgU (generally between USD 130 and 260/kgU).

Two basic categories of resources were defined:

- Reasonably Assured Resources (same as defined in 1965).
- Estimated Additional Resources, which referred to uranium in addition to Reasonably Assured Resources, that was expected to occur, mostly on the basis of direct geological evidence in:
  – Extensions of well-explored deposits;
  – Little-explored deposits;
  – Undiscovered deposits believed to exist along a well-defined geological trend with known deposits.

  Estimated Additional Resources were considered as having potential for later conversion to Reasonably Assured Resources as the result of further exploration efforts.

A third category, Speculative Resources, was established for potential uranium that was thought to exist mostly on the basis of indirect evidence and geological extrapolations, in deposits discoverable with existing exploration techniques. The location of deposits envisaged in this category could generally be specified only as being somewhat within a given region or geological trend. The existence and size of such resources are highly speculative. Speculative Resources were distinct from Reasonably Assured and Estimated Additional Resources because of the very low confidence which could be placed on the amounts reported. Table 6.4A summarises the resource classification used in the 1979 Red Book. Table 6.5A compares the 1979 classification with classification used elsewhere.

Table 6.4A. NEA/IAEA resource classification system – 1979 Red Book

Exploitable at costs higher than USD 130/kgU	Reasonably Assured Resources	Estimated Additional Resources	Speculative Resources
Exploitable at costs between USD 80-130/kgU	Reasonably Assured Resources	Estimated Additional Resources	Speculative Resources
Exploitable at costs up to USD 80/kgU	Reasonably Assured Resources	Estimated Additional Resources	

Table 6.5A. Approximate correlations of terms used in major resource classification systems

NEA/IAEA	Reasonably Assured		Estimated Additional		Speculative
Australia	Reasonably Assured		Estimated Additional		
Canada (EMR)	Measured	Indicated	Inferred	Prognosticated	Speculative
France	Reserves I	Reserves II	Perspectives I	Perspectives II	
South Africa	Reasonably Assured		Estimated Additional		
United States (DOE)	Reserves		Probable Potential Resources		Possible and Speculative Potential Resources

In 1983, in order to improve the comparability of the data provided by national authorities, changes were made to the resource terminology and cost definitions used in previous Red Books.

Resources categories according to confidence level

Prior to 1982, uranium resources were reported according to three categories, Reasonably Assured Resources, Estimated Additional Resources and Speculative Resources. In 1983, the Estimated Additional Resources category (EAR) was replaced by two new categories, Estimated Additional Resources Category I (EAR-I) and Estimated Additional Resources Category II (EAR-II). In general terms, the "discovered" part of the resources in the previous EAR category were to be reported as EAR-I, and the "undiscovered" part in EAR-II. Following are the definitions of resources based on levels of confidence.

Reasonably Assured Resources (RAR) refers to uranium that occurs in known mineral deposits of such size, grade and configuration that it could be recovered within the given production cost ranges, with currently proven mining and processing technology. Estimates of tonnage and grade are based on specific sample data and measurements of the deposits and on knowledge of deposit characteristics. Reasonably Assured Resources have a high assurance of existence and in the cost category below USD 80/kgU are considered as reserves for the purposes of the Red Book.

Estimated Additional Resources – Category I (EAR-I) refers to uranium in addition to RAR that is expected to occur, mostly on the basis of direct geological evidence, in extensions of well-explored deposits, and in deposits in which geological continuity has been established but where specific data and measurements of the deposits and knowledge of the deposits' characteristics are considered to be inadequate to classify the resource as RAR. Such deposits can be delineated and the uranium subsequently recovered, all within the given cost ranges. Estimates of tonnage and grade are based on such sampling as is available and on the knowledge of the deposit characteristics as determined in the best known parts of the deposit or in similar deposits. Less reliance can be placed on the estimates in this category than on those for RAR.

Estimated Additional Resources – Category II (EAR-II) refers to uranium in addition to EAR-I that is expected to occur in deposits believed to exist in well-defined geological trends or areas of mineralisation with known deposits. Such deposits can be discovered, delineated and the uranium subsequently recovered, all within the given cost ranges. Estimates of tonnage and grade are based primarily on the knowledge of deposit characteristics in known deposits within the respective trends or areas and on such sampling, geological, geophysical or geochemical evidence as may be available. Less reliance can be placed on the estimates in this category than on those for EAR-I.

Speculative Resources (SR) refers to uranium, in addition to Estimated Additional Resources – Category II, that is thought to exist mostly on the basis of indirect evidence and geological extrapolations, in deposits discoverable with existing exploration techniques. The location of deposits envisaged in this category could generally be specified only as being somewhere within a given region or geological trend. As the term implies, the existence and size of such resources are highly speculative.

Resources according to cost categories

Cost categories used to report Red Book uranium resources are: <USD 80/kgU, USD 80 to 130/kgU and <USD 130/kgU. High Cost Conventional Resources recoverable at USD 130 to 260/kgU was also reported in the 1983 Red Book, but this category has since been discontinued. When estimating the cost of production for assigning resources within these cost categories, account is taken of the following costs:

The direct costs of mining, transporting and processing the uranium ore.

- The costs of associated environmental and waste management.
- The costs of maintaining non-operating production units where applicable.
- The capital cost of providing new production units where applicable.
- The cost of financing, including any unamortized costs where applicable.
- Indirect costs such as office overheads, taxes and royalties where applicable.
- Future exploration and development costs wherever required for further ore delineation to the stage where it is ready to be mined.

Sunk costs such as exploration are not normally taken into consideration.

Table 6.6A summarises the NEA/IAEA resource classification system used in 1983. Table 6.7A compares the 1983 NEA/IAEA classification with other major resource classification systems.

Table 6.6A. NEA/IAEA classification system – 1983 Red Book

Exploitable at costs between USD 130–260/kgU	Reasonably Assured Resources	Estimated Additional Resources I	Estimated Additional Resources II	Speculative Resources
Exploitable at costs between USD 80–130/kgU	Reasonably Assured Resources	Estimated Additional Resources I	Estimated Additional Resources II	Speculative Resources
Exploitable at costs up to USD 80/kgU	Reasonably Assured Resources	Estimated Additional Resources I	Estimated Additional Resources II	

Table 6.7A. Correlation between the resource categories defined above and other classification systems

NEA/IAEA	Reasonably Assured		Estimated Additional I	Estimated Additional II	Speculative
Australia	Reasonably Assured		Estimated Additional	Undiscovered	
Canada EMR	Measured	Indicated	Inferred	Prognosticated	Speculative
France	Reserves I	Reserves II	Perspectives I	Perspectives II	
South Africa	Reasonably Assured		Estimated Additional I	Estimated Additional II	
United States DOE	Reserves		Probable Potential Resources		Possible and Speculative Potential Resources

Since the modifications that were introduced in the 1983 Red Book, only minor changes have been made to the NEA/IAEA resource terminology and definitions. In 2003, however, a new lower-cost category, i.e. resources recoverable at <USD 40/kgU, had been introduced, in order to reflect a cost range more relevant to current uranium market prices. Table 6.8A shows the 2003 NEA/IAEA resource classification system. Table 6.9A shows a correlation of the major worldwide resource classification systems from the 2003 Red Book.

Table 6.8A. NEA/IAEA classification system – 2003 Red Book

Exploitable at costs between USD 80–130/kgU	Reasonably Assured Resources	Estimated Additional Resources I	Estimated Additional Resources II	Speculative Resources
Exploitable at costs between USD 40–80/kgU	Reasonably Assured Resources	Estimated Additional Resources I	Estimated Additional Resources II	Speculative Resources
Exploitable at costs up to USD 40/kgU	Reasonably Assured Resources	Estimated Additional Resources I	Estimated Additional Resources II	Speculative Resources

Table 6.9A. Approximate correlations of terms used in major resources classification systems – 2003 Red Book

	KNOWN CONVENTIONAL RESOURCES			UNDISCOVERED CONVENTIONAL RESOURCES		
NEA/IAEA	REASONABLY ASSURED		ESTIMATED ADDITIONAL-I	ESTIMATED ADDITIONAL-II	SPECULATIVE	
Australia	DEMONSTRATED: MEASURED	DEMONSTRATED: INDICATED	INFERRED	UNDISCOVERED		
Canada (NRCan)	MEASURED	INDICATED	INFERRED	PROGNOSTICATED	SPECULATIVE	
United States (DOE)	REASONABLY ASSURED		ESTIMATED ADDITIONAL		SPECULATIVE	
Russian Federation, Kazakhstan, Ukraine, Uzbekistan	A + B	C 1	C 2	P1	P2	P3
UNFC*	EF1		EF2	EF3	EF4	

* UNFC (United Nations Framework Classification) correlation with NEA/IAEA and national classifications systems was still under consideration, when added to the 2003 edition of the Red Book.

In the 2005 edition of the Red Book the titles of the resource categories were modified to make them more in line with the major systems in use and to align them for possible future alignment with the UNFC. The EAR-I category was re-titled Inferred Resources and the EAR-II category was re-titled Prognosticated Resources. The definitions were left unchanged.

EVOLUTION OF REASONABLY ASSURED RESOURCES (1965-2003)
(1 000 tU)

Year of publication	<USD 26/kgU (USD 5-10/lbU$_3$O$_8$)	USD 26-39/kgU (USD 10-15/lbU$_3$O$_8$)	<USD 40/kg U (<USD 15/lbU$_3$O$_8$)	USD 40-80/kgU (USD 15-30/lbU$_3$O$_8$)	
1965	494	527	1 021	> 430	
1967	539	496	1 035	556	
1969	539	–	–	–	
1970	645	578	1 223	–	
1973	866	680	1 546	–	
1976	–	–	1 080	730	
1977	–	–	–	–	
1979	–	–	–	–	
1982	–	–	–	–	
1983	–	–	–	–	
1986	–	–	–	–	
1988	–	–	–	–	
1989	–	–	–	–	
1991	–	–	–	–	
1993	–	–	–	–	
1995	–	–	> 544	> 426	a
1997	–	–	> 666	> 555	a
1999	–	–	> 916	> 531	a
2001	–	–	> 1 534	> 557	a
2003	–	–	1 731	> 575	a

Year of publication	<USD 80/kgU (<USD 30/lbU$_3$O$_8$)		USD 80-130/kgU (USD 30-50/lbU$_3$O$_8$)		<USD 130/kgU <USD 50/lbU$_3$O$_8$)		USD 130-260/kgU (USD 50-100/lbU$_3$O$_8$)	<USD 260/kgU (<USD 100/lbU$_3$O$_8$)
1965	> 1 451		–		> 1 451		–	–
1967	1 590		–		1 590		–	–
1969	–		–		–		–	–
1970	1 223		–		–		–	–
1973	1 546		–		1 546		–	–
1976	1 810		–		1 810		–	–
1977	1 650		540		2 190		–	–
1979	1 850		740		2 590		–	–
1982	1 747		546		2 293		–	–
1983	1 425	a	575	a	2 000	a	–	–
1986	1 609	a	641	a	2 250	a	> 417	2 667
1988	1 555	a	678	a	2 233	a	–	–
1989	1 546	a	655	a	2 201	a	> 400	2 601
1991	1 389	a	610	a	1 999	a	–	–
1993	1 424	a	659	a	2 038	a	–	–
1995	2 124	a	827	a	2 951	a	–	–
1997	2 340	a	718	a	3 220	a	–	–
1999	2 270	a	660	a	2 964	a	–	–
2001	2 242	a	590	a	2 853	a	–	–
2003	2 458	a	662	a	3 169	a	–	–

–   Not applicable..

*   From 1965 to 1993, resources in WOCA only. From 1995 onwards, resources of total world gradually are reported.

a) Total adjusted. Since 1983, RAR are given as recoverable resources (mining and milling losses deducted).

EVOLUTION OF REASONABLY ASSURED RESOURCES BY COUNTRY (1965-2003)
(1 000 tU)*

Red Book edition Countries	1965		1967		1970		1973	
	<USD 80/kgU	<USD 130/kgU	<USD 80/kgU	<USD 130/kgU	<USD 40/kgU	<USD 130/kgU	<USD 40/kgU	<USD 130/kgU
Algeria								
Angola			11.6	–				
Argentina	7.7	–	27.0	–	16.2	–	16.9	–
Australia	14.8	–	14.9	–	23.8	–	100.5	–
Austria			20.0	–				
Botswana								
Brazil					0.8	–	0.7	–
Bulgaria								
Canada	338.8	–	331.1	–	278.7	–	307.0	–
Central African Republic					8.0	–	8.0	–
Chile								
China								
Congo / Zaire	4.6	–	4.6	–			1.8	
Czech Republic								
Denmark (Greenland)	3.9	–	3.9	–	3.9	–	5.6	–
Finland							1.3	–
France	32.3	–	38.5	–	41.6	–	56.6	–
Gabon	3.9	–	3.1	–	10.4	–	20.0	–
Germany	NA	–	NA	–	NA	–	NA	
Greece								
Hungary								
India	12.7	–	20.8	–	2.3		2.3	–
Indonesia								
Iran, Islamic Republic of								
Italy	NA	–	24.2	–	1.2	–	1.2	–
Japan	2.0	–	3.1	–	5.5	–	7.0	–
Kazakhstan								
Korea, Republic of								
Malawi								
Mexico					1.9	–	1.9	–
Mongolia								
Morocco	19.3	–	19.3	–				
Namibia								
Niger			19.3	–	30.0	–	50.0	–
Peru								
Philippines								
Portugal	5.4	–	16.2	–	7.4	–	7.4	–
Romania								
Russian Federation								
Slovenia / Yugoslavia	NA	–	NA	–	NA	–	6.0	–
Somalia								
South Africa	107.8	–	250.3	–	196.4	–	264.0	–
Spain	8.5	–	23.1	–	16.2	–	16.2	–
Sweden	385.0	–	385.0	–	269.5	–	270.0	–
Thailand								
Turkey	NA	–	NA	–	NA	–	2.7	–
Ukraine								
United Kingdom								
United States	396.6	–	500.5	–	300.3	–	400.0	–
Uzbekistan								
Vietnam								
Zimbabwe								
Total	1 343 1	–	1 716.2	–	1 214.0	–	1 547.1	–
Adjusted Total (a)								

EVOLUTION OF REASONABLY ASSURED RESOURCES BY COUNTRY (1965-2003)
(1 000 tU)*

Red Book edition Countries	1976 <USD 80/kgU	1976 <USD 130/kgU	1977 <USD 80/kgU	1977 <USD 130/kgU	1979 <USD 80/kgU	1979 <USD130/kgU	1982 <USD 80/kgU	1982 <USD 130/kgU
Algeria	28.0	–	28.0	28.0	28.0	28.0	26.0	26.0
Angola								
Argentina	20.6	–	17.8	41.8	23.0	28.1	25.0	30.3
Australia	243.0	–	289.0	296.0	290.0	299.0	294.0	317.0
Austria			1.8	1.8	1.8	1.8	0.0	0.3
Botswana					0 0	0 4		
Brazil	10.4	–	18.2	18.2	74.2	74.2	119.1	119.1
Bulgaria								
Canada	166.0	–	167.0	182.0	215.0	235.0	230.0	258.0
Central African Republic	8.0	–	8.0	8.0	18.0	18.0	18.0	18.0
Chile			0.0	0.0	0.0	0.0	0.0	
China								
Congo / Zaire	1.8	–	1.8	1.8	1.8	1.8	1.8	1.8
Czech Republic								
Denmark (Greenland)	6.0	–	0.0	5.8	0.0	27.0	0.0	27.0
Finland	1.9	–	1.3	3.2	0.0	2.7	0.0	3.4
France	55.0	–	37.0	51.8	39.6	55.3	59.3	74.9
Gabon	20.0	–	20.0	20.0	37.0	37.0	19.4	21.6
Germany	1.0	–	1.5	2.0	4.0	4.5	1.0	5.0
Greece							1.4	5.4
Hungary								
India	29.2	–	29.8	29.8	29.8	29.8	32.0	32.0
Indonesia								
Iran, Islamic Republic of								
Italy	1.2	–	1.2	1.2	0.0	1.2	0.0	2.4
Japan	7.7	–	7.7	7.7	7.7	7.7	7.7	7.7
Kazakhstan								
Korea, Republic of	2.4	–	0.0	3.0	0.0	4.4	0.0	11.0
Malawi								
Mexico	6.0	–	4.7	4.7	6.0	6.0	2.9	2.9
Mongolia								
Morocco								
Namibia					117.0	133.0	119.0	135.0
Niger	50.0	–	160.0	160.0	160.0	160.0	160.0	160.0
Peru								
Philippines			0.3	0.3	0.3	0.3		
Portugal	6.9	–	6.8	8.3	6.7	8.2	6.7	8.2
Romania								
Russian Federation								
Slovenia / Yugoslavia	6.5	–	4.5	6.5	4.5	6.5		
Somalia			0.0	6.2	0.0	6.6	0.0	6.6
South Africa	276.0	–	306.0	348.0	247.0	391.0	247.0	356.0
Spain	103.5	–	6.8	6.8	9.8	9.8	12.5	16.4
Sweden	300.0	–	1.0	301.0	0.0	301.0	0.0	38.0
Thailand								
Turkey	3.1	–	4.1	4.1	2.4	3.9	2.5	4.6
Ukraine								
United Kingdom	1.8	–	0.0	0.0	0.0	0.0	0.0	0.0
United States	454.0	–	523.0	643.0	531.0	708.0	362.0	605.0
Uzbekistan								
Vietnam								
Zimbabwe								
Total	1 810.0	–	1 647.3	2 191.0	1 854.6	2 446.6	1 747.3	2 293.6
Adjusted Total (a)								

EVOLUTION OF REASONABLY ASSURED RESOURCES BY COUNTRY (1965-2003)
(1 000 tU)*

Red Book edition Countries	1983		1986		1988		1989	
	<USD 80/kgU	<USD 130/kgU	<USD 80/kgU	<USD 130/kgU	<USD 80/kgU	<USD 130/kgU	<USD 80/kgU	<USD 130/kgU
Algeria	26.0	26.0	26.0	26.0	26.0	26.0	26.0	26.0
Angola								
Argentina	18.8	23.3	15.4	18.9	9.3	11.9	9.1	11.7
Australia	314.0	336.0	463.0	526.0	462.0	518.0	480.0	538.0
Austria	0.0	0.3						
Botswana								
Brazil	163.3	163.3	163.3	163.3	163.1	163.1	162.7	162.7
Bulgaria								
Canada	176.0	185.0	155.0	214.0	153.0	249.0	139.0	235.0
Central African Republic	18.0	18.0	8.0	16.0	8.0	16.0	8.0	16.0
Chile	0.0	2.3	0.0	0.0				
China								
Congo / Zaire	1.8	1.8	1.8	1.8	1.8	1.8	1.8	1.8
Czech Republic								
Denmark (Greenland)	0.0	27.0	0.0	27.0	0.0	27.0	0.0	27.0
Finland	0.0	3.4	0.0	1.5	0.0	1.5	0.0	1.5
France	56.2	67.5	56.0	67.1	53.8	65.2	46.7	58.9
Gabon	18.7	23.3	16.7	21.4	14.9	19.6	13.0	17.7
Germany	0.9	5.1	0.9	4.7	0.8	4.8	0.8	4.8
Greece	0.4	0.4	0.4	0.4	0.4	0.4	0.3	0.3
Hungary								
India	31.7	42.6	35.1	46.1	34.7	45.7	41.1	47.3
Indonesia							0.0	1.0
Iran, Islamic Republic of								
Italy	2.9	2.9	4.8	4.8	4.8	4.8	4.8	4.8
Japan	7.7	7.7	7.7	7.7	0.0	6.6	0.0	6.6
Kazakhstan								
Korea, Republic of	0.0	10.0	0.0	10.0			0.0	11.8
Malawi								
Mexico	2.9	2.9	4.5	7.7	4.5	7.7	4.5	7.7
Mongolia								
Morocco								
Namibia	119.0	135.0	104.0	120.0	97.3	113.3	90.9	106.9
Niger	160.0	160.0	180.0	182.2	173.7	175.9	173.7	175.9
Peru	0.5	0.5	0.5	0.5	0.0	1.5	0.0	1.8
Philippines								
Portugal	6.7	8.2	6.8	8.2	7.1	8.5	7.3	8.7
Romania								
Russian Federation								
Slovenia / Yugoslavia								
Somalia	0.0	6.6	0.0	6.6	0.0	6.6	0.0	6.6
South Africa	191.0	313.0	256.6	358.7	247.1	349.2	317.0	418.5
Spain	15.7	20.2	26.7	32.9	26.7	32.9	16.8	35.0
Sweden	2.0	39.0	2.0	39.0	2.0	39.0	2.0	4.0
Thailand								
Turkey	2.5	4.6	2.1	3.9	0.0	3.9	0.0	3.9
Ukraine								
United Kingdom								
United States	131.3	407.2	131.3	398.1	124.0	398.0	111.3	377.5
Uzbekistan								
Vietnam								
Zimbabwe								
Total	1 468.0	2 043.1	1 668.6	2 314.5	1 614.9	2 297.8	1 656.8	2 319.3
Adjusted Total (a)	1 425.0	2 000.0	1 609.0	2 250.0	1 555.0	2 233.0	1 546.0	2 201.0

EVOLUTION OF REASONABLY ASSURED RESOURCES BY COUNTRY (1965-2003)
(1 000 tU)*

Red Book edition Countries	1991		1993		1995		1997	
	<USD 80/kgU	<USD 130/kgU	<USD 80/kgU	<USD 130/kgU	<USD 80/kgU	<USD 130/kgU	<USD 80/kgU	<USD 130/kgU
Algeria	26.0	26.0	26.0	26.0	26.0	26.0	26.0	26.0
Angola								
Argentina	8.7	10.9	4.6	7.3	3.4	5.7	4.6	8.8
Australia	469.0	529.0	462.0	517.0	633.0	710.0	622.0	715.0
Austria								
Botswana								
Brazil	162.0	162.0	162.0	162.0	162.0	162.0	162.0	162.0
Bulgaria						7.9	7.8	7.8
Canada	146.0	214.0	277.0	397.0	270.0	381.0	331.0	331.0
Central African Republic	8.0	16.0	8.0	16.0	8.0	16.0	8.0	16.0
Chile								
China								
Congo / Zaire	1.8	1.8	1.8	1.8	1.8	1.8	1.8	1.8
Czech Republic			15.9	22.3	11.8	31.2	6.6	30.2
Denmark (Greenland)	0.0	27.0	0.0	27.0	0.0	27.0	0.0	27.0
Finland	0.0	1.5	0.0	1.5	0.0	1.5	0.0	1.5
France	23.8	39.5	19.9	33.7	16.0	25.0	13.5	22.4
Gabon	11.0	15.7	9.8	14.4	10.0	10.0	6.0	6.0
Germany	0.6	4.6	0.0	3.0	0.0	3.0	0.0	3.0
Greece	0.3	0.3	0.3	0.3	1.0	1.0	1.0	1.0
Hungary	1.6	3.1	0.6	1.1	0.4	0.7	0.4	0.4
India						NA		52.1
Indonesia	4.3	4.3	0.0	5.4	0.0	6.3	0.0	6.3
Iran, Islamic Republic of								
Italy	4.8	4.8	4.8	4.8	4.8	4.8	4.8	4.8
Japan	0.0	6.6	0.0	6.6	0.0	6.6	0.0	6.6
Kazakhstan					439.5	598.7	439.2	601.3
Korea, Republic of	0.0	11.8	0.0	11.8	0.0	11.8		
Malawi								
Mexico	0.0	1.7	0.0	1.7	0.0	1.7	0.0	1.7
Mongolia					62.0	62.0	61.6	61.6
Morocco								
Namibia	84.8	100.8	80.6	96.6	160.6	191.8	156.1	187.4
Niger	166.1	172.7	159.2	165.8	57.4	87.1	70.0	70.0
Peru	1.8	1.8	1.8	1.8	1.8	1.8	1.8	1.8
Philippines								
Portugal	7.3	8.7	7.3	8.7	7.3	8.9	7.3	8.9
Romania					0.0	5.5	0.0	6.9
Russian Federation							145.0	145.0
Slovenia / Yugoslavia	0.0	0.9	0.0	1.8	0.0	1.8	2.2	2.2
Somalia	0.0	6.6	0.0	6.6	0.0	6.6	0.0	6.6
South Africa	247.6	344.4	144.4	240.8	204.7	258.6	218.3	269.8
Spain	17.9	39.0	17.9	39.0	9.2	11.5	4.7	12.2
Sweden	2.0	4.0	2.0	4.0	0.0	4.0	0.0	4.0
Thailand			0.0	0.0				
Turkey	0.0	9.1	9.1	9.1	9.1	9.1	9.1	9.1
Ukraine					42.6	81.0	45.6	84.0
United Kingdom			0.0	0.0				
United States	101.9	356.0	114.0	369.0	113.0	366.0	110.0	361.0
Uzbekistan							66.2	83.7
Vietnam							NA	1.3
Zimbabwe	1.8	1.8	1.8	1.8	1.8	1.8	1.8	1.8
Total	1 499.0	2 126.4	1 530.7	2 205.8	2 265.1	3 137.1	2 534.4	3 349.9
Adjusted Total (a)	1 389.0	1 999.0	1 424.0	2 083.0	2 124.0	2 951.0	2 340.0	3 220.0

EVOLUTION OF REASONABLY ASSURED RESOURCES BY COUNTRY (1965-2003)
(1 000 tU)*

Red Book edition Countries	1999 <USD 80/kgU	1999 <USD 130/kgU	2001 <USD 80/kgU	2001 <USD 130/kgU	2003 <USD 80/kgU	2003 <USD 130/kgU
Algeria	26.0	26.0	26.0	26.0	19.5	19.5
Angola						
Argentina	5.2	7.5	5.1	7.1	4.9	7.1
Australia	607.0	716.0	667.0	697.0	702.0	735.0
Austria						
Botswana						
Brazil	162.0	162.0	162.0	162.0	86.2	86.2
Bulgaria	7.8	7.8	7.8	7.8	5.9	5.9
Canada	326.4	326.4	314.6	314.6	333.8	333.8
Central African Republic	8.0	16.0	8.0	16.0	6.0	12.0
Chile					NA	0.6
China					35.1	35.1
Congo / Zaire	1.8	1.8	1.8	1.8	1.4	1.4
Czech Republic	4.1	7.0	2.4	2.4	0.8	0.8
Denmark (Greenland)	0.0	27.0	0.0	27.0	0.0	20.3
Finland	0.0	1.5	0.0	1.5	0.0	1.1
France	12.5	14.2	0.2	0.2	0.0	0.0
Gabon	4.8	4.8	4.8	4.8	0.0	4.8
Germany	0.0	3.0	0.0	3.0	0.0	3.0
Greece	1.0	1.0	1.0	1.0	1.0	1.0
Hungary	0.0	0.0	0.0	0.0	0.0	0.0
India	NA	NA	NA	NA	NA	41.0
Indonesia	0.0	6.3	0.0	6.8	0.3	4.6
Iran, Islamic Republic of	0.0	0.5	0.0	0.5	0.0	0.4
Italy	4.8	4.8	4.8	4.8	4.8	4.8
Japan	0.0	6.6	0.0	6.6	0.0	6.6
Kazakhstan	436.6	598.7	432.8	594.8	384.6	530.5
Korea, Republic of					0.0	0.0
Malawi	11.7	11.7	11.7	11.7	8.8	8.8
Mexico	0.0	1.7	0.0	1.7	0.0	1.3
Mongolia	61.6	61.6	61.6	61.6	46.2	46.2
Morocco						
Namibia	149.3	180.5	143.9	175.1	139.3	170.5
Niger	71.1	71.1	29.6	29.6	102.2	102.2
Peru	1.8	1.8	1.8	1.8	1.2	1.2
Philippines						
Portugal	7.5	7.5	7.5	7.5	7.5	7.5
Romania	0.0	6.6	0.0	4.6	0.0	3.3
Russian Federation	140.9	140.9	138.0	138.0	124.1	143.0
Slovenia / Yugoslavia	2.2	2.2	2.2	2.2	2.2	2.2
Somalia	0.0	6.6	0.0	6.6	0.0	5.0
South Africa	232.9	292.8	231.1	291.0	231.7	315.3
Spain	3.1	6.7	2.5	4.9	2.5	4.9
Sweden	0.0	4.0	0.0	4.0	0.0	4.0
Thailand	0.0	0.0	0.0	0.0	0.0	0.0
Turkey	9.1	9.1	9.1	9.1	6.8	6.8
Ukraine	42.6	81.0	42.6	81.0	34.6	64.7
United Kingdom						
United States	106.0	355.0	104.0	348.0	102.0	345.0
Uzbekistan	65.6	83.1	90.1	115.4	61.5	79.6
Vietnam	0.0	1.3	0.0	1.3	0.0	1.0
Zimbabwe	1.8	1.8	1.8	1.8	1.4	1.4
Total	2 515.3	3 266.0	2 515.6	3 182.5	2 458.2	3 169.2
Adjusted Total (a)	2 274.0	2 964.0	2 242.5	2 853.3	2 458.2	3 169.2

* Numbers in *italics* represent *in situ* resources, not taking into account mining and milling losses.

(a) Total adjusted by the Secretariat to account for mining and milling losses.

NA Not available.

– Not applicable.

228

EVOLUTION OF ESTIMATED ADDITIONAL RESOURCES
AND ESTIMATED ADDITIONAL RESOURCES CATEGORY-I (1965-2003)
(1 000 tU)

Year of publication	<USD 26/kgU USD 5-10/lbU$_3$0$_8$	USD 26-39/kgU USD 10-15/lbU$_3$O$_8$	<USD 40/kgU <USD 15/lbU$_3$O$_8$	USD 40-80/kgU USD 15-30/lbU$_3$O$_8$	<USD 80/kgU <USD 30/lbU$_3$O$_8$
1965	524	387	911	847	1 758
1967	552	428	980	1 081	2 061
1969	no estimates of EAR in all cost ranges				
1970	677	660	1 337	–	1 337
1973	1 191	821	2 012	–	2 012
1976	–	1 000	1 000	680	1 680
1977	–	–	–	–	1 510
1979	–	–	–	–	1 480
1982	–	–	–	–	1 605
1983	–	–	–	–	885
1986	–	–	–	–	897
1988	–	–	–	–	891
1989	–	–	–	–	773
1991	–	–	–	–	791
1993	–	–	–	–	670
1995	–	–	> 204	> 191	637
1997	–	–	> 257	> 158	745
1999	–	–	> 338	> 145	728
2001	–	–	> 552	> 187	865
2003	–	–	793	275	1 079

Year of publication	USD 80-130/kgU USD 30-50/lbU$_3$O$_8$	<USD 130/kgU <USD 50/lbU$_3$O$_8$	USD 130-260/kgU USD 50-100/lbU$_3$O$_8$	<USD 260/kgU <USD 100/lbU$_3$O$_8$
1965	–	1 758	–	–
1967	–	2 061	–	–
1969	no estimates of EAR in all cost ranges			
1970	–	> 1 337	–	–
1973	–	> 2 012	–	–
1976	–	1 680	–	–
1977	590	2 100	–	–
1979	970	2 450	–	–
1982	1 115	2 720	–	–
1983	305	1 190	–	–
1986	406	1 303	444	1 747
1988	425	1 316	–	–
1989	391	1 164	185	1 349
1991	335	1 126	–	–
1993	296	966	–	–
1995	263	900	–	–
1997	244	1 079	–	–
1999	250	990	–	–
2001	225	1 080	–	–
2003	321	1 419	–	–

Notes: In 1983, Estimated Additional Resources have been subdivided into Category I and Category II (EAR-I and EAR-II).
Since 1983, EAR-I are given as recoverable resources (mining and milling losses deducted).
– Not applicable.

ESTIMATED ADDITIONAL RESOURCES (EAR) (1965-2003)
(1 000 tU)

Red Book edition Countries	1965		1967		1970	
	< USD 80/kg U	< USD 130/kgU	< USD 80/kgU	< USD 130/kgU	< USD 40/kgU	< USD 130/kgU
Algeria	NA	–	NA	–	NA	–
Angola	11.55	–	11.55	–	11.55	–
Argentina	20.79	–	97.02	–	42.4	–
Australia	NA	–	3.39	–	9.335	–
Austria	NA	–	NA	–	NA	–
Bolivia	NA	–	NA	–	NA	–
Botswana	NA	–	NA	–	NA	–
Brazil (a)	NA	–	NA	–	0.8	–
Bulgaria	NA	–	NA	–	NA	–
Cameroon	NA	–	NA	–	NA	–
Canada	508.2	–	585.2	–	307.9	–
Central African Republic	NA	–	NA	–	8	–
Chile	NA	–	NA	–	NA	–
China	NA	–	NA	–	NA	–
Congo / Zaire	NA	–	NA	–	NA	–
Czech Republic	NA	–	NA	–	NA	–
Denmark (Greenland)	NA	–	NA	–	NA	–
Egypt	NA	–	NA	–	NA	–
Finland	NA	–	NA	–	NA	–
France	29.26	–	23.1	–	30.935	–
Gabon	NA	–	2.7	–	10	–
Germany	NA	–	NA	–	NA	–
Greece	NA	–	NA	–	NA	–
Hungary	NA	–	NA	–	NA	–
India	NA	–	47.74	–	0.77	–
Indonesia	NA	–	NA	–	NA	–
Iran, Islamic Republic of	NA	–	NA	–	NA	–
Italy	NA	–	NA	–	NA	–
Japan	NA	–	NA	–	NA	–
Kazakhstan	NA	–	NA	–	NA	–
Korea, Republic of	NA	–	NA	–	NA	–
Madagascar	NA	–	NA	–	NA	–
Malawi	NA	–	NA	–	NA	–
Mexico	NA	–	NA	–	NA	–
Mongolia	NA	–	NA	–	NA	–
Morocco	NA	–	NA	–	NA	–
Namibia	NA	–	NA	–	NA	–
Niger	NA	–	10.1	–	39	–
Peru	NA	–	NA	–	NA	–
Philippines	NA	–	NA	–	NA	–
Portugal	14.63	–	21.56	–	17.55	–
Romania	NA	–	NA	–	NA	–
Russian Federation	NA	–	NA	–	NA	–
Slovenia / Yugoslavia	NA	–	NA	–	NA	–
Somalia	NA	–	NA	–	NA	–
South Africa	NA	–	92.4	–	38.45	–
Spain (b)	223.3	–	215.6	–	NA	–
Sweden	192.5	–	192.5	–	38.5	–
Thailand	NA	–	NA	–	NA	–
Turkey	NA	–	NA	–	NA	–
Ukraine	NA	–	NA	–	NA	–
United Kingdom	NA	–	NA	–	NA	–
United States (c)	743.05	–	762.3	–	621	–
Uzbekistan	NA	–	NA	–	NA	–
Vietnam	NA	–	NA	–	NA	–
Zimbabwe	NA	–	NA	–	NA	–
Total	1 743	–	2 065	–	1 176	–
Adjusted Total (d)	–	–	–	–	–	–

ESTIMATED ADDITIONAL RESOURCES (EAR) (1965-2003)
(1 000 tU)

Red Book edition Countries	1973		1976		1977	
	<USD 40/kgU	<USD 130/kgU	<USD 80/kgU	<USD 130/kgU	<USD 80/kgU	<USD 130/kgU
Algeria	NA	–	NA	–	50	50
Angola	13	–	NA	–	NA	NA
Argentina	37	–	39	–	0	0
Australia	107.5	–	80	–	44	49
Austria	NA	–	NA	–	0	0
Bolivia	NA	–	NA	–	0	0,5
Botswana	NA	–	NA	–	NA	NA
Brazil (a)	72.5	–	8.8	–	8.2	8.2
Bulgaria	NA	–	NA	–	NA	NA
Cameroon	NA	–	NA	–	NA	NA
Canada	409	–	419	–	392	656
Central African Republic	8	–	8	–	8	8
Chile	NA	–	NA	–	5.1	5.1
China	NA	–	NA	–	NA	NA
Congo / Zaire	1.7	–	1.7	–	1.7	1.7
Czech Republic	NA	–	NA	–	NA	NA
Denmark (Greenland)	10	–	10	–	0	8.7
Egypt	NA	–	NA	–	NA	NA
Finland	NA	–	NA	–	0	0
France	49.3	–	40	–	24.1	44.1
Gabon	10	–	10	–	5	10
Germany	NA	–	4	–	3	3.5
Greece	NA	–	NA	–	NA	NA
Hungary	NA	–	NA	–	NA	NA
India	0.8	–	23.3	–	23.7	23.7
Indonesia	NA	–	NA	–	NA	NA
Iran, Islamic Republic of	NA	–	NA	–	NA	NA
Italy	NA	–	1	–	1	1
Japan	NA	–	NA	–	0	0
Kazakhstan	NA	–	NA	–	NA	NA
Korea, Republic of	NA	–	NA	–	0	0
Madagascar	NA	–	NA	–	0	2
Malawi	NA	–	NA	–	NA	NA
Mexico	NA	–	NA	–	2.4	2.4
Mongolia	NA	–	NA	–	NA	NA
Morocco	NA	–	NA	–	NA	NA
Namibia	NA	–	NA	–	NA	NA
Niger	30	–	30	–	53	53
Peru	NA	–	NA	–	NA	NA
Philippines	NA	–	NA	–	0	0
Portugal	15.9	–	NA	–	0.9	0.9
Romania	NA	–	NA	–	NA	NA
Russian Federation	NA	–	NA	–	NA	NA
Slovenia / Yugoslavia	10	–	15.2	–	5	20.5
Somalia	NA	–	NA	–	0	3 4
South Africa	34	–	74	–	34	72
Spain (b)	NA	–	106.8	–	8.5	8.5
Sweden	40	–	NA	–	3	3
Thailand	NA	–	NA	–	NA	NA
Turkey	NA	–	0.4	–	0	0
Ukraine	NA	–	NA	–	NA	NA
United Kingdom	NA	–	4	–	0	7.4
United States (c)	839	–	812	–	838	1 053
Uzbekistan	NA	–	NA	–	NA	NA
Vietnam	NA	–	NA	–	NA	NA
Zimbabwe	NA	–	NA	–	NA	NA
Total	1 688	–	1 687	–	1 511	2 096
Adjusted Total (d)	–	–	–	–	–	–

ESTIMATED ADDITIONAL RESOURCES (EAR) (1965-2003)
(1 000 tU)

Red Book edition Countries	1979		1982		1983	
	<USD 80/kgU	<USD 130/kgU	<USD 80/kgU	<USD 130/kgU	<USD 80/kgU	<USD 130/kgU
Algeria	0	5.5	0	0	NA	NA
Angola	NA	NA	NA	NA	NA	NA
Argentina	3.8	9.1	3.8	13.4	7	7
Australia	47	53	264	285	369	394
Austria	0	0	0.7	1.7	0.7	1.7
Bolivia	0	0.5	NA	NA	NA	NA
Botswana	0	0	NA	NA	NA	NA
Brazil (a)	90.1	90.1	81.2	81.2	92.4	92.4
Bulgaria	NA	NA	NA	NA	NA	NA
Cameroon	NA	NA	NA	NA	0	1.2
Canada	370	728	358	760	181	229
Central African Republic	0	0	0	0	NA	NA
Chile	5.1	5.1	0	6.7	0	2.3
China	NA	NA	NA	NA	NA	NA
Congo / Zaire	1.7	1.7	1.7	1.7	1.7	1.7
Czech Republic	NA	NA	NA	NA	NA	NA
Denmark (Greenland)	0	16	0	16	0	16
Egypt	0	5	0	5	0	5
Finland	0	0.5	0	0.5	NA	NA
France	26.2	46.2	28.4	46.5	26.6	32.85
Gabon	0	0	0	9.9	1.3	9.6
Germany	7	7.5	1.5	8.5	1.3	8.2
Greece	NA	NA	2	7.3	6	6
Hungary	NA	NA	NA	NA	NA	NA
India	0.9	23.7	0.9	25.1	4.8	19.3
Indonesia	NA	NA	NA	NA	NA	NA
Iran, Islamic Republic of	NA	NA	NA	NA	NA	NA
Italy	0	2	0	2	NA	NA
Japan	0	0	0	0	NA	NA
Kazakhstan	NA	NA	NA	NA	NA	NA
Korea, Republic of	0	0	NA	NA	NA	NA
Madagascar	0	2	NA	NA	NA	NA
Malawi	NA	NA	NA	NA	NA	NA
Mexico	2.4	2.4	3.5	6.1	3.5	6.1
Mongolia	NA	NA	NA	NA	NA	NA
Morocco	NA	NA	NA	NA	NA	NA
Namibia	30	53	30	53	30	53
Niger	53	53	53	53	53	53
Peru	NA	NA	NA	NA	NA	NA
Philippines	0	0	NA	NA	NA	NA
Portugal	2.5	2.5	2.5	2.5	1	1
Romania	NA	NA	NA	NA	NA	NA
Russian Federation	NA	NA	NA	NA	NA	NA
Slovenia / Yugoslavia	5	20.5	NA	NA	NA	NA
Somalia	0	3.4	0	3.4	0	3.4
South Africa	54	139	84	175	99	147
Spain (b)	8.5	8.5	8.5	8.5	5	5
Sweden	0	3	0	44	0.3	43.3
Thailand	NA	NA	NA	NA	NA	NA
Turkey	0	0	0	0	NA	NA
Ukraine	NA	NA	NA	NA	NA	NA
United Kingdom	0	7.4	0	7.4	NA	NA
United States (c)	773	1 158	681	1 097	30.4	82.6
Uzbekistan	NA	NA	NA	NA	NA	NA
Vietnam	NA	NA	NA	NA	NA	NA
Zimbabwe	NA	NA	NA	NA	NA	NA
Total	1 480	2 447	1 605	2 720	914	1 221
Adjusted Total (d)	–	–	–	–	885	1 190

ESTIMATED ADDITIONAL RESOURCES (EAR) (1965-2003)
(1 000 tU)

Red Book edition Countries	1986 <USD 80/kgU	1986 <USD 130/kgU	1988 <USD 80/kgU	1988 <USD 130/kgU	1989 <USD80/kgU	1989 <USD 130/kgU
Algeria	NA	NA	NA	NA	NA	NA
Angola	NA	NA	NA	NA	NA	NA
Argentina	7.7	7.9	0.8	3.9	0.84	3.98
Australia	251	377	257	384	262	393
Austria	0.7	1.7	0.7	1.7	0.7	1.7
Bolivia	NA	NA	NA	NA	NA	NA
Botswana	NA	NA	NA	NA	NA	NA
Brazil (a)	92.39	92.39	92.39	92.39	92.39	92.39
Bulgaria	NA	NA	NA	NA	NA	NA
Cameroon	NA	NA	NA	NA	NA	NA
Canada	105	197	112	211	109	204
Central African Republic	NA	NA	NA	NA	NA	NA
Chile	NA	0.3	0	0.3	0	0.3
China	NA	NA	NA	NA	NA	NA
Congo / Zaire	1.7	1.7	1.7	1.7	1.7	1.7
Czech Republic	NA	NA	NA	NA	NA	NA
Denmark (Greenland)	NA	16	0	16	0	16
Egypt	NA	NA	NA	NA	NA	NA
Finland	NA	2 9	0	2 9	NA	NA
France	26.83	45.17	21.19	38.07	20	36
Gabon	1.3	9.6	1.3	9.6	1.3	9.6
Germany	1.6	7.3	1.6	7.3	1.6	7.3
Greece	6	6	6	6	6	6
Hungary	NA	NA	NA	NA	NA	NA
India	2.12	16.61	2.12	16.61	4.08	17.32
Indonesia	NA	NA	0	7.31	0	6.7
Iran, Islamic Republic of	NA	NA	NA	NA	NA	NA
Italy	NA	1.3	0	1.3	0	1.3
Japan	NA	NA	NA	NA	NA	NA
Kazakhstan	NA	NA	NA	NA	NA	NA
Korea, Republic of	NA	NA	NA	NA	0	3
Madagascar	NA	NA	NA	NA	NA	NA
Malawi	NA	NA	NA	NA	NA	NA
Mexico	NA	2.98	0	2.98	0	2.98
Mongolia	NA	NA	NA	NA	NA	NA
Morocco	NA	NA	NA	NA	NA	NA
Namibia	30	53	30	53	30	53
Niger	283.6	300.3	283.6	300.3	283.6	300.3
Peru	5	5	0	1.8	0	1.86
Philippines	NA	NA	a	NA	NA	NA
Portugal	1.25	1.25	1 45	1.45	1.45	1.45
Romania	NA	NA	NA	NA	NA	NA
Russian Federation	NA	NA	NA	NA	NA	NA
Slovenia / Yugoslavia	NA	NA	NA	NA	NA	NA
Somalia	NA	3.4	0	3.4	0	3.4
South Africa	97.5	124.6	97.5	124.6	72.6	110.2
Spain (b)	9	9	9	9	0	9
Sweden	2	46	1	46.3	1	6.3
Thailand	NA	NA	NA	NA	NA	NA
Turkey	NA	3.2	0	3.2	0	3 2
Ukraine	NA	NA	NA	NA	NA	NA
United Kingdom	NA	NA	NA	NA	NA	NA
United States (c)	*	*	*	*	*	*
Uzbekistan	NA	NA	NA	NA	NA	NA
Vietnam	NA	NA	NA	NA	NA	NA
Zimbabwe	NA	NA	NA	NA	NA	NA
Total	925	1 332	919	1 346	888	1 292
Adjusted Total (d)	897	1 303	891	1 316	773	1 164

ESTIMATED ADDITIONAL RESOURCES (EAR) (1965-2003)
(1 000 tU)

Red Book edition Countries	1991 <USD 80/kgU	1991 <USD 130/kgU	1993 <USD 80/kgU	1993 <USD 130/kgU	1995 <USD 80/kgU	1995 <USD 130/kgU
Algeria	NA	NA	NA	NA	NA	NA
Angola	NA	NA	NA	NA	NA	NA
Argentina	0.54	2.49	2.3	2.6	2.1	3.25
Australia	264	390	272	394	154	194
Austria	0.7	1.7	0.7	1.7	0.7	1.7
Bolivia	NA	NA	NA	NA	NA	NA
Botswana	NA	NA	NA	NA	NA	NA
Brazil (a)	*94*	*94*	*94*	*94*	*100.2*	*100.2*
Bulgaria	NA	NA	NA	NA	8.4	8.4
Cameroon	NA	NA	NA	NA	NA	NA
Canada	149	229	31	74	30	73
Central African Republic	NA	NA	NA	NA	NA	NA
Chile	NA	NA	NA	NA	NA	NA
China	NA	NA	NA	NA	NA	NA
Congo / Zaire	*1.7*	*1.7*	*1.7*	*1.7*	*1.7*	*1.7*
Czech Republic	NA	NA	1.35	21.35	1.66	19.45
Denmark (Greenland)	0	16	0	16	0	16
Egypt	NA	NA	NA	NA	NA	NA
Finland	NA	NA	NA	NA	NA	NA
France	4.2	8.1	3.55	6.73	1.95	2.14
Gabon	1.3	9.6	1.3	9.6	5.86	5.86
Germany	1.6	7.3	0	4	0	4
Greece	*6*	*6*	*6*	*6*	*6*	*6*
Hungary	*9.1*	*18.25*	*1.32*	*16.66*	*1.25*	*15.64*
India	NA	NA	NA	NA	NA	NA
Indonesia	NA	NA	0	2.15	NA	*1.67*
Iran, Islamic Republic of	NA	NA	NA	NA	NA	NA
Italy	0	1.3	0	1.3	0	1.3
Japan	NA	NA	NA	NA	NA	NA
Kazakhstan	NA	NA	NA	NA	*195.9*	*259.3*
Korea, Republic of	0	3	0	3	0	3
Madagascar	NA	NA	NA	NA	NA	NA
Malawi	NA	NA	NA	NA	NA	NA
Mexico	0	*0.7*	0	*0.7*	0	*0.7*
Mongolia	NA	NA	NA	NA	21	21
Morocco	NA	NA	NA	NA	NA	NA
Namibia	30	53	30	53	90.82	107.52
Niger	*295.77*	*305.77*	*295.77*	*305.77*	6	6
Peru	*1.72*	*1.86*	*1.72*	*1.86*	*1.86*	*1.86*
Philippines	NA	NA	NA	NA	NA	NA
Portugal	1.45	1.45	1.45	1.45	1.45	1.45
Romania	NA	NA	NA	NA	0	6.04
Russian Federation	NA	NA	NA	NA	NA	NA
Slovenia / Yugoslavia	0	5	5	5	5	5
Somalia	0	3.4	0	3.4	0	3.4
South Africa	51.8	82.6	34.72	54 72	55.84	75.24
Spain (b)	0	9	4.2	4.2	10.69	13.46
Sweden	1	6.3	1	6.3	0	6
Thailand	NA	NA	NA	NA	NA	NA
Turkey	NA	NA	NA	NA	NA	NA
Ukraine	NA	NA	NA	NA	20	50
United Kingdom	NA	NA	0	0	NA	NA
United States (c)	*	*	*	*	*	*
Uzbekistan	NA	NA	NA	NA	NA	NA
Vietnam	0	0.2	0	0.2	*0.49*	*0.54*
Zimbabwe	NA	NA	NA	NA	NA	NA
Total	914	1 258	789	1 091	723	1 015
Adjusted Total (d)	791	1 126	670	966	637	900

ESTIMATED ADDITIONAL RESOURCES (EAR) (1965-2003)
(1 000 tU)

Red Book edition Countries	1997		1999		2001	
	<USD 80/kgU	<USD 130/kgU	<USD 80/kgU	<USD 130/kgU	<USD 80/kgU	<USD 130/kgU
Algeria	0.7	1.7	0 7	1.7	NA	NA
Angola	NA	NA	NA	NA	NA	NA
Argentina	0.9	3.11	2.38	2.45	2.38	8.56
Australia	136	180	147	194	196	233
Austria	NA	NA	NA	NA	NA	NA
Bolivia	NA	NA	NA	NA	NA	NA
Botswana	NA	NA	NA	NA	NA	NA
Brazil (a)	100.2	100.2	100.2	100.2	100 2	100.2
Bulgaria	8.4	8.4	8.4	8.4	8.4	8.4
Cameroon	NA	NA	NA	NA	NA	NA
Canada	99	99	106.59	106.59	122.39	122.39
Central African Republic	NA	NA	NA	NA	NA	NA
Chile	NA	NA	NA	NA	NA	NA
China	NA	NA	NA	NA	NA	NA
Congo / Zaire	1.7	1.7	1.7	1.7	1.7	1.7
Czech Republic	1.18	18.96	1.11	22.66	0.31	0.31
Denmark (Greenland)	0	16	0	16	0	16
Egypt	NA	NA	NA	NA	NA	NA
Finland	NA	NA	NA	NA	NA	NA
France	1.21	1.4	0.55	0.55	0	11.74
Gabon	1	1	1	1	1	1
Germany	0	4	0	4	0	4
Greece	6	6	6	6	6	6
Hungary	0	15.41	0	18.4	0	18.4
India	NA	24.25	NA	NA	NA	NA
Indonesia	NA	1.67	NA	1.67	0	1.7
Iran, Islamic Republic of	NA	NA	NA	0.88	0	0.88
Italy	0	1.3	0	1.3	0	1 3
Japan	NA	NA	NA	NA	NA	NA
Kazakhstan	195.9	259.3	195.6	259.3	195.9	259.3
Korea, Republic of	NA	NA	NA	NA	NA	NA
Madagascar	NA	NA	NA	NA	NA	NA
Malawi	NA	NA	NA	NA	NA	NA
Mexico	0	0.7	0	0.7	0	0.7
Mongolia	21	21	21	21	21	21
Morocco	NA	NA	NA	NA	NA	NA
Namibia	90.82	107.52	90.82	107.51	90.82	107.51
Niger	1.2	1.2	0	18.58	25.53	25.53
Peru	1.86	1.86	1.86	1.86	1.86	1.86
Philippines	NA	NA	NA	NA	NA	NA
Portugal	1.45	1.45	NA	1.45	0	1.45
Romania	NA	8.95	NA	8.95	NA	4.69
Russian Federation	36.5	36.5	36.5	36.5	36.5	36.5
Slovenia / Yugoslavia	5	10	5	10	5	10
Somalia	0	3.4	0	3.4	0	3.4
South Africa	66.1	87.8	66.8	76.4	66.8	76.4
Spain (b)	NA	8.19	0	7.54	0	6.38
Sweden	0	6	0	6	0	6
Thailand	NA	NA	NA	0.01	NA	0.01
Turkey	NA	NA	NA	NA	NA	NA
Ukraine	17	47	20	50	20	50
United Kingdom	NA	NA	NA	NA	NA	NA
United States (c)	*	*	NA	NA	NA	NA
Uzbekistan	39.36	46.5	39.85	46.99	46.8	56.71
Vietnam	0.49	6.74	0.49	6.74	1.1	6.74
Zimbabwe	NA	NA	NA	NA	NA	NA
Total	833	1 138	854	1 150	950	1 210
Adjusted Total (d)	745	1 079	728	990	865	1 080

ESTIMATED ADDITIONAL RESOURCES (EAR) (1965-2003)
(1 000 tU)

Red Book edition Countries	2003 <USD 80/kgU	2003 <USD 130/kgU
Algeria	NA	NA
Angola	NA	NA
Argentina	2.86	8.56
Australia	287	323
Austria	NA	NA
Bolivia	NA	NA
Botswana	NA	NA
Brazil (a)	57.14	57.14
Bulgaria	6.3	6.3
Cameroon	NA	NA
Canada	104.71	104.71
Central African Republic	NA	NA
Chile	NA	0.885
China	14.69	14.69
Congo / Zaire	1.275	1.275
Czech Republic	0.09	0.09
Denmark (Greenland)	0	12
Egypt	NA	NA
Finland	NA	NA
France	0	9.51
Gabon	0	1
Germany	0	4
Greece	6	6
Hungary	0	13.8
India	NA	18.935
Indonesia	0	1.155
Iran, Islamic Republic of	0	0.7
Italy	0	1.3
Japan	NA	NA
Kazakhstan	237.78	317.16
Korea, Republic of	NA	NA
Madagascar	NA	NA
Malawi	NA	NA
Mexico	0	0.525
Mongolia	15.75	15.75
Morocco	NA	NA
Namibia	73.56	87.085
Niger	125.377	125.377
Peru	1.265	1.265
Philippines	NA	NA
Portugal	1.45	1.45
Romania	0	3.608
Russian Federation	34.26	121.22
Slovenia / Yugoslavia	5	10
Somalia	0	2.55
South Africa	66.94	80.34
Spain (b)	0	6.38
Sweden	0	6
Thailand	0	0.005
Turkey	NA	NA
Ukraine	4.735	11.41
United Kingdom	NA	NA
United States (c)	NA	NA
Uzbekistan	31.76	38.84
Vietnam	0.82	5.435
Zimbabwe	NA	NA
Total	1 079	1 419
Adjusted Total (d)	1 079	1 419

Notes:

Numbers *in italics* represent *in situ* losses, not taking into account mining and milling resources.
a) 1973 total includes non-conventional phosphate resources.
b) 1967 total includes non-conventional lignite and quartzite resources.
c) 1973 total includes non-conventional phosphate resources.
d) Total adjusted by the Secretariat to account for mining and milling losses.
NA Not available.
– Not applicable.

Appendix 6.6

OTHER KNOWN RESOURCES
(REPORTED RESOURCES NOT CORRESPONDING
WITH DEFINITIONS OF RAR AND EAR-I)

Year of publication	<USD 80/kgU <USD 30/lbU$_3$O$_8$	USD 80-130/kgU USD 30-50/lbU$_3$O$_8$	<USD 130/kgU <USD 50/lbU$_3$O$_8$	Cost range unassigned	Total
1991	465	221	829	–	–
1993	718	–	1 021	395	1 436
1995	215	–	295	364	659

– Not applicable.

Appendix 6.7

TOTAL KNOWN CONVENTIONAL RESOURCES (RAR+EAR/EAR-I) (1965-2003)
(1 000 tU)

Year of publication	<USD 26/kgU (USD 5-10/lbU_3O_8)	<USD 40/kgU (<USD 15/lbU_3O_8)	<USD 80/kgU (<USD 30/lbU_3O_8)	<USD 130/kgU (<USD 50/lbU_3O_8)	<USD 260/kgU (<USD 100/lbU_3O_8)	Total (including cost range unassigned)
1965	1 018	1 932	3 209	3 209	3 209	–
1967	1 091	2 015	3 651	3 651	3 651	–
1970	1 322	2 563	–	–	–	–
1973	2 057	3 558	–	–	–	–
1976	–	2 080	3 490	3 490	–	–
1977	–	–	3 160	4 290	–	–
1979	–	–	3 330	5 040	–	–
1982	–	–	3 352	5 013	–	–
1983	–	–	2 310	3 190	–	–
1986	–	–	2 506	3 553	4 414	–
1988	–	–	2 446	3 549	–	–
1989	–	–	2 319	3 365	3 950	–
1991	–	–	2 645	3 954	–	–
1993	–	–	2 812	4 025	–	5 461
1995	–	> 748	2 976	4 146	–	4 805
1997	–	> 923	3 085	4 299	–	–
1999	–	> 1 254	2 998	3 954	–	–
2001	–	> 2 080	3 107	3 933	–	–
2003	–	2 524	3 537	4 588	–	–

–   Not applicable.

Appendix 6.8

EVOLUTION OF UNDISCOVERED RESOURCES (1982-2003)

Year of publication	Estimated Additional Resources Category-II* (tU)					Speculative Resources (tU)	
	<USD 80/kgU	USD 80-130/kgU	<USD 130/kg U	USD 130-260/kgU	<USD260/kgU	<USD 130/kg U	Total (including cost range unassigned)
	<USD 30/lbU$_3$0$_8$	(USD 30-50/lbU$_3$0$_8$)	<USD 50/lbU$_3$0$_8$	(USD 50-100/lbU$_3$0$_8$)	(<USD 100/lbU$_3$0$_8$)	<USD 50/lbU$_3$0$_8$)	
1982	–	–	–	–	–	–	9 900-22 100**
1983	657	474	1 341	–	–	6 260-7 760	9 900-24 600**
1985	601	529	1 613	634	2 247	3 757-5 057	5 803-7 103
1987	663	536	1 681	–	–	3 033-4 333	6 364-7 664
1989	675	524	1 686	649	2 335	3 969-5 269	6 813-8 113
1991	1 304	792	2 108	–	–	4 850-6 150	9 662-10 962
1993	984	1 137	2 430	–	–	2 005	9 977-11 277
1995	1 404	–	2 448	–	–	3 796	8 479
1997	1 466	–	2 266	–	–	4 453	9 709
1999	1 460	–	2 295	–	–	3 043	9 164
2001	1 480	–	2 332	–	–	4 438	9 939
2003	1 475	–	2 255	–	–	4 437	7 539

–   Not applicable.

*   The United States did not separate EAR into EAR-I and EAR-II. All EAR of the USA was reported as EAR-II.

**   IUREP estimates.

EVOLUTION OF KNOWN CONVENTIONAL RESOURCES
IN SELECTED COUNTRIES (1965-2003)
(< USD 80/kgU)

Australia	1965	1966	1967	1968	1969	1970	1971	1972	1973	1974
RAR <USD 80/kgU (tU)	14 800		11 500			23 800			100 500	
EAR-I <USD 80/kgU (tU)	–		3 100			9 500			107 500	
KCR <USD 80/kgU (tU)	> 14 800		14 600			33 300			208 000	
Production (tU)	285	166	166	165	254	254	0	0	0	0
Change in KCR (<USD 80/kgU)			200			18 700			174 700	
Change in KCR + production (tU)			651			19 285			174 954	

Australia	1975	1976	1977	1978	1979	1980	1981	1982	1983	1984
RAR <USD 80/kgU (tU)	243 000		289 000		290 000		294 000		314 000	
EAR-I <USD 80/kgU (tU)	80 000		44 000		47 000		264 000		369 000	
KCR <USD 80/kgU (tU)	323 000		333 000		337 000		558 000		683 000	
Production (tU)	0	359	356	516	706	1 561	2 860	4 453	3 218	4 390
Change in KCR (<USD 80/kgU)	115 000		10 000		4 000		221 000		125 000	
Change in KCR + production (tU)	115 000		10 359		4 872		223 267		132 313	

Australia	1985	1986	1987	1988	1989	1990	1991	1992	1993	1994
RAR <USD 80/kgU (tU)	463 000		462 000		480 000		469 000		462 000	
EAR-I <USD 80/kgU (tU)	251 000		257 000		262 000		264 000		272 000	
KCR <USD 80/kgU (tU)	714 000		719 000		742 000		733 000		734 000	
Production (tU)	3 252	4 154	3 780	3 534	3 657	3 519	3 759	2 318	2 238	2 200
Change in KCR (<USD 80/kgU)	31 000		5 000		23 000		-9 000		1 000	
Change in KCR + production (tU)	38 608		12 406		30 314		-1 824		7 077	

Australia	1995	1996	1997	1998	1999	2000	2001	2002	2003	
RAR <USD 80/kgU (tU)	633 000		622 000		607 000		667 000		702 000	
EAR-I <USD 80/kgU (tU)	154 000		136 000		147 000		196 000		287 000	
KCR <USD 80/kgU (tU)	787 000		758 000		754 000		863 000		989 000	
Production (tU)	3 700	4 959	5 479	4 894	5 984	7 579	7 720	6 854	7 573	
Change in KCR (<USD 80/kgU)	53 000		-29 000		-4 000		109 000		126 000	
Change in KCR + production (tU)	57 438		-20 341		6 373		122 563		140 574	

EVOLUTION OF KNOWN CONVENTIONAL RESOURCES
IN SELECTED COUNTRIES (1965-2003)
(< USD 80/kgU)

Canada	1965	1966	1967	1968	1969	1970	1971	1972	1973	1974
RAR <USD 80/kgU (tU)	339 000		331 000			279 000			307 000	
EAR-I <USD 80/kgU (tU)	508 000		585 000			308 000			409 000	
KCR <USD 80/kgU (tU)	847 000		916 000			587 000			716 000	
Production (tU)	3 418	3 025	3 234	3 200	3 430	3 520	3 830	4 000	3 710	3 420
Change in KCR (<USD 80/kgU)			200			-329 000			129 000	
Change in KCR + production (tU)			6 643			-319 136			140 350	

Canada	1975	1976	1977	1978	1979	1980	1981	1982	1983	1984
RAR <USD 80/kgU (tU)	166 000		167 000		215 000		230 000		176 000	
EAR-I <USD 80/kgU (tU)	419 000		392 000		370 000		358 000		181 000	
KCR <USD 80/kgU (tU)	585 000		559 000		585 000		588 000		357 000	
Production (tU)	3 560	4 850	5 790	6 800	6 820	7 150	7 720	8 080	7 140	11 169
Change in KCR (<USD 80/kgU)	-131 000		-26 000		26 000		3 000		-231 000	
Change in KCR + production (tU)	-123 870		-17 590		38 590		16 970		-215 200	

Canada	1985	1986	1987	1988	1989	1990	1991	1992	1993	1994
RAR <USD 80/kgU (tU)	155 000		153 000		139 000		146 000		277 000	
EAR-I <USD 80/kgU (tU)	105 000		112 000		109 000		149 000		31 000	
KCR <USD 80/kgU (tU)	260 000		265 000		248 000		295 000		308 000	
Production (tU)	10 880	11 720	12 440	12 393	11 323	8 729	8 160	9 297	9 155	9 647
Change in KCR (<USD 80/kgU)	-97 000		5 000		-17 000		47 000		13 000	
Change in KCR + production (tU)	-78 691		27 600		7 833		67 052		30 457	

Canada	1995	1996	1997	1998	1999	2000	2001	2002	2003
RAR <USD 80/kgU (tU)	270 000		331 000		326 000		315 000		334 000
EAR-I <USD 80/kgU (tU)	30 000		99 000		107 000		122 000		105 000
KCR <USD 80/kgU (tU)	300 000		430 000		433 000		437 000		439 000
Production (tU)	10 473	11 706	12 031	10 922	8 214	10 683	12 522	11 607	10 457
Change in KCR (<USD 80/kgU)	-8 000		130 000		3 000		4 000		2 000
Change in KCR + production (tU)	10 802		152 179		25 953		22 897		26 129

EVOLUTION OF KNOWN CONVENTIONAL RESOURCES
IN SELECTED COUNTRIES (1965-2003)
(< USD 80/kgU)

France	1965	1966	1967	1968	1969	1970	1971	1972	1973	1974
RAR <USD 80/kgU (tU)	32 300		38 500			41 600			56 600	
EAR-I <USD 80/kgU (tU)	29 300		23 100			31 200			49 300	
KCR <USD 80/kgU (tU)	61 600		61 600			72 800			105 900	
Production (tU)	1 032	1 423	1 078	1 377	1 180	1 250	1 250	1 545	1 616	1 673
Change in KCR (<USD 80/kgU)			0			11 200			33 100	
Change in KCR + production (tU)			2 455			14 835			37 145	

France	1975	1976	1977	1978	1979	1980	1981	1982	1983	1984
RAR <USD 80/kgU (tU)	55 000		37 000		39 600		59 300		56 200	
EAR-I <USD 80/kgU (tU)	40 000		24 100		26 200		28 400		26 600	
KCR <USD 80/kgU (tU)	95 000		61 100		65 800		87 700		82 800	
Production (tU)	1 731	1 871	2 097	2 183	2 362	2 634	2 552	2 859	3 271	3 168
Change in KCR (<USD 80/kgU)	-10 900		-33 900		4 700		21 900		-4 900	
Change in KCR + production (tU)	-7 611		-30 298		8 980		26 896		511	

France	1985	1986	1987	1988	1989	1990	1991	1992	1993	1994
RAR <USD 80/kgU (tU)	56 000		53 800		46 700		23 800		19 900	
EAR-I <USD 80/kgU (tU)	26 800		21 200		20 000		4 200		3 600	
KCR <USD 80/kgU (tU)	82 800		75 000		66 700		28 000		23 500	
Production (tU)	3 189	3 248	3 376	3 394	3 241	2 841	2 477	2 149	1 730	1 053
Change in KCR (<USD 80/kgU)	0		-7 800		-8 300		-38 700		-4 500	
Change in KCR + production (tU)	6 439		-1 363		-1 530		-32 618		126	

France	1995	1996	1997	1998	1999	2000	2001	2002	2003
RAR <USD 80/kgU (tU)	16 000		13 500		12 460		190		0
EAR-I <USD 80/kgU (tU)	1 950		1 210		550		0		0
KCR <USD 80/kgU (tU)	17 950		14 710		13 010		190		0
Production (tU)	1 016	930	572	452	416	296	184	18	0
Change in KCR (<USD 80/kgU)	-5 550		-3 240		-1 700		-12 820		-190
Change in KCR + production (tU)	-2 767		-1 294		-676		-12 108		12

EVOLUTION OF KNOWN CONVENTIONAL RESOURCES
IN SELECTED COUNTRIES (1965-2003)
(< USD 80/kgU)

Namibia	1965	1966	1967	1968	1969	1970	1971	1972	1973	1974
RAR <USD 80/kgU (tU)										
EAR-I <USD 80/kgU (tU)										
KCR <USD 80/kgU (tU)										
Production (tU)										
Change in KCR (<USD 80/kgU)										
Change in KCR + production (tU)										

Namibia	1975	1976	1977	1978	1979	1980	1981	1982	1983	1984
RAR <USD 80/kgU (tU)					117 000		119 000		119 000	
EAR-I <USD 80/kgU (tU)					30 000		30 000		30 000	
KCR <USD 80/kgU (tU)					147 000		149 000		149 000	
Production (tU)		654	2 340	2 697	3 840	4 042	3 971	3 776	3 719	3 700
Change in KCR (<USD 80/kgU)							2 000		0	
Change in KCR + production (tU)							9 882		7 747	

Namibia	1985	1986	1987	1988	1989	1990	1991	1992	1993	1994
RAR <USD 80/kgU (tU)	104 000		97 000		91 000		85 000		81 000	
EAR-I <USD 80/kgU (tU)	30 000		30 000		30 000		30 000		30 000	
KCR <USD 80/kgU (tU)	134 000		127 000		121 000		115 000		111 000	
Production (tU)	3 400	3 470	3 540	3 511	3 077	3 211	2 450	1 660	1 679	1 895
Change in KCR (<USD 80/kgU)	-15 000		-7 000		-6 000		-6 000		-4 000	
Change in KCR + production (tU)	-7 581		-130		1 051		288		110	

Namibia	1995	1996	1997	1998	1999	2000	2001	2002	2003
RAR <USD 80/kgU (tU)	161 000		156 000		149 000		144 000		139 000
EAR-I <USD 80/kgU (tU)	91 000		91 000		91 000		91 000		74 000
KCR <USD 80/kgU (tU)	252 000		247 000		240 000		235 000		213 000
Production (tU)	2 016	2 447	2 905	2 780	2 690	2 715	2 239	2 333	2 036
Change in KCR (<USD 80/kgU)	141 000		-5 000		-7 000		-5 000		-22 000
Change in KCR + production (tU)	144 574		-537		-1 315		405		-17 428

EVOLUTION OF KNOWN CONVENTIONAL RESOURCES
IN SELECTED COUNTRIES (1965-2003)
(< USD 80/kgU)

Niger	1965	1966	1967	1968	1969	1970	1971	1972	1973	1974
RAR <USD 80/kgU (tU)			18 000			30 000			50 000	
EAR-I <USD 80/kgU (tU)			10 000			39 000			30 000	
KCR <USD 80/kgU (tU)			28 000			69 000			80 000	
Production (tU)							430	867	948	1 117
Change in KCR (<USD 80/kgU)									11 000	
Change in KCR + production (tU)									12 297	

Niger	1975	1976	1977	1978	1979	1980	1981	1982	1983	1984
RAR <USD 80/kgU (tU)	50 000		160 000		160 000		160 000		160 000	
EAR-I <USD 80/kgU (tU)	30 000		53 000		53 000		53 000		53 000	
KCR <USD 80/kgU (tU)	80 000		213 000		213 000		213 000		213 000	
Production (tU)	1 306	1 460	1 609	2 060	3 620	4 120	4 363	4 259	3 426	3 276
Change in KCR (<USD 80/kgU)	0		133 000		0		0		0	
Change in KCR + production (tU)	2 065		135 766		3 669		7 740		8 622	

Niger	1985	1986	1987	1988	1989	1990	1991	1992	1993	1994
RAR <USD 80/kgU (tU)	180 000		174 000		174 000		166 000		159 000	
EAR-I <USD 80/kgU (tU)	284 000		284 000		284 000		296 000		296 000	
KCR <USD 80/kgU (tU)	464 000		458 000		458 000		462 000		455 000	
Production (tU)	3 181	3 110	2 970	2 965	2 962	2 839	2 963	2 965	2 914	2 975
Change in KCR (<USD 80/kgU)	251 000		-6 000		0		4 000		-7 000	
Change in KCR + production (tU)	257 702		291		5 935		9 801		-1 072	

Niger	1995	1996	1997	1998	1999	2000	2001	2002	2003
RAR <USD 80/kgU (tU)	57 000		70 000		71 000		30 000		102 000
EAR-I <USD 80/kgU (tU)	6 000		1 000		0		26 000		125 000
KCR <USD 80/kgU (tU)	63 000		71 000		71 000		56 000		227 000
Production (tU)	2 974	3 329	3 487	3 714	2 907	2 914	2 919	3 080	3 150
Change in KCR (<USD 80/kgU)	-392 000		8 000		0		-15 000		171 000
Change in KCR + production (tU)	-386 111		14 303		7 201		-9 179		176 999

EVOLUTION OF KNOWN CONVENTIONAL RESOURCES
IN SELECTED COUNTRIES (1965-2003)
(< USD 80/kgU)

South Africa	1965	1966	1967	1968	1969	1970	1971	1972	1973	1974
RAR <USD 80/kgU (tU)	108 000		250 000			204 000			264 000	
EAR-I <USD 80/kgU (tU)	–		92 000			31 000			34 000	
KCR <USD 80/kgU (tU)	> 108 000		342 000			235 000			298 000	
Production (tU)	2 262	2 530	3 080	2 985	3 080	3 167	3 220	3 197	2 735	2 711
Change in KCR (<USD 80/kgU)			234 000			-107 000			63 000	
Change in KCR + production (tU)			238 792			-97 855			72 584	

South Africa	1975	1976	1977	1978	1979	1980	1981	1982	1983	1984
RAR <USD 80/kgU (tU)	276 000		306 000		247 000		247 000		191 000	
EAR-I <USD 80/kgU (tU)	74 000		34 000		54 000		84 000		99 000	
KCR <USD 80/kgU (tU)	350 000		340 000		301 000		331 000		290 000	
Production (tU)	2 488	2 758	3 360	3 961	4 797	6 146	6 131	5 816	6 060	5 732
Change in KCR (<USD 80/kgU)	52 000		-10 000		-39 000		30 000		-41 000	
Change in KCR + production (tU)	57 446		-4 754		-31 679		40 943		-29 053	

South Africa	1985	1986	1987	1988	1989	1990	1991	1992	1993	1994
RAR <USD 80/kgU (tU)	257 000		247 000		317 000		248 000		144 000	
EAR-I <USD 80/kgU (tU)	98 000		98 000		73 000		52 000		35 000	
KCR <USD 80/kgU (tU)	355 000		345 000		390 000		300 000		179 000	
Production (tU)	4 880	4 602	3 963	3 800	2 943	2 460	1 712	1 669	1 699	1 671
Change in KCR (<USD 80/kgU)	65 000		-10 000		45 000		-90 000		-121 000	
Change in KCR + production (tU)	76 792		-518		52 763		-84 597		-117 619	

South Africa	1995	1996	1997	1998	1999	2000	2001	2002	2003
RAR <USD 80/kgU (tU)	205 000		218 000		233 000		231 000		232 000
EAR-I <USD 80/kgU (tU)	56 000		66 000		67 000		67 000		67 000
KCR <USD 80/kgU (tU)	261 000		284 000		300 000		298 000		299 000
Production (tU)	1 421	1 436	1 100	965	927	798	878	824	758
Change in KCR (<USD 80/kgU)	82 000		23 000		16 000		-2 000		1 000
Change in KCR + production (tU)	85 370		25 857		18 065		-275		2 702

EVOLUTION OF KNOWN CONVENTIONAL RESOURCES
IN SELECTED COUNTRIES (1965-2003)
(< USD 80/kgU)

United States	1965	1966	1967	1968	1969	1970	1971	1972	1973	1974
RAR <USD 80/kgU (tU)	289 000		293 000			300 000			400 000	
EAR-I <USD 80/kgU (tU)	512 000		743 000			624 000			769 000	
KCR <USD 80/kgU (tU)	801 000		1 036 000			924 000			1 169 000	
Production (tU)	8 033	8 146	8 657	9 515	8 931	9 928	9 442	9 924	10 182	8 868
Change in KCR (<USD 80/kgU)			235 000			-112 000			245 000	
Change in KCR + production (tU)			251 179			-84 897			274 294	

United States	1975	1976	1977	1978	1979	1980	1981	1982	1983	1984
RAR <USD 80/kgU (tU)	454 000		523 000		531 000		362 000		131 000	
EAR-I <USD 80/kgU (tU)	818 000		838 000		773 000		681 000		30 000	
KCR <USD 80/kgU (tU)	1 272 000		1 361 000		1 304 000		1 043 000		161 000	
Production (tU)	8 924	9 806	11 493	14 221	14 414	16 811	14 799	10 335	8 138	5 724
Change in KCR (<USD 80/kgU)	103 000		89 000		-57 000		-261 000		-882 000	
Change in KCR + production (tU)	122 050		107 730		-31 286		-229 775		-856 866	

United States	1985	1986	1987	1988	1989	1990	1991	1992	1993	1994
RAR <USD 80/kgU (tU)	131 000		124 000		111 000		102 000		114 000	
EAR-I <USD 80/kgU (tU)	0		0		0		0		0	
KCR <USD 80/kgU (tU)	131 000		124 000		111 000		102 000		114 000	
Production (tU)	4 352	5 195	4 997	5 050	5 324	3 420	3 060	2 171	1 178	1 289
Change in KCR (<USD 80/kgU)	-30 000		-7 000		-13 000		-9 000		12 000	
Change in KCR + production (tU)	-16 138		2 547		-2 953		-256		17 231	

United States	1995	1996	1997	1998	1999	2000	2001	2002	2003
RAR <USD 80/kgU (tU)	113 000		110 000		106 000		104 000		102 000
EAR-I <USD 80/kgU (tU)	0		0		0		0		0
KCR <USD 80/kgU (tU)	113 000		110 000		106 000		104 000		102 000
Production (tU)	2 324	2 431	2 170	1 810	1 773	1 522	1 015	902	769
Change in KCR (<USD 80/kgU)	-1 000		-3 000		-4 000		-2 000		-2 000
Change in KCR + production (tU)	1 467		1 755		-20		1 295		-83

–   Not applicable.

EVOLUTION OF KNOWN CONVENTIONAL RESOURCES
IN SELECTED COUNTRIES (1965-2003)
(< USD 130/kgU)

Australia	1965	1966	1967	1968	1969	1970	1971	1972	1973	1974
RAR <USD 130/kgU (tU)	14 800		11 500			23 800			100 500	
EAR-I <USD 130/kgU (tU)	–		3 100			9 500			107 500	
KCR <USD 130/kgU (tU)	14 800		14 600			33 300			208 000	
production (tU)	285	166	166	165	254	254	0	0	0	0
Change in resource estimates			-200			18 700			174 700	
Change in KCR + production			251			19 119			174 954	

Australia	1975	1976	1977	1978	1979	1980	1981	1982	1983	1984
RAR <USD 130/kgU (tU)	243 000		296 000		290 000		317 000		336 000	
EAR-I <USD 130/kgU (tU)	80 000		49 000		53 000		283 000		394 000	
KCR <USD 130/kgU (tU)	323 000		345 000		343 000		600 000		730 000	
production (tU)	0	359	356	516	706	1 561	2 860	4 453	3 218	4 390
Change in resource estimates	115 000		22 000		-2 000		257 000		130 000	
Change in KCR + production	115 000		22 359		-1 128		259 267		137 313	

Australia	1985	1986	1987	1988	1989	1990	1991	1992	1993	1994
RAR <USD 130/kgU (tU)	526 000		518 000		538 000		529 000		517 000	
EAR-I <USD 130/kgU (tU)	377 000		384 000		393 000		390 000		394 000	
KCR <USD 130/kgU (tU)	903 000		902 000		931 000		919 000		911 000	
production (tU)	3 252	4 154	3 780	3 534	3 657	3 519	3 759	2 318	2 238	2 200
Change in resource estimates	173 000		-1 000		29 000		-12 000		-8 000	
Change in KCR + production	181 843		6 406		36 314		-4 824		-1 923	

Australia	1995	1996	1997	1998	1999	2000	2001	2002	2003
RAR <USD 130/kgU (tU)	710 000		715 000		716 000		697 000		735 000
EAR-I <USD 130/kgU (tU)	194 000		180 000		194 000		233 000		323 000
KCR <USD 130/kgU (tU)	904 000		895 000		910 000		930 000		1 058 000
production (tU)	3 700	4 959	5 479	4 894	5 984	7 579	7 720	6 854	7 573
Change in resource estimates	-7 000		-9 000		15 000		20 000		128 000
Change in KCR + production	-2 562		-341		25 373		33 563		142 574

EVOLUTION OF KNOWN CONVENTIONAL RESOURCES
IN SELECTED COUNTRIES (1965-2003)
(< USD 130/kgU)

Brazil	1965	1966	1967	1968	1969	1970	1971	1972	1973	1974
RAR <USD 130/kgU (tU)						800			700	
EAR-I <USD 130/kgU (tU)						800			2 500	
KCR <USD 130/kgU (tU)						1 600			3 200	
production (tU)										
Change in resource estimates						1 600			1 600	
Change in KCR + production						1 600			1 600	

Brazil	1975	1976	1977	1978	1979	1980	1981	1982	1983	1984
RAR <USD 130/kgU (tU)	10 400		18 200		74 200		119 100		163 300	
EAR-I <USD 130/kgU (tU)	8 800		8 200		90 100		81 200		92 400	
KCR <USD 130/kgU (tU)	19 200		26 400		164 300		200 300		255 700	
production (tU)							4	242	189	117
Change in resource estimates	16 000		7 200		137 900		36 000		55 400	
Change in KCR + production	16 000		7 200		137 900		36 000		55 646	

Brazil	1985	1986	1987	1988	1989	1990	1991	1992	1993	1994
RAR <USD 130/kgU (tU)	163 280		163 050		162 710		162 000		162 000	
EAR-I <USD 130/kgU (tU)	92 390		92 390		92 000		94 000		94 000	
KCR <USD 130/kgU (tU)	255 670		255 440		254 710		256 000		256 000	
production (tU)	115	115	0	18	35	5	0	0	24	106
Change in resource estimates	-30		-230		-730		1 290		0	
Change in KCR + production	-30		0		-712		1 330		0	

Brazil	1995	1996	1997	1998	1999	2000	2001	2002	2003
RAR <USD 130/kgU (tU)	162 000		162 000		162 000		162 000		86 190
EAR-I <USD 130/kgU (tU)	100 200		100 200		100 200		100 200		57 140
KCR <USD 130/kgU (tU)	262 200		262 200		262 200		262 200		143 330
production (tU)	106	0	0	0	0	11	58	272	230
Change in resource estimates	6 200		0		0		0		-118 870
Change in KCR + production	6 330		106		0		11		-118 540

EVOLUTION OF KNOWN CONVENTIONAL RESOURCES
IN SELECTED COUNTRIES (1965-2003)
(< USD 130/kgU)

Canada	1965	1966	1967	1968	1969	1970	1971	1972	1973	1974
RAR <USD 130/kgU (tU)	338 800		331 100			278 700			307 000	
EAR-I <USD 130/kgU (tU)	508 200		585 200			308 000			409 000	
KCR <USD 130/kgU (tU)	847 000		916 300			586 700			716 000	
production (tU)	3 418	3 025	3 234	3 200	3 430	3 520	3 830	4 000	3 710	3 420
Change in resource estimates			69 300			-329 600			129 300	
Change in KCR + production			75 743			-319 736			142 630	

Canada	1975	1976	1977	1978	1979	1980	1981	1982	1983	1984
RAR <USD 130/kgU (tU)	166 000		182 000		235 000		258 000		185 000	
EAR-I <USD 130/kgU (tU)	419 000		656 000		728 000		760 000		229 000	
KCR <USD 130/kgU (tU)	585 000		838 000		963 000		1 018 000		414 000	
production (tU)	3 560	4 850	5 790	6 800	6 820	7 150	7 720	8 080	7 140	11 169
Change in resource estimates	-131 000		253 000		125 000		55 000		-604 000	
Change in KCR + production	-123 870		261 410		137 590		68 970		-588 200	

Canada	1985	1986	1987	1988	1989	1990	1991	1992	1993	1994
RAR <USD 130/kgU (tU)	214 000		249 000		235 000		214 000		397 000	
EAR-I <USD 130/kgU (tU)	197 000		211 000		204 000		229 000		74 000	
KCR <USD 130/kgU (tU)	411 000		460 000		439 000		443 000		471 000	
production (tU)	10 880	11 720	12 440	12 393	11 323	8 729	8 160	9 297	9 155	9 647
Change in resource estimates	-3 000		49 000		-21 000		4 000		28 000	
Change in KCR + production	15 309		71 600		3 833		24 052		45 457	

Canada	1995	1996	1997	1998	1999	2000	2001	2002	2003
RAR <USD 130/kgU (tU)	381 000		331 000		326 420		314 560		333 834
EAR-I <USD 130/kgU (tU)	73 000		99 000		106 590		122 390		104 710
KCR <USD 130/kgU (tU)	454 000		430 000		433 010		436 950		438 544
production (tU)	10 473	11 706	12 031	10 922	8 214	10 683	12 522	11 607	10 457
Change in resource estimates	-17 000		-24 000		3 010		3 940		1 594
Change in KCR + production	1 802		-1 821		25 963		22 837		25 723

EVOLUTION OF KNOWN CONVENTIONAL RESOURCES
IN SELECTED COUNTRIES (1965-2003)
(< USD 130/kgU)

France	1965	1966	1967	1968	1969	1970	1971	1972	1973	1974
RAR <USD 130/kgU (tU)	32 300		38 500			41 600			56 600	
EAR-I <USD 130/kgU (tU)	29 300		23 100			31 200			49 300	
KCR <USD 130/kgU (tU)	61 600		61 600			72 800			105 900	
production (tU)	1 032	1 423	1 078	1 377	1 180	1 250	1 250		1 545	1 616
Change in resource estimates			0			11 200			33 100	
Change in KCR + production			2 455			13 757			35 600	

France	1975	1976	1977	1978	1979	1980	1981	1982	1983	1984
RAR <USD 130/kgU (tU)	55 000		51 800		55 300		74 900		67 500	
EAR-I <USD 130/kgU (tU)	40 000		44 100		46 200		46 500		32 900	
KCR <USD 130/kgU (tU)	95 000		95 900		101 500		121 400		100 400	
production (tU)	1 673	1 731	1 871	2 097	2 183	2 362	2 634	2 552	2 859	3 271
Change in resource estimates	-10 900		900		5 600		19 900		-21 000	
Change in KCR + production	-7 739		4 304		9 568		24 445		-15 814	

France	1985	1986	1987	1988	1989	1990	1991	1992	1993	1994
RAR <USD 130/kgU (tU)	67 060		65 150		58 900		39 500		33 650	
EAR-I <USD 130/kgU (tU)	45 170		38 070		36 000		8 100		6 730	
KCR <USD 130/kgU (tU)	112 230		103 220		94 900		47 600		40 380	
production (tU)	3 168	3 189	3 248	3 376	3 394	3 241	2 841	2 477	2 149	1 730
Change in resource estimates	11 830		-9 010		-8 320		-47 300		-7 220	
Change in KCR + production	17 960		-2 653		-1 696		-40 665		-1 902	

France	1995	1996	1997	1998	1999	2000	2001	2002	2003
RAR <USD 130/kgU (tU)	24 950		22 360		14 240		190		0
EAR-I <USD 130/kgU (tU)	2 140		1 400		550		11 740		9 510
KCR <USD 130/kgU (tU)	27 090		23 760		14 790		11 930		9 510
production (tU)	1 053	1 016	930	572	452	416	296	184	18
Change in resource estimates	-13 290		-3 330		-8 970		-2 860		-2 420
Change in KCR + production	-9 411		-1 261		-7 468		-1 992		-1 940

EVOLUTION OF KNOWN CONVENTIONAL RESOURCES
IN SELECTED COUNTRIES (1965-2003)
(< USD 130/kgU)

Namibia	1965	1966	1967	1968	1969	1970	1971	1972	1973	1974
RAR <USD 130/kgU (tU)	–		–				–		–	
EAR-I <USD 130/kgU (tU)	–		–				–		–	
KCR <USD 130/kgU (tU)	–		–				–		–	
production (tU)	0	0	0	0	0	0	0	0	0	0
Change in resource estimates										
Change in KCR + production										

Namibia	1975	1976	1977	1978	1979	1980	1981	1982	1983	1984
RAR <USD 130/kgU (tU)	–		–		–		135 000		135 000	
EAR-I <USD 130/kgU (tU)	–		–		53 000		53 000		53 000	
KCR <USD 130/kgU (tU)	–		–		> 53000		188 000		188 000	
production (tU)	0	654	2 340	2 697	3 840	4 042	3 971	3 776	3 719	3 700
Change in resource estimates							135 000		0	
Change in KCR + production							142 882		7 747	

Namibia	1985	1986	1987	1988	1989	1990	1991	1992	1993	1994
RAR <USD 130/kgU (tU)	120 000		113 300		106 900		100 750		96 640	
EAR-I <USD 130/kgU (tU)	53 000		53 000		53 000		53 000		53 000	
KCR <USD 130/kgU (tU)	173 000		166 300		159 900		153 750		149 640	
production (tU)	3 400	3 470	3 540	3 511	3 077	3 211	2 450	1 660	1 679	1 895
Change in resource estimates	-15 000		-6 700		-6 400		-6 150		-4 110	
Change in KCR + production	-7 581		170		651		138		0	

Namibia	1995	1996	1997	1998	1999	2000	2001	2002	2003
RAR <USD 130/kgU (tU)	191 820		187 360		180 510		175 100		170 532
EAR-I <USD 130/kgU (tU)	107 520		107 520		107 510		107 510		87 085
KCR <USD 130/kgU (tU)	299 340		294 880		288 020		282 610		257 617
production (tU)	2 016	2 447	2 905	2 780	2 690	2 715	2 239	2 333	2 037
Change in resource estimates	149 700		-4 460		-6 860		-5 410		-24 993
Change in KCR + production	153 274		3		-1 175		-5		-20 421

EVOLUTION OF KNOWN CONVENTIONAL RESOURCES
IN SELECTED COUNTRIES (1965-2003)
(< USD 130/kgU)

Niger	1965	1966	1967	1968	1969	1970	1971	1972	1973	1974
RAR <USD 130/kgU (tU)	–		19 250			30 030			50 000	
EAR-I <USD 130/kgU (tU)	–		–			–			–	
KCR <USD 130/kgU (tU)	–		19 250			30 030			50 000	
production (tU)	0	0	0	0	0	0	430	867	948	1 117
Change in resource estimates						10 780			19 970	
Change in KCR + production						10 780			21 267	

Niger	1975	1976	1977	1978	1979	1980	1981	1982	1983	1984
RAR <USD 130/kgU (tU)	50 000		160 000		160 000		160 000		160 000	
EAR-I <USD 130/kgU (tU)	–		53 000		53 000		53 000		53 000	
KCR <USD 130/kgU (tU)	50 000		213 000		213 000		213 000		213 000	
production (tU)	1 306	1 460	1 609	2 060	3 620	4 120	4 363	4 259	3 426	3 276
Change in resource estimates	0		163 000		0		0		0	
Change in KCR + production	2 065		165 766		3 669		7 740		8 622	

Niger	1985	1986	1987	1988	1989	1990	1991	1992	1993	1994
RAR <USD 130/kgU (tU)	182 200		175 910		175 910		172 720		165 820	
EAR-I <USD 130/kgU (tU)	300 300		300 300		300 300		305 770		305 770	
KCR <USD 130/kgU (tU)	482 500		476 210		476 210		478 490		471 590	
Production (tU)	3 181	3 110	2 970	2 965	2 962	2 839	2 963	2 965	2 914	2 975
Change in resource estimates	269 500		-6 290		0		2 280		-6 900	
Change in KCR + production	276 202		1		5 935		8 081		-972	

Niger	1995	1996	1997	1998	1999	2000	2001	2002	2003
RAR <USD 130/kgU (tU)	87 100		69 960		71 120		29 600		102 227
EAR-I <USD 130/kgU (tU)	6 000		1 200		18 580		25 530		125 377
KCR <USD 130/kgU (tU)	93 100		71 160		89 700		55 130		227 604
Production (tU)	2 974	3 329	3 487	3 714	2 907	2 914	2 919	3 080	3 157
Change in resource estimates	-378 490		-21 940		18 540		-34 570		172 474
Change in KCR + production	-372 601		-15 637		25 741		-28 749		178 473

EVOLUTION OF KNOWN CONVENTIONAL RESOURCES
IN SELECTED COUNTRIES (1965-2003)
(< USD 130/kgU)

South Africa	1965	1966	1967	1968	1969	1970	1971	1972	1973	1974
RAR <USD 130/kgU (tU)	107 800		250 250			204 050			264 000	
EAR-I <USD 130/kgU (tU)	–		–			–			–	
KCR <USD 130/kgU (tU)	107 800		250 250			204 050			264 000	
Production (tU)	2 262	2 530	3 080	2 985	3 080	3 167	3 220	3 197	2 735	2 711
Change in resource estimates			142 450			-46 200			59 950	
Change in KCR + production			147 242			-37 055			69 534	

South Africa	1975	1976	1977	1978	1979	1980	1981	1982	1983	1984
RAR <USD 130/kgU (tU)	276 000		348 000		391 000		356 000		313 000	
EAR-I <USD 130/kgU (tU)	–		72 000		139 000		175 000		147 000	
KCR <USD 130/kgU (tU)	276 000		420 000		530 000		531 000		460 000	
Production (tU)	2 488	2 758	3 360	3 961	4 797	6 146	6 131	5 816	6 060	5 732
Change in resource estimates	12 000		144 000		110 000		1 000		-71 000	
Change in KCR + production	17 446		149 246		117 321		11 943		-59 053	

South Africa	1985	1986	1987	1988	1989	1990	1991	1992	1993	1994
RAR <USD 130/kgU (tU)	358 700		349 170		418 500		344 400		240 840	
EAR-I <USD 130/kgU (tU)	124 600		124 600		110 200		82 600		54 720	
KCR <USD 130/kgU (tU)	483 300		473 770		528 700		427 000		295 560	
Production (tU)	4 880	4 602	3 963	3 800	2 943	2 460	1 712	1 669	1 699	1 671
Change in resource estimates	23 300		-9 530		54 930		-101 700		-131 440	
Change in KCR + production	35 092		-48		62 693		-96 297		-128 059	

South Africa	1995	1996	1997	1998	1999	2000	2001	2002	2003
RAR <USD 130/kgU (tU)	258 560		269 800		292 800		291 000		315 330
EAR-I <USD 130/kgU (tU)	75 240		87 800		76 400		76 400		80 340
KCR <USD 130/kgU (tU)	333 800		357 600		369 200		367 400		395 670
Production (tU)	1 421	1 436	1 100	965	927	798	878	828	763
Change in resource estimates	38 240		23 800		11 600		-1 800		28 270
Change in KCR + production	41 610		26 657		13 665		-75		29 976

EVOLUTION OF KNOWN CONVENTIONAL RESOURCES IN SELECTED COUNTRIES (1965-2003) (< USD 130/kgU)

United States	1965	1966	1967	1968	1969	1970	1971	1972	1973	1974
RAR <USD 130/kgU (tU)	288 800		292 600			300 300			400 000	
EAR-I <USD 130/kgU (tU)	512 100		743 100			623 700			769 000	
KCR <USD 130/kgU (tU)	800 900		1 035 700			924 000			1 169 000	
Production (tU)	8 033	8 146	8 657	9 515	8 931	9 928	9 442	9 924	10 182	8 868
Change in resource estimates			234 800			-111 700			245 000	
Change in KCR + production			250 979			-93 254			264 370	

United States	1975	1976	1977	1978	1979	1980	1981	1982	1983	1984
RAR <USD 130/kgU (tU)	454 000		743 000		708 000		605 000		407 200	
EAR-I <USD 130/kgU (tU)	812 000		1 053 000		1 158 000		1 097 000		82 600	
KCR <USD 130/kgU (tU)	1 266 000		1 796 000		1 866 000		1 702 000		489 800	
Production (tU)	8 924	9 806	11 493	14 221	14 414	16 811	14 799	10 335	8 138	5 724
Change in resource estimates	97 000		530 000		70 000		-164 000		-1 212 200	
Change in KCR + production	116 050		548 730		95 714		-132 775		-1 187 066	

United States	1985	1986	1987	1988	1989	1990	1991	1992	1993	1994
RAR <USD 130/kgU (tU)	398 100		398 000		377 500		356 000		369 000	
EAR-I <USD 130/kgU (tU)										
KCR <USD 130/kgU (tU)	398 100		398 000		377 500		356 000		369 000	
Production (tU)	4 352	5 195	4 997	5 050	5 324	3 420	3 060	2 171	1 178	1 289
Change in resource estimates	-91 700		-100		-20 500		-21 500		13 000	
Change in KCR + production	-77 838		9 447		-10 453		-12 756		18 231	

United States	1995	1996	1997	1998	1999	2000	2001	2002	2003
RAR <USD 130/kgU (tU)	366 000		361 000		355 000		348 000		345 000
EAR-I <USD 130/kgU (tU)									
KCR <USD 130/kgU (tU)	366 000		361 000		355 000		348 000		345 000
Production (tU)	2 324	2 431	2 170	1 810	1 773	1 522	1 015	902	769
Change in resource estimates	-3 000		-5 000		-6 000		-7 000		-3 000
Change in KCR + production	-533		-245		-2 020		-3 705		-1 083

– Not applicable.

Appendix 7.1

WORLD URANIUM PRODUCTION (1945-2003)[a]

Country	1945	1946	1947	1948	1949	1950	1951	1952	1953	1954
Argentina	0	0	0	0	0	0	0	0	0	0
Australia	0	0	0	0	0	0	0	0	0	59
Belgium	0	0	0	0	0	0	0	0	0	0
Brazil	0	0	0	0	0	0	0	0	0	0
Bulgaria	0	50	50	50	50	100	100	100	100	100
Canada	200	238	239	115	188	169	168	187	283	678
China	0	0	0	0	0	0	0	0	0	0
Congo, D.R.	300	300	500	1 000	1 600	2 220	2 470	2 190	2 200	2 100
Czech Republic	x	x	x	x	x	x	x	x	x	x
Czechoslovakia	0	18	49	103	147	281	527	816	1 162	1 535
Finland	0	0	0	0	0	0	0	0	0	0
France	0	0	0	0	0	0	0	0	0	50
Gabon	0	0	0	0	0	0	0	0	0	0
Germany, F.R.	0	0	0	0	0	0	0	0	0	0
GDR	0	17	150	321	766	1 224	1 675	2 199	3 094	3 967
Hungary	0	0	0	0	0	0	0	0	0	0
India	0	0	0	0	0	0	0	0	0	0
Japan	0	0	0	0	0	0	0	0	0	0
Kazakhstan	x	x	x	x	x	x	x	x	x	x
Madagascar	0	0	0	0	0	0	0	0	0	0
Mexico	0	0	0	0	0	0	0	0	0	0
Mongolia (b)	0	0	0	0	0	0	0	0	0	0
Namibia	0	0	0	0	0	0	0	0	0	0
Niger	0	0	0	0	0	0	0	0	0	0
Pakistan	0	0	0	0	0	0	0	0	0	0
Poland	0	0	0	40	40	40	42	42	42	42
Portugal	0	0	0	0	0	0	5	94	112	108
Romania	0	0	0	0	0	0	0	500	1 000	1 500
Russian Federation	x	x	x	x	x	x	x	x	x	x
South Africa	0	0	0	0	0	0	0	154	1 154	1 538
Spain	0	0	0	0	0	0	0	0	0	0
Sweden	0	0	0	0	0	0	0	0	0	0
Ukraine	x	x	x	x	x	x	x	x	x	x
United States	0	0	52	78	136	353	589	672	895	1 308
USSR*	7	50	129	182	278	417	650	1 000	1 750	2 100
Uzbekistan	x	x	x	x	x	x	x	x	x	x
Yugoslavia	0	0	0	0	0	0	0	0	0	0
Total	507	673	1 169	1 889	3 205	4 804	6 226	7 954	11 792	15 085
OECD										
WOCA	500	538	791	1 193	1 924	2 742	3 227	3 203	4 532	5 733
Non-WOCA	7	135	378	696	1 281	2 062	2 994	4 657	7 148	9 244

WORLD URANIUM PRODUCTION (1945-2003)[a]

Country	1955	1956	1957	1958	1959	1960	1961	1962	1963	1964
Argentina	0	0	0	0	0	0	4	4	8	30
Australia	100	250	321	467	860	935	1 199	1 049	919	283
Belgium	0	0	0	0	0	0	0	0	0	0
Brazil	0	0	0	0	0	0	0	0	0	0
Bulgaria	150	150	200	200	300	300	350	350	350	350
Canada	725	1 754	5 105	10 311	12 227	9 807	7 417	6 485	6 425	5 604
China	0	0	0	0	0	0	0	500	500	500
Congo, D.R.	2 100	2 000	2 000	1 810	1 780	910	120	0	0	0
Czech Republic	x	x	x	x	x	x	x	x	x	x
Czechoslovakia	2 077	2 430	2 745	2 903	2 987	3 036	2 846	2 941	2 859	2 878
Finland	0	0	0	11	2	11	17	0	0	0
France	50	150	300	737	816	958	1 067	1 048	1 045	1 001
Gabon	0	0	0	0	0	0	313	390	450	450
Germany, F.R.	0	0	0	0	0	0	8	11	3	16
GDR	4 522	5 157	5 278	5 302	5 345	5 356	5 991	6 371	6 730	6 983
Hungary	0	50	200	300	400	500	500	500	500	550
India	0	50	50	50	50	50	50	50	100	100
Japan	0	0	0	0	0	0	0	0	0	0
Kazakhstan	x	x	x	x	x	x	x	x	x	x
Madagascar	0	0	50	70	80	80	70	85	90	130
Mexico	0	0	0	0	0	0	0	0	0	0
Mongolia (b)	0	0	0	0	0	0	0	0	0	0
Namibia	0	0	0	0	0	0	0	0	0	0
Niger	0	0	0	0	0	0	0	0	0	0
Pakistan	0	0	0	0	0	0	0	0	0	0
Poland	42	42	42	42	41	41	41	41	40	0
Portugal	109	112	109	120	127	108	102	21	8	34
Romania	1 500	2 000	2 000	2 000	1 900	1 900	0	0	0	0
Russian Federation	x	x	x	x	x	x	x	x	x	x
South Africa	2 539	3 462	4 385	4 805	4 954	4 930	4 207	3 865	3 479	3 415
Spain	0	0	0	0	0	0	40	40	40	40
Sweden	0	0	0	0	0	0	8	8	8	8
Ukraine	x	x	x	x	x	x	x	x	x	x
United States	2 142	4 583	6 525	9 568	12 493	13 568	13 346	13 084	10 937	9 113
USSR*	2 300	2 700	3 000	3 100	3 400	3 800	3 900	4 000	4 100	4 200
Uzbekistan	x	x	x	x	x	x	x	x	x	x
Yugoslavia	0	0	0	0	0	0	0	0	0	0
Total	18 356	24 890	32 310	41 796	47 762	46 290	41 596	40 843	38 591	35 685
OECD							21 988	20 697	18 466	15 816
WOCA	7 656	12 361	18 845	27 949	33 389	31 357	27 968	26 140	23 512	20 224
Non-WOCA	10 591	12 529	13 465	13 847	14 373	14 933	13 628	14 703	15 079	15 461

WORLD URANIUM PRODUCTION (1945-2003)[a]

Country	1985	1986	1987	1988	1989	1990	1991	1992	1993	1994
Argentina	126	173	95	142	52	9	18	123	126	80
Australia	3 252	4 154	3 780	3 534	3 657	3 519	3 759	2 318	2 238	2 200
Belgium	37	41	44	43	43	39	38	36	34	40
Brazil	115	115	0	18	35	5	0	0	24	106
Bulgaria	650	650	650	650	620	405	240	150	100	70
Canada	10 880	11 720	12 440	12 393	11 323	8 729	8 160	9 297	9 155	9 647
China	800	800	800	344	800	800	800	955	780	480
Congo, D.R.	0	0	0	0	0	0	0	0	0	0
Czech Republic	x	x	x	x	x	x	x	x	955	541
Czechoslovakia	2 623	2 578	2 560	2 468	2 407	2 142	1 779	1 539	x	x
Finland	0	0	0	0	0	0	0	0	0	0
France	3 189	3 248	3 376	3 394	3 241	2 841	2 477	2 149	1 730	1 053
Gabon	940	900	800	929	868	709	678	589	556	650
Germany, F.R.	31	26	58	41	17	2 972	1 207	232	116	47
GDR	4 470	4 086	4 059	3 924	3 800	x	x	x	x	x
Hungary	570	570	570	576	530	490	417	430	380	413
India	200	200	200	200	200	230	200	150	148	155
Japan	7	6	5	0	0	0	0	0	0	0
Kazakhstan	x	x	x	x	x	x	x	2 802	2 700	2 240
Madagascar	0	0	0	0	0	0	0	0	0	0
Mexico	0	0	0	0	0	0	0	0	0	0
Mongolia (b)	0	0	0	0	94	89	101	105	54	72
Namibia	3 400	3 470	3 540	3 511	3 077	3 211	2 450	1 660	1 679	1 895
Niger	3 181	3 110	2 970	2 965	2 962	2 839	2 963	2 965	2 914	2 975
Pakistan	30	30	30	30	30	30	30	23	23	23
Poland	0	0	0	0	0	0	0	0	0	0
Portugal	118	111	142	159	129	111	28	29	32	24
Romania	220	290	275	260	250	210	160	120	120	120
Russian Federation	x	x	x	x	x	x	x	2 640	2 697	2 541
South Africa	4 880	4 602	3 963	3 800	2 943	2 460	1 712	1 669	1 699	1 671
Spain	201	215	223	228	227	213	196	187	184	256
Sweden	0	0	0	0	0	0	0	0	0	0
Ukraine	x	x	x	x	x	x	x	1 000	1 000	900
United States	4 352	5 195	4 997	5 050	5 324	3 420	3 060	2 171	1 178	1 289
USSR* (c)	15 900	15 900	15 900	16 000	15 000	14 500	12 000	x	x	x
Uzbekistan	x	x	x	x	x	x	x	2 680	2 600	2 015
Yugoslavia	30	59	72	80	86	53	x	x	x	x
Zambia**										
Total	60 202	62 249	61 549	60 739	57 715	50 026	42 473	36 019	33 222	31 503
OECD	22 067	24 716	25 065	24 842	23 961	21 844	18 925	16 419	14 667	14 556
WOCA	34 969	37 375	36 735	36 517	34 214	31 390	26 976			
Non-WOCA	25 233	24 874	24 814	24 222	23 501	18 636	15 497			

WORLD URANIUM PRODUCTION (1945-2003)[a]

Country	1995	1996	1997	1998	1999	2000	2001	2002	2003	1945-2003
Argentina	65	16	30	7	4	0	0	0	0	2 631
Australia	3 700	4 959	5 479	4 894	5 984	7 579	7 720	6 854	7 573	113 304
Belgium	25	28	27	15	0	0	0	0	0	680
Brazil	106	0	0	0	0	11	56	272	230	1 645
Bulgaria	0	0	0	0	0	0	0	0	0	16 735
Canada	10 473	11 706	12 031	10 922	8 214	10 683	12 522	11 607	10 455	374 548
China	*500*	*560*	*570*	*590*	*700*	*700*	*700*	*730*	*730*	27 689
Congo, D.R.	*0*	*0*	*0*	*0*	*0*	*0*	*0*	*0*	*0*	25 600
Czech Republic	600	604	603	610	612	507	456	465	452	6 405
Czechoslovakia	x	x	x	x	x	x	x	x	x	102 244
Finland	0	0	0	0	0	0	0	0	0	41
France	1 016	930	572	452	416	296	184	18	9	75 965
Gabon	652	568	470	725	0	0	0	0	0	25 403
Germany, F.R.	35	39	28	30	29	28	27	221	150	5 859
GDR	x	x	x	x	x	x	x	x	x	213 380
Hungary	210	200	200	10	10	10	10	10	4	21 080
India	*155*	*207*	*207*	*207*	*207*	*207*	*230*	*230*	*230*	7 963
Japan	0	0	0	0	0	0	0	0	0	84
Kazakhstan	1 630	1 210	1 090	1 270	1 560	1 870	2 114	2 826	3 327	24 639
Madagascar	0	0	0	0	0	0	0	0	0	785
Mexico	0	0	0	0	0	0	0	0	0	49
Mongolia (b)	20	0	0	0	0	0	0	0	0	535
Namibia	2 016	2 447	2 905	2 780	2 690	2 715	2 239	2 333	2 037	78 794
Niger	2 974	3 329	3 487	3 714	2 907	2 914	2 919	3 080	3 157	91 186
Pakistan	*23*	*23*	*23*	*23*	*23*	*23*	*46*	*38*	*40*	961
Poland	*0*	*0*	*0*	*0*	*0*	*0*	*0*	*0*	*0*	660
Portugal	18	15	17	19	10	13	4	0		715
Romania	120	105	107	132	89	86	*85*	*90*	*90*	17 989
Russian Federation	2 160	2 605	2 580	2 530	2 610	2 760	3 090	2 850	3 073	32 136
South Africa	1 421	1 436	1 100	965	927	798	878	828	763	157 618
Spain	255	255	255	255	255	255	30	37		156
Sweden	0	0	0	0	0	0	0	0	0	91
Ukraine	*800*	*800*	*800*	*750*	*750*	*750*	*750*	*800*	*800*	9 900
United States	2 324	2 431	2 170	1 810	1 773	1 522	1 015	902	769	356 485
USSR* (c)	x	x	x	x	x	x	x	x	x	377 613
Uzbekistan	1 644	1 459	1 764	1 926	2 159	2 028	1 945	1 859	1 603	23 682
Yugoslavia	x	x	x	x	x	x	x	x	x	380
Zambia**										102
Total	32 942	35 932	36 515	34 636	31 929	35 755	37 020	36 050	35 492	2 204 732
OECD	17 846	21 167	21 382	19 017	17 303	20 893	21 968	20 114	19 412	
WOCA										
Non-WOCA										

Notes:

Numbers in **bold italics** are data from a Red Book that were estimated by the Secretariat.

X   National entity not in existence or politically redefined.

*   Numbers modified after VETROV *et al.*, "Raw Material Branch of Nuclear Industry", Russian Nuclear Industry, 2000, Moscow.

** Zambian production is estimated using sources other than the Red Book with production estimated to have occurred prior to 1965.

(a) Data prior to 1965 is new data received from member countries and sources other than the Red Book.

(b) Ore mined at the Dornod mine (design capacity of 2 million tonnes or ore/year, 2 400 tU/yr) was processed by the Priargunsky Mining and Processing Combinate in Krasnokamensk in Russia.

(c) Production includes production in former Soviet Socialist Republics of Estonia, Kazakhstan, Kyrgystan, Russian Federation, Turkmenistan, Ukraine and Uzbekistan.

Appendix 7.2

URANIUM PRODUCTION CAPACITY OF EXISTING PRODUCTION CENTRES
(1968-2003) (tU/YEAR)

	1968	1969	1970	1971	1972	1973	1974	1975	1976
Argentina	35	50	50	50	50	50	50	50	50
Australia	1 117	1 117	1 117	700	700	700	700	700	400
Belgium	0	0	0	0	0	0	0	0	0
Brazil	0	0	0	0	0	0	0	0	0
Bulgaria	*500*	*500*	*500*	*500*	*500*	*500*	*500*	*500*	*600*
Canada	8 470	8 800	9 625	5 690	4 600	4 600	4 600	4 600	5 860
China	*500*	*500*	*500*	*500*	*500*	*500*	*500*	*500*	*800*
Czechoslovakia (a)	3 000	3 000	3 000	3 000	3 000	3 000	2 750	2 750	2 750
Czech Republic	X	X	X	X	X	X	X	X	X
France	1 540	1 770	1 770	1 800	1 800	1 800	1 800	1 800	2 200
Gabon	600	600	600	600	600	600	600	600	800
Germany	125	125	125	125	125	125	125	125	125
GDR (a)	*7 500*	*7 500*	*7 500*	*7 500*	*7 500*	*7 500*	*7 500*	*7 500*	*7 500*
Hungary	*600*	*600*	*600*	*600*	*600*	*600*	*600*	*600*	*600*
India	*150*	*150*	*200*	*200*	*200*	*200*	*200*	*200*	*200*
Japan	0	30	30	30	30	30	30	30	30
Kazakhstan	x	x	x	x	x	x	x	x	x
Mexico	30	30	30	30	30	30	30	80	80
Mongolia (b)	0	0	0	0	0	0	0	0	0
Namibia	0	0	0	0	0	0	0	0	3 700
Niger	0	0	0	750	750	750	750	1 500	1 500
Pakistan	*0*	*0*	*0*	*30*	*30*	*30*	*30*	*30*	*30*
Portugal	112	112	112	112	112	112	115	115	85
Romania	*0*	*0*	*0*	*0*	*0*	*0*	*0*	*0*	*0*
Russian Federation	x	x	x	x	x	x	x	x	x
South Africa	3 080	3 160	4 130	4 130	4 130	4 130	2 700	2 600	3 410
Spain	55	55	55	60	60	115	144	144	170
Sweden	120	120	120	120	120	0	0	0	0
Ukraine	x	x	x	x	x	x	x	x	x
United States	12 320	12 320	14 630	14 600	14 600	14 600	12 000	15 000	15 000
USSR (a) (c)	*10 000*	*10 000*	*10 000*	*10 000*	*10 000*	*15 000*	*15 000*	*15 000*	*15 000*
Uzbekistan	x	x	x	x	x	x	x	x	x
Yugoslavia	0	0	0	0	0	0	0	0	0
Total World	49 854	50 539	54 694	51 127	50 037	54 972	50 724	54 424	60 890

URANIUM PRODUCTION CAPACITY OF EXISTING PRODUCTION CENTRES
(1968-2003) (tU/YEAR)

	1977	1978	1979	1980	1981	1982	1983	1984	1985
Argentina	130	280	185	200	180	240	340	340	360
Australia	400	500	600	600	2 600	4 500	3 400	3 800	4 500
Belgium	0	0	0	45	45	45	45	45	45
Brazil	0	0	0	0	420	420	420	420	420
Bulgaria	*600*	*600*	*600*	*600*	*600*	*600*	*600*	*600*	*700*
Canada	6 100	6 450	6 900	7 200	8 400	9 500	7 300	10 500	11 500
China	*800*	*900*	*900*	*900*	*900*	*900*	*900*	*900*	*900*
Czechoslovakia (a)	2 750	2 750	2 750	3 550	3 550	3 550	3 550	3 550	3 550
Czech Republic	X	X	X	X	X	X	X	X	X
France	2 200	2 850	2 950	3 450	3 706	3 900	3 900	3 900	3 900
Gabon	800	1 200	1 000	1 000	1 000	1 200	1 500	1 500	1 500
Germany	125	125	125	125	125	125	125	125	125
GDR (a)	*7 500*	*7 500*	*7 500*	*7 500*	*7 500*	*7 500*	*7 500*	*7 500*	*7 500*
Hungary	*600*	*600*	*600*	*600*	*600*	*600*	*600*	*600*	*600*
India	*200*	*200*	*200*	*200*	*200*	*200*	*200*	*200*	*200*
Japan	30	30	30	30	30	30	9	9	9
Kazakhstan	x	x	x	x	x	x	x	x	x
Mexico	80	80	80	80	80	80	0	0	0
Mongolia (b)	0	0	0	0	0	0	0	0	0
Namibia	3 700	3 700	3 845	4 100	3 939	3 923	3 806	3 806	3 700
Niger	1 609	2 400	3 350	4 300	4 500	4 500	4 500	4 500	4 600
Pakistan	*30*	*30*	*30*	*30*	*30*	*30*	*30*	*30*	*30*
Portugal	85	85	85	95	107	126	115	115	119
Romania	*0*	*100*	*100*	*100*	*100*	*100*	*150*	*200*	*250*
Russian Federation	x	x	x	x	x	x	x	x	x
South Africa	6 700	8 800	5 240	6 500	6 700	7 200	6 134	6 134	5 500
Spain	191	191	339	678	145	110	150	150	200
Sweden	0	0	0	0	0	0	0	0	0
Ukraine	x	x	x	x	x	x	x	x	x
United States	14 700	19 300	19 000	20 900	17 100	16 900	8 100	8 100	8 200
USSR (a) (c)	*16 500*	*16 500*	*16 500*	*16 500*	*16 500*	*16 500*	*16 500*	*16 500*	*16 500*
Uzbekistan	x	x	x	x	x	x	x	x	x
Yugoslavia	0	0	0	0	0	0	0	0	102
Total World	65 830	75 171	72 909	79 283	79 057	82 779	69 874	73 524	75 010

URANIUM PRODUCTION CAPACITY OF EXISTING PRODUCTION CENTRES
(1968-2003) (tU/YEAR)

	1986	1987	1988	1989	1990	1991	1992	1993	1994
Argentina	360	205	205	205	205	205	150	150	150
Australia	4 500	4 250	5 200	5 400	4 800	5 400	4 200	4 200	4 200
Belgium	45	45	45	45	45	45	45	45	45
Brazil	420	420	420	420	420	420	420	420	420
Bulgaria	*700*	*700*	*700*	*700*	*700*	*300*	*300*	*300*	*300*
Canada	12 000	12 000	12 100	11 400	11 400	12 200	9 300	8 650	9 710
China	*900*	*900*	*900*	*900*	*900*	*900*	*1 000*	*1 000*	*1 000*
Czechoslovakia (a)	3 550	3 550	3 550	3 550	3 550	3 550	2 050	x	x
Czech Republic	X	X	X	X	X	X	X	1 100	1 000
France	3 900	3 900	3 900	3 900	3 920	3 920	3 070	2 570	2 570
Gabon	1 500	1 500	1 500	1 500	1 500	1 500	1 500	1 500	1 500
Germany	125	125	125	125	4 500	4 500	0	0	0
GDR (a)	*7 500*	*7 500*	*7 500*	*7 500*	x	x	x	x	x
Hungary	*600*	*600*	*600*	*600*	*600*	*600*	*600*	*600*	*600*
India	*200*	*200*	*200*	*200*	*200*	*230*	*200*	*200*	*200*
Japan	9	9	0	0	0	0	0	0	0
Kazakhstan	x	x	x	x	x	x	*3 000*	*3 000*	*3 000*
Mexico	0	0	0	0	0	0	0	0	0
Mongolia (b)	0	0	0	0	0	0	0	0	0
Namibia	3 700	3 500	3 500	3 500	3 500	3 500	3 500	3 500	3 500
Niger	4 600	4 600	3 800	3 800	3 700	3 700	3 400	3 400	3 800
Pakistan	*30*	*30*	*30*	*30*	*30*	*30*	*30*	*30*	*30*
Portugal	170	170	170	170	170	170	50	50	50
Romania	*300*	*300*	*300*	*300*	*300*	*300*	*300*	*300*	*300*
Russian Federation	x	x	x	x	x	x	3 500	3 500	3 500
South Africa	5 500	5 500	4 000	4 000	2 500	2 500	1 900	1 900	1 900
Spain	220	220	225	225	225	205	187	270	800
Sweden	0	0	0	0	0	0	0	0	0
Ukraine	x	x	x	x	x	x	*1 000*	*1 000*	*1 000*
United States	9 900	9 700	10 000	10 000	10 000	10 000	10 000	10 000	10 000
USSR (a) (c)	*16 500*	*16 500*	*16 500*	*16 500*	*16 500*	*16 500*	x	x	x
Uzbekistan	x	x	x	x	x	x	*3 000*	*3 000*	*2 500*
Yugoslavia	102	102	102	102	102	x	x	x	x
Total World	77 331	76 526	75 572	75 072	69 767	70 675	52 702	50 685	52 075

URANIUM PRODUCTION CAPACITY OF EXISTING PRODUCTION CENTRES
(1968-2003) (tU/YEAR)

	1995	1996	1997	1998	1999	2000	2001	2002	2003
Argentina	150	120	120	120	120	120	120	120	120
Australia	4 200	4 200	5 000	5 500	5 700	8 200	9 400	9 400	9 400
Belgium	45	45	45	45	0	0	0	0	0
Brazil	420	0	0	0	150	340	340	340	340
Bulgaria	0	0	0	0	0	0	0	0	0
Canada	9 840	10 820	12 950	14 250	17 150	11 830	14 300	16 290	14 890
China	*1 000*	*1 000*	740	740	740	740	700	700	850
Czechoslovakia (a)	x	x	x	x	x	x	x	x	x
Czech Republic	1 000	1 000	680	680	680	680	660	550	450
France	1 500	1 500	760	600	600	600	0	0	0
Gabon	1 500	1 500	1 500	1 500	1 500	0	0	0	0
Germany	0	0	0	0	0	0	0	0	0
GDR (a)	x	x	x	x	x	x	x	x	x
Hungary	*600*	*600*	*600*	*600*	0	0	0	0	0
India	*230*	*230*	*220*	*220*	*210*	*210*	*210*	*210*	*230*
Japan	0	0	0	0	0	0	0	0	0
Kazakhstan	2 300	2 500	1 500	1 600	2 000	2 500	3 000	3 000	3 000
Mexico	0	0	0	0	0	0	0	0	0
Mongolia (b)	0	0	0	0	0	0	0	0	0
Namibia	3 000	3 000	3 000	3 000	4 000	4 000	4 000	4 000	4 000
Niger	3 500	3 500	3 800	3 800	2 910	2 910	2 910	2 960	3 800
Pakistan	*30*	*30*	*30*	*30*	*30*	*30*	*30*	*30*	*65*
Portugal	50	50	50	50	170	170	170	170	170
Romania	*300*	*300*	*300*	*300*	*300*	*300*	*100*	*100*	*100*
Russian Federation	3 500	3 500	3 500	3 500	3 500	3 500	3 700	3 700	3 700
South Africa	1 900	1 900	1 900	1 900	1 700	1 700	1 157	1 319	1 270
Spain	800	800	255	255	255	255	255	255	255
Sweden	0	0	0	0	0	0	0	0	0
Ukraine	*1 000*	*1 000*	*1 000*	*1 000*	*1 000*	*1 000*	*1 000*	*1 000*	*1 000*
United States	10 585	10 670	10 670	10 670	10 670	9 510	9 510	8 000	4 000
USSR (a) (c)	x	x	x	x	x	x	x	x	x
Uzbekistan	*2 500*	*2 500*	2 050	2 500	2 300	2 300	2 300	2 300	2 300
Yugoslavia	x	x	x	x	x	x	x	x	x
Total World	49 950	50 765	50 670	52 860	55 685	50 895	53 862	54 444	49 940

Notes:

Numbers in **bold italics** are data from a Red Book that were estimated by the Secretariat.

(a) Data provided by member state from non-Red Book sources.

(b) Ore mined at the Dornod mine (design capacity of 2 million tonnes or ore/year, 2 400 tU/yr) was processed by the Priargunsky Mining and Processing Combinate in Krasnokamensk in Russia.

(c) Production capacity includes production centres in former Soviet Socialist Republics of Estonia, Kazakhstan, Kyrgystan, Russian Federation, Turkmenistan, Ukraine and Uzbekistan.

Appendix 7.3

EMPLOYMENT IN EXISTING PRODUCTION CENTRES
(1980-2003)

Country	1980	1981	1982	1983	1984	1985	1986	1987	1988	1989	1990	1991
Argentina	700	700	650	650	650	630	630	600	590	500	340	250
Australia	500	500	500	500	480	460	460	460	621	484	1 183	1 189
Belgium	NA	NA	NA	NA	30	5	5	5	5	5	5	5
Brazil	NA	NA	NA	NA	780	724	637	579	564	568	521	463
Bulgaria	NA	NA	NA	NA	NA	NA	NA	NA	NA	NA	NA	NA
Canada	6 100	NA	4 830	5 850	5 810	5 333	5 080	4 825	4 730	4 280	2 495	2 195
China	NA	NA	NA	NA	NA	NA	NA	NA	NA	NA	10 000	9 500
Czechoslovakia/Czech Republic	NA	NA	NA	NA	NA	31 700	31 800	32 100	31 600	30 400	12 100	9 300
France	NA	NA	NA	NA	3 682	3 508	3 247	3 145	3 089	2 786	2 276	1 773
Gabon	NA	NA	NA	NA	NA	NA	NA	NA	NA	NA	NA	NA
Germany (a)	85	80	75	75	27 636	27 893	27 481	26 835	26 450	25 189	15 710	7 488 c
Hungary	NA	NA	NA	NA	7 000	7 454	7 493	7 364	7 090	6 483	4 798	2 240
India	NA	NA	NA	NA	NA	NA	NA	NA	NA	NA	NA	NA
Kazakhstan	NA	NA	NA	NA	NA	NA	NA	NA	NA	NA	NA	NA
Namibia	NA	NA	NA	NA	NA	NA	NA	NA	NA	NA	NA	NA
Niger	NA	NA	NA	NA	3 550 *	3 552	3 541	3 295	3 243	3 203	3 173	2 562
Portugal	614	598	573	549	535	519	506	487	460	450	231	217
Romania	NA	NA	NA	NA	NA	NA	NA	NA	NA	NA	NA	NA
Russian Federation	NA	NA	NA	NA	NA	NA	NA	NA	NA	NA	NA	NA
Slovenia/Yugoslavia	NA	NA	NA	NA	420 *	420	460	470	490	490	440	200
South Africa	NA	NA	NA	NA	NA	NA	NA	NA	NA	NA	NA	NA
Spain	252	355	348	347	325	311	349	348	348	309	309	240
United States (b)	19 920	13 680	8 970	5 620	3 600	2 450	2 120	2 000	2 140	1 580	1 335	1 016
Uzbekistan	NA	NA	NA	NA	NA	NA	NA	NA	NA	NA	NA	NA

Country	1992	1993	1994	1995	1996	1997	1998	1999	2000	2001	2002	2003
Argentina	220	220	180	120	100	80	80	80	70	62	60	60
Australia	376	405	412	413	464	468	501	565	527	550	502	655
Belgium	5	5	5	5	5	6	6	6	5	5	4	0
Brazil	430	410	408	390	305	280	180	110	48	128	128	140
Bulgaria	13 000	8 000	NA	NA	NA	NA	NA	NA	NA	NA	NA	NA
Canada	1 310	1 320	1 370	1 350	1 155	1 105	1 134	1 076	1 026	973	972	965
China	9 500	9 300	9 100	8 000	8 500	8 500	8 500	8 500	8 500	8 200	8 000	7 700
Czechoslovakia/Czech Republic	6 600	5 900	5 400	4 500	3 600	3 580	3 410	3 300	2 887	2 641	2 507	2 426
France	1 368	824	496	468	441	141	144	NA	NA	NA	NA	NA
Gabon	207	193	263	276	259	150	NA	NA	15	15	15	NA
Germany (a)	6 093 c	4 895 c	4 613 c	4 400 c	4 200 c	3 980 c	3 615 c	3 149 c	3 115 c	3 004 c	2 691 c	2 444 c
Hungary	1 855	1 755	1 766	1 250	1 300	900	0	0	0	0	0	0
India	3 780	3 898	3 898	NA	NA	4 000	4 000	4 000	4 000	4 200	4 200	4 200
Kazakhstan	11 800	10 550	8 050	6 850	6 000	5 100	4 800	4 600	4 000	4 000	3 770	3 870
Namibia	1 266	1 240	1 246	1 246	1 189	1 254	1 104	1 009	902	785	782	NA
Niger	2 340	2 118	2 104	2 109	2 070	2 033	2 012	1 830	1 680	1 607	1 558	1 606
Portugal	94	52	46	52	56	57	61	54	47	30	11	0
Romania	NA	NA	6 500	6 000	5 000	4 550	3 300	2 800	2 150	2 000 *	2 000 *	2 000 *
Russian Federation	NA	15 900	14 400	14 000	13 000	12 900	1 2 800	12 700	12 500	12 325	12 800	12 785
Slovenia/Yugoslavia	150	145	145	140	115	105	NA	NA	79	69	48	45
South Africa	NA	NA	NA	NA	NA	NA	160	160	160	150	150	150
Spain	232	186	185	183	178	172	148	135	134	58	56	56
United States (b)	682	380 d	452	535 d	689 d	793 d	911 d	649 d	401 d	245 d	277 d	204 d
Uzbekistan	NA	NA	6 688	7 378	8 201	8 230	8 165	7 734	7 331	7 300	8 370	8 460

Notes:

NA Not available.
*   Data from a Red Book that were a Secretariat estimate.
(a) Includes former GDR.
(b) Employment from 1984 through 1990 for exploration, mining and milling.
(c) Employment for decommissioning.
(d) Employment for decommissioning not included.

Appendix 8.1

URANIUM INVENTORIES BY RED BOOK (1976-2003)
(tonnes natural U equivalent)

Countries	1976				
	Natural uranium	Enriched uranium	Depleted uranium	Reprocessed uranium	Reported inventories
Argentina					
Australia					1 750
Belgium					NA
Brazil					
Canada					5 580
Czech Republic					
Finland					
France					NA
Germany, F.R.					1 370
Italy					5
Japan					
Korea, Republic of					
Lithuania					
Mexico					40
Netherlands					
Niger					
Portugal					350
Slovak Republic					
Slovenia					
South Africa					
Spain					
Sweden					200
Switzerland					
Turkey					
United Kingdom (a)	13 011	1 032	6 113		20 156
United States					73 300
Yugoslavia					
Total					~103 000

Countries	1977				
	Natural uranium	Enriched uranium	Depleted uranium	Reprocessed uranium	Reported inventories
Argentina					
Australia					1 750
Belgium					
Brazil					
Canada					5 500
Czech Republic					
Finland					
France					
Germany, F.R.					4 080
Italy					
Japan					
Korea, Republic of					
Lithuania					
Mexico					42
Netherlands					
Niger					
Portugal					570
Slovak Republic					
Slovenia					
South Africa					
Spain					1 775
Sweden					
Switzerland					
Turkey					
United Kingdom (a)	12 018	1 154	6 030		19 202
United States					84 900
Yugoslavia					
Total					117 819

URANIUM INVENTORIES BY RED BOOK (1976-2003)
(tonnes natural U equivalent)

Countries	1979				
	Natural uranium	Enriched uranium	Depleted uranium	Reprocessed uranium	Reported inventories
Argentina					
Australia					
Belgium					
Brazil					
Canada					5 500
Czech Republic					
Finland					
France					
Germany, F.R.					2 950
Italy					1 828
Japan					
Korea, Republic of					1 000
Lithuania					
Mexico					
Netherlands					
Niger					
Portugal					725
Slovak Republic					
Slovenia					
South Africa					
Spain					
Sweden					200
Switzerland					
Turkey					
United Kingdom (a)	13 634	1 191	5 458		20 283
United States					108 100
Yugoslavia					
Total					140 586

Countries	1983				
	Natural uranium	Enriched uranium	Depleted uranium	Reprocessed uranium	Reported inventories
Argentina	50				50
Australia	1 744				1 744
Belgium					
Brazil	50				50
Canada					
Czech Republic					
Finland	600	713			1 313
France					
Germany, F.R.	6 954	7 210	2 750		16 914
Italy	2 200	5 150			7 350
Japan					
Korea, Republic of	1 520	90			1 610
Lithuania					
Mexico					
Netherlands	788	885	1 440		3 113
Niger					
Portugal	631				631
Slovak Republic					
Slovenia					
South Africa					
Spain					
Sweden					
Switzerland					
Turkey					
United Kingdom (a)	NA	NA	5 000		5 000
United States	77 800	42 600	12 000		132 400
Yugoslavia					
Total	92 337	56 648	21 190		~170 000

URANIUM INVENTORIES BY RED BOOK (1976-2003)
(tonnes natural U equivalent)

Countries	1986				
	Natural uranium	Enriched uranium	Depleted uranium	Reprocessed uranium	Reported inventories
Argentina	300				300
Australia	1 743				1 743
Belgium					1 800 *
Brazil					110 *
Canada					
Czech Republic					
Finland					350 *
France					18 000 *
Germany, F.R.	9 154	6 000	252		15 406
Italy	800	3 700	520		5 020
Japan					
Korea, Republic of	1 300				1 300
Lithuania					
Mexico					
Netherlands	1 600		215		1 815
Niger					
Portugal	589				
Slovak Republic					
Slovenia					
South Africa					
Spain					
Sweden					
Switzerland					
Turkey					
United Kingdom	NA	NA	4 000		4 000
United States	74 300	54 900	13 000		142 200
Yugoslavia					
Total	89 786	64 600	17 987		~192 000

Countries	1988				
	Natural uranium	Enriched uranium	Depleted uranium	Reprocessed uranium	Reported inventories
Argentina					
Australia					1 744
Belgium					1 920 *
Brazil					
Canada					
Czech Republic					
Finland	455	445			900
France					14 800 *
Germany, F.R.					16 030
Italy	1 672	1 885	145	417	4 119
Japan					10 900 *
Korea, Republic of					
Lithuania					
Mexico					
Netherlands					
Niger	550				550
Portugal	606				606
Slovak Republic					
Slovenia					
South Africa					
Spain					
Sweden					780
Switzerland					
Turkey					
United Kingdom			6 600		6 600
United States	70 000	49 900	15 900		135 800
Yugoslavia					
Total	73 283	52 230	22 645	417	194 749

URANIUM INVENTORIES BY RED BOOK (1976-2003)
(tonnes natural U equivalent)

Countries	1989				
	Natural uranium	Enriched uranium	Depleted uranium	Reprocessed uranium	Reported inventories
Argentina	400				400
Australia	1 900				1 900
Belgium					1 480 *
Brazil					110 *
Canada					
Czech Republic					
Finland	560	480			1 040
France					18 600 *
Germany, F.R.	10 500	2 918	3 000		16 418
Italy	897	2 308	3	134	3 342
Japan					
Korea, Republic of	300		12		312
Lithuania					
Mexico					
Netherlands	70	18	520	16	624
Niger					
Portugal	675				675
Slovak Republic					
Slovenia					
South Africa					
Spain					
Sweden		780			780
Switzerland					570 *
Turkey					
United Kingdom	NA	NA	2 150	3 300	5 450
United States	69 500	30 800	16 900	NA	117 200
Yugoslavia	85	76			
Total	84 887	37 380	22 585	3 450	~169 000

Countries	1991				
	Natural uranium	Enriched uranium	Depleted uranium	Reprocessed uranium	Reported inventories
Argentina	260				260
Australia	1 900				1 900
Belgium					1 900 *
Brazil					110 *
Canada					
Czech Republic					
Finland	600	465			1 065
France					21 600*
Germany, F.R.	12 432	2 500	-	-	14 932
Italy					
Japan					
Korea, Republic of	540	300	5		845
Lithuania					
Mexico					
Netherlands	429	18	825	6	1 278
Niger					
Portugal	656				656
Slovak Republic					
Slovenia					
South Africa					
Spain					
Sweden		780			780
Switzerland					570 *
Turkey					
United Kingdom					
United States	56 500	29 900	NA	NA	86 400
Yugoslavia					
Total	73 317	33 963	830	6	~132 000

URANIUM INVENTORIES BY RED BOOK (1976-2003)
(tonnes natural U equivalent)

Countries	1993				
	Natural uranium	Enriched uranium	Depleted uranium	Reprocessed uranium	Reported inventories
Argentina	100	-	-	-	100
Australia	1 900				1 900
Belgium					1 900 *
Brazil					110 *
Canada	-	-	-	-	
Czech Republic	1 500	-	-	-	1 500
Finland	815	1 065			1 880
France					23 700 *
Germany, F.R.	1 610	897	3 324	-	5 831
Italy					
Japan					
Korea, Republic of	686	518			1 204
Lithuania	-	336	-	-	336
Mexico	147	382	261	-	790
Netherlands	407	12	950	10	1 379
Niger					
Portugal	631				631
Slovak Republic	800	28			828
Slovenia					
South Africa					
Spain					
Sweden		780			780
Switzerland					530 *
Turkey					
United Kingdom					
United States (b)	50 700	20 800	NA	NA	71 500
Yugoslavia					
Total	59 296	24 818	4 535	10	~115 000

Countries	1995				
	Natural uranium	Enriched uranium	Depleted uranium	Reprocessed uranium	Reported inventories
Argentina	100				100
Australia					
Belgium					2 060*
Brazil	300	270	800		1 370
Canada					
Czech Republic					> 2 225
Finland	800	1 100			1 900
France					26 700 *
Germany, F.R.					
Italy					
Japan					
Korea, Republic of					1 850*
Lithuania					
Mexico					
Netherlands					375
Niger	123				
Portugal					
Slovak Republic		220			
Slovenia					
South Africa					
Spain					
Sweden	780*				780*
Switzerland					530*
Turkey					
United Kingdom					
United States (b)	52 474	18 013			70 487
Yugoslavia					
Total	53 797	19 603	129 600		>109 177

(tonnes natural U equivalent)

Countries	1997				
	Natural uranium	Enriched uranium	Depleted uranium	Reprocessed uranium	Reported inventories
Argentina	188				188
Australia					
Belgium					2 100*
Brazil					NA
Canada					
China, P. R.					
Czech Republic	> 2 700				> 2 700
Finland	700	1 120			1 850
France					25 800 *
Germany, F.R.					
Italy					
Japan					
Korea, Republic of					2 760 *
Lithuania		200 *			200 *
Mexico					170 *
Netherlands					
Niger	340				
Portugal					
Russian Federation					
Slovak Republic		220			220
Slovenia					
South Africa					
Spain					
Sweden	780 *				780 *
Switzerland					
Turkey					
United Kingdom					
United States					52 910
Yugoslavia					
Total (c)	4 708	1 540			89 678

Countries	1999				
	Natural uranium	Enriched uranium	Depleted uranium	Reprocessed uranium	Reported inventories
Argentina		16			16
Australia					
Belgium					2 100 *
Brazil					
Canada					
China, P. R.			2 000***		
Czech Republic					> 2 700
Finland	730	1 120			1 850
France			190 000***		27 800 *
Germany, F.R.					
Italy					
Japan			10 000***		
Korea, Republic of			200***		2456*
Lithuania					
Mexico					
Netherlands					
Niger					
Portugal					
Russian Federation			460 000***		
Slovak Republic		108			
Slovenia					
South Africa					
Spain					
Sweden					
Switzerland					480*
Turkey					
United Kingdom			30 000***		8 400
United States	39 080	23 240	480 000***	NA	> 62 320
Yugoslavia					
Total (c)	39 810	24 484	1 188 200***		>108 122

URANIUM INVENTORIES BY RED BOOK (1976-2003)
(tonnes natural U equivalent)

Countries	2001				
	Natural uranium	Enriched uranium	Depleted uranium	Reprocessed uranium	Reported inventories
Argentina	110				110
Australia					NA
Belgium					2 100 *
Brazil	81	0	0	0	81
Canada					
China	NA	NA	560	NA	560
Czech Republic	> 2 500	0	0	0	> 2 500
Finland					500*
France					25 700*
Germany, F.R.					
Italy					
Japan	NA	NA	2800	NA	2 800
Korea, Republic of	800	1 800	0	0	2 600
Lithuania		120*			120*
Mexico					
Netherlands					
Niger					
Portugal					
Slovak Republic					
Slovenia					
South Africa					
Spain					
Sweden					
Switzerland					375*
Turkey	2				2
United Kingdom					
United States	39 480	24 140	134400	NA	198 020
Total	42973	26060	137760		>235 468

Countries	2003				
	Natural uranium	Enriched uranium	Depleted uranium	Reprocessed uranium	Reported inventories
Argentina	110	0	NA	NA	110
Australia	NA	0	NA	NA	n.a
Belgium	NA	NA	NA	NA	n.a
Brazil	20	NA	NA	NA	> 20
Canada	NA	0	NA	NA	n.a
China	NA	NA	NA	NA	
Czech Republic	2 000	NA	NA	NA	> 2 000
Finland	NA	NA	NA	NA	> 290 *
France	NA	NA	NA	NA	25 700 *
Germany, F.R.	NA	NA	NA	NA	n.a
Italy	0	0	NA	NA	-
Japan	NA	NA	NA	NA	n.a
Korea, Republic of	1 100	2 100	NA	NA	3 200
Lithuania	0	140	NA	NA	140
Mexico	300	0	NA	NA	300
Netherlands	0	0	NA	NA	
Niger	NA	NA	NA	NA	
Portugal	286	0	NA	NA	286
Slovak Republic	0	0	NA	NA	
Slovenia	0	0	NA	NA	
South Africa	0	0	NA	NA	
Spain	NA	380	NA	NA	> 380
Sweden	NA	NA	NA	NA	n.a
Switzerland	NA	NA	NA	NA	NA
Turkey	2	0	NA	NA	2
United Kingdom	NA	NA	NA	NA	n.a
United States	37 845	20 820	NA	NA	58 665
Total	41 663	23 440	NA	NA	~115 000

URANIUM INVENTORIES BY RED BOOK (1976-2003)
(tonnes natural U equivalent)

Notes:

NA Not available.
* Estimated inventory (natural and enriched), according to stock policy.
** Estimated inventories at the end of the year by the Uranium Institute.
*** OECD/IAEA, *Management of Depleted Uranium*, Paris, 2001.
(a) Data from source other than a Red Book.
(b) Data for 1995 comes from US Energy Information Administration, *Uranium Industry Annual 1996*.
(c) Total for 1999 depleted uranium includes 16 000 tU held by URENCO in Germany, Netherlands and United Kingdom.

Stock of depleted uranium (end of 1999)

Country	tU
China	2 000
France	190 000
Japan	10 000
Korea, Republic of	200
Russian Federation	460 000
United Kingdom	30 000
United States	480 000
Urenco (Germany, Netherlands, United Kingdom)	16 000
Total	1 188 200

Source: *Management of Depleted Uranium* (OECD/IAEA 2001).

Appendix 8.2

URANIUM INVENTORY POLICIES BY RED BOOK

The following are uranium stock policies for individual countries as they were stated in the Red Book that was published in the year indicated. The policies stated in this summary may well have changed in response to changing perceptions of uranium supply and demand relationships. In particular, many utilities that have maintained as much as three years of inventory have shortened this time frame as the uranium supply industry has become more broadly diversified.

Belgium

1986: Utilities maintain uranium stocks equivalent to two years forward requirements.

Brazil:

1986: A minimum of one year forward requirements, plus a 10% safety margin, are to be held in national stockpiles.

Canada

1983: In the past, Canada has implemented domestic stockpiling programmes, to maintain a nucleus of production and employment during periods of depressed markets. The last such programme was completed in 1974. A portion of the uranium accumulated under these past programmes was sold on the export market in the early-1970s, the remainder, 5 570 tU, was transferred to Eldorado Nuclear Ltd in March 1981, to increase the Crown's equity investment in that company. No further uranium stockpiling programmes are currently contemplated.

1989: Canada's three nuclear utilities seek to maintain an on-site inventory of finished fuel bundles, which in no case is equivalent to less than six months' forward requirements. Additional inventory includes, in varying proportions, finished fuel bundles, "in-process" material, and those quantities required to feed the refining process. In all cases, overall quantities do not normally exceed one year's forward requirements. No national stockpiles currently exist or are contemplated.

Czech Republic

2003: Stocks in the form of natural uranium are held by the government and by DIAMO, sp (Producer). Utilities (CEZ, a.s.) do not hold strategic inventories in the form of natural uranium. CEZ's general policy is to keep uranium in the form of fabricated fuel and to maintain optimum quantities of uranium in processing, in order to approximately cover its annual needs.

Finland

1986: Utilities maintain uranium stocks equivalent to more than one year of forward requirements. The government maintains no stock.

1989: Utilities maintain uranium stocks equivalent to more than one year of forward requirements. In addition, TVO Power Company owns uranium in different processing stages abroad, equivalent to one year's forward requirements. The government maintains no stock.

2003: The nuclear power utilities maintain reserves of fuel assemblies sufficient for 7-12 months use.

France

1986: *Electricité de France* possesses emergency uranium stocks, the minimum level of which has been fixed at the equivalent of three years' consumption, to face possible supply interruptions.

Germany

1983: A uranium stockpile, equal to approximately two years of requirements is held by the government and the utilities.

Japan

1989: Japanese utilities have inventories equivalent to several years of forward uranium requirements, mainly in fabrication facilities and nuclear plant sites.

Korea, Republic of

1986: At least one-year of forward reactor consumption is maintained as a strategic inventory. About one half of the stock is stored as natural uranium in overseas conversion facilities and the remainder is stored as enriched uranium at the local fabrication facilities.

1991: KEPCO is required to carry stocks in form of natural and enriched uranium of six months forward demand.

Lithuania

1997: A six-month stock of enriched fuel is generally maintained by the utility.

2003: Policy is to maintain sufficient stockpile for three months power plant operation.

Mexico

1991: It is the policy of the National Electricity Utility (CFE) to carry stocks equivalent to two years of forward reactor consumption.

1997: Policy is to maintain one to two reloads of natural uranium at an enrichment facility. The policy has been not to have stockpiles of enriched uranium or fabricated fuel.

Netherlands

1989: The policy is to maintain approximately five years forward reactor consumption.

1997: The natural uranium stocks were disposed of by 31 December 1995. Since then, the Netherlands have held no further stocks.

Spain

2001: In 2000, the legislation that regulated enriched uranium strategic inventories was changed in a way that the minimum strategic inventories, which total until 2000 were financed by the Spanish Government, have been reduced from 1.5 years requirements to less than a year and are now being financed by the Spanish utilities.

2003: A strategic uranium inventory of at least 382 tU (450 tU$_3$O$_8$), maintained as enriched uranium, should be held jointly by the utilities that own nuclear power plants. Additional inventories could be maintained depending on uranium market conditions.

Sweden

1986: Stockpile should enable reactors to operate 22 months without further supply.

1998: The Swedish Parliament decided to replace the previous obligation that utilities had to keep a stockpile of enriched uranium corresponding to the production of 35 TWh with a reporting mechanism.

Switzerland

1991: Switzerland reported a one year requirement in the form of finished fuel assemblies.

United Kingdom

1983: The UK policy is to maintain a stockpile equivalent to two years forward uranium requirements, provided adequate supply diversification has been activated. In addition, it is the policy to hold a minor proportion of such stocks in the enriched form.

2003: The UK uranium stockpile practices are the responsibility of the individual bodies concerned. Actual stock levels are commercially confidential.

United States

1983: The United States has no specific policy regarding the maintenance of national strategic stockpiles of uranium. A large stockpiles exists which will be used to meet needs for the toll enrichment operation of the government uranium enrichment plants, other government uses, and as an emergency source of supply for the industry.

The US government will maintain working inventories at its enrichment plant to ensure that enriched uranium is delivered on time. A 45-day working inventory of natural uranium (6 700 tU as UF$_6$) is needed to assure the smooth operation of the enrichment plants.

A 90-day enriched uranium working inventory (equivalent to 11 500 t of natural uranium) is required to assure enrichment service customers that their deliveries can be met on contracted schedules.

The enriched uranium stockpile could be used as an emergency source of supply for the nuclear industry. Emergency commercial use of material from US government stocks would normally be under a lease rather than through outright sales.

1986: The United States has no specific policy regarding the maintenance of national strategic stockpiles of uranium. However the government will maintain a 45-day working inventory for its enrichment plants (about 3 000 tU) and a 120-day enriched working inventory for customers of enrichment services (about 7 000 – 8 000 t of natural uranium). Excess enriched uranium stockpiles are intended to be reduced over the next three years.

1993: Under the Energy Policy Act of 1992, a National Strategic Uranium reserve was established under the direction and control of the Secretary of Energy. The resource consists of natural uranium and uranium equivalent contained in stockpiles or inventories currently held by the United States for defence purposes. The use of the reserve is to be restricted to military and research for a period of six years from the passage of the 1992 Energy Policy Act. The Department of Energy's stockpile of enrichment tailings is to be restricted to military uses for a period of six years from the passage of the act.

Utilities: Not all utilities report having a policy regarding maintaining certain levels for forward coverage of requirements. At the end of 1992, utilities held stocks of about 35 200 tU as natural and enriched uranium, which amounted to more than two years of requirements.

Switzerland

1989: The policy of nuclear plant operating companies is to maintain one to two years supply in form of fresh fuel assemblies.

Yugoslavia

1989: The required utility stock level is 10.6 tonnes of enriched U (4.3% ^{235}U).

OECD PUBLICATIONS, 2, rue André-Pascal, 75775 PARIS CEDEX 16
PRINTED IN FRANCE
(66 2006 09 1 P) ISBN 92-64-02806-4 – No. 55271 2006